AF388944

Nanomaterials for Ion Battery Applications

Nanomaterials for Ion Battery Applications

Editor

Jaehyun Hur

MDPI • Basel • Beijing • Wuhan • Barcelona • Belgrade • Manchester • Tokyo • Cluj • Tianjin

Editor
Jaehyun Hur
Department of Chemical and
Biological Engineering
Gachon University
Gyeoggi
Korea, South

Editorial Office
MDPI
St. Alban-Anlage 66
4052 Basel, Switzerland

This is a reprint of articles from the Special Issue published online in the open access journal *Nanomaterials* (ISSN 2079-4991) (available at: www.mdpi.com/journal/nanomaterials/special_issues/nano_battery).

For citation purposes, cite each article independently as indicated on the article page online and as indicated below:

LastName, A.A.; LastName, B.B.; LastName, C.C. Article Title. *Journal Name* **Year**, *Volume Number*, Page Range.

ISBN 978-3-0365-4770-1 (Hbk)
ISBN 978-3-0365-4769-5 (PDF)

Contents

About the Editor

Jaehyun Hur

Jaehyun Hur achieved his B.S. in Applied Chemical Engineering at Seoul National University, Seoul, Republic of Korea, in 2000. He achieved a Ph.D. at the Department of Chemical Engineering in 2008 from Purdue University, West Lafayette, IN, USA, in 2008. He worked as a research scientist in Samsung Advanced Institute of Technology (SAIT) until 2014, where he developed various secondary batteries. He is currently an Associate Professor in the Department of Chemical and Biological Engineering at Gachon University. His research is focused on the synthesis of various nanomaterials toward the development of novel materials for secondary batteries. He has published over 100 papers in his research area.

nanomaterials

Editorial

Nanomaterials for Ion Battery Applications

Jaehyun Hur

Department of Chemical and Biological Engineering, Gachon University, Seongnam 13120, Gyeonggi, Korea; jhhur@gachon.ac.kr; Tel.: +82-31-750-5593

Citation: Hur, J. Nanomaterials for Ion Battery Applications. *Nanomaterials* **2022**, *12*, 2293. https://doi.org/10.3390/nano12132293

Received: 14 June 2022
Accepted: 30 June 2022
Published: 4 July 2022

Publisher's Note: MDPI stays neutral with regard to jurisdictional claims in published maps and institutional affiliations.

Nanomaterials offer opportunities to improve battery performance in terms of energy density and electrochemical reaction kinetics owing to a significant increase in the effective surface area of electrodes and reduced ion diffusion pathways. Nevertheless, a large number of unwanted secondary reactions in nanomaterials, which lead to the rapid deterioration in prolonged cycling, are important issues to be overcome. Therefore, the judicious design of nanoarchitecture is one of the essential themes in battery research. The Special Issue of "Nanomaterials for Ion Battery Applications" of *Nanomaterials* covers the recent advancements in nanotechnologies and nanomaterials for various ion batteries including Li-ion batteries (LIBs), Li-O$_2$ batteries, and multivalent aqueous batteries. Seeking facile, inexpensive, and scalable processes to synthesize new nanomaterials and nanoarchitectures into high-performance ion batteries is another important research topic in this Special Issue.

The development of new, high-performance anode materials for LIBs was intensively reported in this Special Issue. Nguyen et al. demonstrated various important roles of layered-structure transition-metal chalcogenide (TMC) materials for LIBs. Although MoS$_2$ is one of the most widely studied TMC materials for LIB anodes, they showed new nanostructures and functions of MoS$_2$. First, they demonstrated that few-layer MoS$_2$ covered SnO$_2$ nanoparticles (NPs) to achieve the highly reversible and high-capacity anode material for LIBs [1]. Few-layer 1T MoS$_2$ nanosheets prepared using a conventional liquid chemical exfoliation method (lithium intercalation in butyllithium-dissolved hexane) was coated on the SnO$_2$ NPs via layer-by-layer fashion at the air–water interface. The coverage of few-layer MoS$_2$ on SnO$_2$ contributed to the reduced volume expansion of active SnO$_2$ NPs, which improved the reversibility of lithiation/delithiation. After optimizing the number of MoS$_2$ layers, they found single-layer MoS$_2$-coated SnO$_2$ NPs exhibited good cycling performance. Second, few-layer 1T MoS$_2$ was demonstrated as an active material for the high-performance LIB anode by the same group [2]. Because of the cumbersome restacking of few-layer MoS$_2$ during the electrochemical reactions, they introduced Ag NPs in the MoS$_2$-dispersed solution via a simple reduction of AgNO$_3$ with NaBH$_4$. The uniform distribution of Ag NPs (hundreds of nanometers) on few-layer MoS$_2$ was facilitated by the 3-mercaptopropionic acid. The resulting Ag@MoS$_2$ electrode exhibited good cycling performance (510 mAh g^{-1} after 100 cycles) and rate performance. Third, a WS$_2$/W$_2$C composite material was demonstrated as a great candidate TMC composite material for the LIB anode [3]. Although the flower-like WS$_2$/W$_2$C nanostructure possesses high surface-to-volume ratio, it showed unstable cycling behavior. The in situ incorporation of carbon nanotubes (CNTs) in WS$_2$/W$_2$C highly improved electrical conductivity without significant agglomeration of CNTs. The obtained performance of the WS$_2$/W$_2$C-CNT composite anode (650 mAh g^{-1} at 100 mA g^{-1} after 100 cycles) was better than those of most WS$_2$-based anodes with various nanostructures. Metallic alloy is another high-potential anode for LIBs. Huy et al. reported C-decorated InSb alloy (InSb-C) as a potential anode material for LIBs [4]. InSb-C was synthesized via a ball-milling process simply using In, Sb, and C as reacting materials. The effects of the binder and buffering matrix on the InSb anode performance were systematically studied. Owing to the hydroxyl functional groups on InSb particles, the poly(acrylic acid) containing carboxylate groups showed better adhesion and stability as a binder than a conventional poly(vinylidene fluoride) binder. The addition of amorphous C in InSb enhanced mechanical stability and electrical conductivity. The InSb-C delivered a high reversible capacity (878 mAh g^{-1} at 100 mA g^{-1} after 140 cycles) and outstanding rate capability (capacity retention of 98% at 10 A g^{-1} relative to 0.1 A g^{-1}). Semiconducting metal oxides (SMOs) are widely studied materials for high-performance LIB anodes owing to high theoretical capacity and safety. Among various SMOs, Bui et al. reviewed ZnO-based binary and ternary composites for LIB anodes [5]. ZnO possesses various benefits such as high theoretical capacity (978 mAh g^{-1}), low-cost, environmental friendliness, high chemical and mechanical stability, and abundance. However, it generally shows a severe capacity fading during cycling due to the remarkable volume change (228%) and aggregation. This review gave an overview of the ZnO binary and ternary composites that overcame these problems via new designs of materials, structures, and processes, which provides useful guidance for the development of not only ZnO-based anode material, but also other SMO-based anode materials.

Aqueous rechargeable batteries are promising alternatives for LIBs because of safety and facile cell assembly. Among these, aqueous Zn-ion batteries (AZIBs) have emerged as a promising technology owing to high theoretical capacity (820 mAh g^{-1}, 5854 mAh L^{-1}), abundance, low-cost, and appropriate redox potential (-0.76 V vs. standard hydrogen electrode). However, the performance of current AZIBs is not satisfactory due to various issues in both the cathode and anode. Therefore, the development of new materials for the cathode and anode that can overcome such issues are of great importance in this research field. Kim et al. synthesized $NH_4V_4O_{10}$ (NHVO) flower-like layered oxide via a simple and rapid microwave method to use as a high-performance cathode material for AZIBs [6]. Although vanadium-based cathodes have been intensively studied, they have been mainly synthesized using the hydrothermal process. The microwave synthesis of NHVO significantly reduced a synthesis time without sacrificing the performance. The NHVO cathode delivered 277 mAh g^{-1} with 75% capacity retention after 150 cycles at 0.25 A g^{-1} and showed exceptional high coulombic efficiencies. Meanwhile, the utilization of TMCs as promising cathode materials in AZIBs, as well as other aqueous multivalent metal-ion batteries, was extensively reviewed by Huy et al. [7]. The interlayer spacing of various TMCs is the key aspect of cathode materials to be successfully used in various multivalent ion-based aqueous batteries. Various strategies including (i) intercalation modification, (ii) defect modification, (iii) hybridization, and (iv) phase modification were explained and future research directions were discussed in this review. On the anode side, one of the most important hurdles is the control of random Zn dendrite growth and hydrogen evolution on Zn metal. Liu et al. proposed that the Cu–Zn alloy can significantly improve the Zn metal performance [8]. The Cu–Zn alloy was fabricated simply by dropping 0.1 M of $CuSO_4$ aqueous solution (100 µL) on Zn foil and allowed to sit for 3 min, followed by washing and drying. A spontaneous replacement reaction changed the Zn foil color from gray to black, indicating the formation of the Cu–Zn alloy. Benefitting from reduced overpotential, the Cu–Zn alloy guided the uniform Zn nucleation, resulting in the remarkable improvement in cycling life of symmetric cells. Furthermore, when coupled with the V_2O_5-PEDOT cathode, the full aqueous battery exhibited stable and reversible cycling behavior over 1000 cycles. The general challenges and strategies of Zn metal anodes in mild aqueous electrolytes are extensively summarized by Huy et al. [9]. The promising strategies were categorized to be (i) Zn metal shielding, (ii) control of Zn nucleation and Zn-ion flux redistribution, and (iii) establishment of uniform electric field. This review presented design guidelines for the development of high-performance AZIBs.

Li-O_2 batteries have drawn significant attention as next-generation energy storage with high energy density (10 times higher than conventional LIBs), low-cost, and green technology. However, the practical viability has been challenging due to sluggish O_2 reaction kinetics. Karuppiah et al. proposed the $La_{0.8}Ce_{0.2}Fe_{0.5}Mn_{0.5}O_3$ perovskite oxide/graphene nanosheet composite (LCFM8255/GNS) as a potential cathode for Li-O_2 batteries, which can resolve such problems [10]. Mesoporous LCFM8255 (high crystallinity and specific surface area)-embedded GNS boosted the electrochemical kinetics owing to its structural benefits. In addition, a high capacity (8475 mAh g^{-1}) and cycle stability were achieved at the current density of 100 mA g^{-1}.

A gel polymer electrolyte (GPE) is one of the solid electrolytes that can overcome the safety issues in organic solvent-based liquid electrolytes. In GPEs, the introduction of inorganic fillers is an effective strategy to achieve high ionic conductance and strong interfacial contact with an electrode. Huy et al. provided an overview of recently reported studies regarding the detailed functions of inorganic fillers including TiO_2, Al_2O_3, SiO_2, ZrO_2, CeO_2, and $BaTiO_3$ added in GPEs for various batteries (Li, Na, Mg, and Zn-ion batteries) [11].

To summarize, this Special Issue covers the progress in various battery systems. The recent studies conducted by many dedicated researchers highlight the importance of innovative nanostructures and new functional materials. Electrochemical performance is highly correlated with these, which can open up new opportunities for battery research.

Funding: This research was supported by the Basic Science Research Program through the National Research Foundation of Korea (NRF) funded by the Ministry of Education (NRF-2021R1F1A1050130) and Basic Science Research Capacity Enhancement Project through Korea Basic Science Institute (National research Facilities and Equipment Center) grant funded by the Ministry of Education (2019R1A6C1010016).

Conflicts of Interest: The authors declare no conflict of interest.

References

1. Nguyen, T.P.; Kim, I.T. Self-assembled few-layered MoS_2 on SnO_2 anode for enhancing lithium-ion storage. *Nanomaterials* **2020**, *10*, 2558. [CrossRef]
2. Nguyen, T.P.; Kim, I.T. Ag nanoparticle-decorated MoS_2 nanosheets for enhancing electrochemical performance in lithium storage. *Nanomaterials* **2021**, *11*, 626. [CrossRef] [PubMed]
3. Nguyen, T.P.; Kim, I.T. In situ growth of W_2C/WS_2 with carbon-nanotube networks for lithium-ion storage. *Nanomaterials* **2022**, *12*, 1003. [CrossRef] [PubMed]
4. Huy, V.P.H.; Kim, I.T.; Hur, J. The effects of the binder and buffering matrix on InSb-based anodes for high-performance rechargeable Li-ion batteries. *Nanomaterials* **2021**, *11*, 3420.
5. Bui, V.K.H.; Pham, T.M.; Hur, J.; Lee, Y.-C. Review of ZnO binary and ternary composite anodes for lithium-ion batteries. *Nanomaterials* **2021**, *11*, 2001. [CrossRef] [PubMed]
6. Kim, S.; Soundharrajan, V.; Sambandam, B.; Mathew, V.; Hwang, J.-Y.; Kim, J. Microwave-assisted rapid synthesis of $NH_4V_4O_{10}$ layered oxide: A high energy cathode for aqueous rechargeable zinc ion batteries. *Nanomaterials* **2021**, *11*, 1905. [CrossRef] [PubMed]
7. Huy, V.P.H.; Ahn, Y.N.; Hur, J. Recent advances in transition metal dichalcogenide cathode materials for aqueous rechargeable multivalent metal-ion batteries. *Nanomaterials* **2021**, *11*, 1517. [CrossRef]
8. Liu, C.; Lu, Q.; Omar, A.; Mikhailova, D. A facile chemical method enabling uniform Zn deposition for improved aqueous Zn-ion batteries. *Nanomaterials* **2021**, *11*, 764. [CrossRef] [PubMed]
9. Huy, V.P.H.; Hieu, L.T.; Hur, J. Zn metal anodes for Zn-ion batteries in mild aqueous electrolytes: Challenges and strategies. *Nanomaterials* **2021**, *11*, 2746.
10. Karuppiah, C.; Wei, C.-N.; Karikalan, N.; Wu, Z.-H.; Thirumalraj, B.; Hsu, L.-F.; Alagar, S.; Piraman, S.; Hung, T.-F.; Li, Y.-J.J.; et al. Graphene nanosheet-wrapped mesoporous $La_{0.8}Ce_{0.2}Fe_{0.5}Mn_{0.5}O_3$ perovskite oxide composite for improved oxygen reaction electro-kinetics and Li-O_2 battery application. *Nanomaterials* **2021**, *11*, 1025. [CrossRef] [PubMed]
11. Huy, V.P.H.; So, S.; Hur, J. Inorganic fillers in composite gel polymer electrolytes for high-performance lithium and non-lithium polymer batteries. *Nanomaterials* **2021**, *11*, 614. [CrossRef]

nanomaterials

Article

In Situ Growth of W$_2$C/WS$_2$ with Carbon-Nanotube Networks for Lithium-Ion Storage

Thang Phan Nguyen and Il Tae Kim *

Department of Chemical and Biological Engineering, Gachon University, Seongnam-si 13120, Korea; phanthang87@gmail.com
* Correspondence: itkim@gachon.ac.kr

Abstract: The combination of W_2C and WS_2 has emerged as a promising anode material for lithium-ion batteries. W_2C possesses high conductivity but the W_2C/WS_2-alloy nanoflowers show unstable performance because of the lack of contact with the leaves of the nanoflower. In this study, carbon nanotubes (CNTs) were employed as conductive networks for in situ growth of W_2C/WS_2 alloys. The analysis of X-ray diffraction patterns and scanning/transmission electron microscopy showed that the presence of CNTs affected the growth of the alloys, encouraging the formation of a stacking layer with a lattice spacing of ~7.2 Å. Therefore, this self-adjustment in the structure facilitated the insertion/desertion of lithium ions into the active materials. The bare W_2C/WS_2-alloy anode showed inferior performance, with a capacity retention of ~300 mAh g^{-1} after 100 cycles. In contrast, the WCNT01 anode delivered a highly stable capacity of ~650 mAh g^{-1} after 100 cycles. The calculation based on impedance spectra suggested that the presence of CNTs improved the lithium-ion diffusion coefficient to 50 times that of bare nanoflowers. These results suggest the effectiveness of small quantities of CNTs on the in situ growth of sulfides/carbide alloys: CNTs create networks for the insertion/desertion of lithium ions and improve the cyclic performance of metal-sulfide-based lithium-ion batteries.

Keywords: WS_2; W_2C; hydrothermal method; carbon nanotubes; lithium-ion batteries

Citation: Nguyen, T.P.; Kim, I.T. In Situ Growth of W$_2$C/WS$_2$ with Carbon-Nanotube Networks for Lithium-Ion Storage. *Nanomaterials* **2022**, *12*, 1003. https://doi.org/10.3390/nano12061003

Academic Editor: Christian M. Julien

Received: 16 February 2022
Accepted: 17 March 2022
Published: 18 March 2022

Publisher's Note: MDPI stays neutral with regard to jurisdictional claims in published maps and institutional affiliations.

1. Introduction

The rise of graphene, transition-metal chalcogenides (TMCs), transition-metal oxides, and layered-structure transition-metal carbides/nitrides (MXenes) shows the significance and potential of 2D-layered nanomaterials, which can be applied in various fields such as displays, energy storage, energy conversion, and electronic devices [1–18]. TMC materials possess high theoretical lithium-storage capacity (~670 mAh g^{-1} with MoS$_2$ and 433 mAh g^{-1} with WS_2). However, the practical showed that a high abnormal capacity was recorded, which can contribute by conversion reaction, the derived solid electrolyte interface (SEI)-layer formation, or the high lithiation process in the interfacial lithium-storage spaces [19–21]. For example, Feng et al. fabricated WS_2 nanoflakes for lithium-ion batteries (LIBs), which delivered a high initial discharge capacity of ~ 1700 mAh g^{-1} at a current of 47.5 mA g^{-1} [22]. Liu et al. synthesized mesoporous WS_2, showing a high initial discharge capacity of ~1300 mAh g^{-1} [23]. However, the TMCs anode material with the conversion reaction could be significantly degraded due to the dissolution of the sulfur into electrolyte, creating a gel-like polymeric layer [24]. Recently, the combination of TMCs with MXenes has received significant attention owing to the tunable bandgap of TMCs, active edge of chalcogenide atoms with high conductivity of MXenes, high stability, and active edge of metal atoms [25–28]. For example, Zhao et al. developed vertical MoS$_2$/Mo$_2$C nanosheets on carbon paper, which maximized the active sites of the active edges and resulted in high electrocatalytic performance in the hydrogen-evolution reaction [29]. Cheng et al. used guar gum as the carbon source for nanoflower MoS$_2$/Mo$_2$C as an efficient sustainable electrocatalyst for the production of hydrogen gas [30]. Faizan et al. fabricated Mo_2C stacked with MoS_2 nanosheets for lithium-storage applications. Li et al. [31] modified the surface of WS_2/W_2C materials with N and S, which improved their electrochemical catalytic properties [28]. Nguyen et al. controlled the growth of W_2C/WS_2 nanoflowers via a hydrothermal method for use as stable anode materials in lithium-ion batteries (LIBs) [32,33]. The carbide MXenes possess high conductivity and stability; however, they are not highly active materials by themselves [6–10]. Their lithium-storage capability is low, and thus they could only be employed as additive materials [34–37]. Meanwhile, carbon nanotubes (CNTs) possess high conductivity and light weight and are popular network materials for enhancing connectivity in electronic applications [38,39]. Therefore, CNTs and derived carbon materials have been widely used as skeleton or network of the active materials for LIBs. For example, Lu et al. used a CNT/MoS_2 composite as a binder-free anode material showing high performance in LIBs [40]. Chen et al. developed a FeS_2/CNT composite material with a neural-network-like structure, which delivered a superior rate and high cycling performance in sodium-ion batteries [41]. The use of TMCs, MXenes, and CNTs in a system could combine their advantage such as physical, chemical stability, high conductivity, and high capability for lithium-storage applications.

In this study, W_2C/WS_2-alloy nanoflowers were fabricated with a small quantity of CNTs as a connective network using a hydrothermal method. The presence of a CNT network is not only effective for the formation of alloy flowers but also improves the electrochemical performance of the as-prepared anode materials in lithium storage. The structural changes and stable performance of W_2C/WS_2 in a CNT network (WCNT) were investigated and discussed.

2. Materials and Methods

2.1. Chemical Materials

Thioacetamide (TAA, C_2H_5NS, 99%), WCl_6 powder (99.9%), multiwalled CNTs (>90%), and polyvinylidene fluoride (PVDF, M_W ~ 534,000) were purchased from Sigma-Aldrich Inc. (St. Louis, MO, USA). Super-P amorphous carbon black (C, approximately 40 nm, 99.99%) and absolute ethanol were purchased from Alpha Aesar, Inc. (Ward Hill, MA, USA). All materials were used as received. WCl_6 was stored in an Ar-filled glove box.

2.2. Synthesis of WCNT

The WCNT was prepared using a modified procedure for W_2C/WS_2 nanoflower synthesis [33]. CNTs were dispersed in ethanol using sonication. Then, 0.6 g WCl_6 was added to 4 mL of the CNT solution with an adjusted weight ratio of CNT: WCl_6 of 5, 10, and 15%. TAA (1.2 g) was dispersed in a separate vessel containing 4 mL of absolute ethanol. The TAA solution was then quickly mixed with the WCl_6/CNT solution and stirred for 5 min. Then, 10 mL of deionized (DI) water was added to the solution, and the mixture was transfer into a 40 mL polypropylene-lined autoclave and heated at 250 °C for 12 h. The obtained powder was washed four times with ethanol and DI water and dried in a vacuum oven at 60 °C. The samples with different quantities of CNTs (5, 10, and 15%) were marked as WCNT01/02/03, respectively.

2.3. Material Characterization

The structures of W_2C/WS_2 and WCNT samples were determined using X-ray diffraction (XRD, D/MAX-2200 Rigaku, Tokyo, Japan) over the 2θ range of 10–70° and transmission electron microscopy (TEM, TECNAI G2F30, FEI Corp., Hilsboro, OR, USA). Their morphologies were analyzed using field emission scanning electron microscopy (FESEM, SIGMA HD, Carl Zeiss, Jena, Germany) at an accelerating voltage of 5 kV. Thermogravimetric analysis (TGA) was measured using a thermal analyzer (Q600 SDT, TA Instruments, New Castle, DE, USA).

2.4. Electrochemical Measurements

The W_2C/WS_2 and WCNT anode materials were evaluated by assembling half-cell LIBs using a coin-type cell (CR 2032, Rotech Inc., Gwangju, Korea) with a lithium reference electrode. The active materials were mixed with carbon super P and PVDF (weight ratio of 70:15:15) in a n-methyl-2-pyrrolidone solution to form a slurry, which was then coated on copper foil using a doctor blade. The working electrodes were dried in a vacuum oven at 70 °C for 24 h to remove the solvent. The anodes were punched into 12 mm circular disks. The loading mass of the active materials was ~1.0–1.3 mg. LIBs were assembled in an Ar-filled glove box using 1 M $LiPF_6$ in ethylene carbonate/diethylene carbonate (EC:DEC = 1:1 by volume) as the electrolyte. Cyclic voltammetry (CV) tests and electrochemical impedance spectroscopy (EIS) were performed using a battery-cycle tester (WBCS3000, WonAtech, Seocho-gu, Seoul, Korea) over the voltage range of 0.01–3.0 V vs. Li/Li^+ and frequency range from 100 kHz to 0.1 Hz, respectively. The cycling stabilities were measured over the voltage range of 0.01–3.00 V using a ZIVE MP1 (WonAtech, Seocho-gu, Seoul, Korea).

3. Results

The morphologies of the W_2C/WS_2 nanoflowers and the WCNT samples are shown in Figure 1. The sizes of nanoflowers range from 100 to 300 nm with many leaves, which consist of 2D nanosheets, as shown in Figure 1a. The presence of CNTs in the samples reduced the number of leaves. All the alloy nanoflowers were wrapped in the CNT network. In addition, at a low concentration of CNTs in the WCNT01 sample, the W_2C/WS_2 nanoflowers grew to a larger size of ~300–400 nm, as illustrated in Figure 1b. At above 10% of CNTs, the size of nanoflowers decreased to 200–300 nm, as shown in Figure 1c,d. At lower quantity of CNTs (2 and 3 wt%), the separate growth of W_2C/WS_2 nanoflowers was found (data not shown), indicating the nonuniformity. Therefore, the minimum content for the effective coverage of W_2C/WS_2 was 5 wt% CNTs. It is noteworthy that the presence of CNTs could act as a seed point for growth of W_2C/WS_2 nanoflowers. In the bare W_2C/WS_2 nanoflower, their leaves were bended around a center. In WCNT samples, the leaf surface was flat, resulting in an increase in the flower size. However, the increased quantity of CNTs could occupy more spaces in solution, which could limit the growth of W_2C/WS_2 flower leaves. Moreover, the high concentration of CNTs could lead to the aggregation in the prepared solution. Therefore, the increased quantity of CNTs in the samples led to the size reduction and the absence of nanoflowers in the frame network. Moreover, the separate growth of W_2C/WS_2 was observed as a result of CNT aggregation, as shown in Figure S1.

Figure 2a shows the XRD patterns and TEM/HR-TEM images of the W_2C/WS_2 alloy flowers and WCNT samples. The XRD patterns of the W_2C/WS_2 alloys were confirmed by the standard W_2C and WS_2 peaks, as reported in previous studies [28,42,43] The (001) and (100) peaks of W_2C are clearly observed. The (002) peak of WS_2 overlapped with the stacking layer peak at ~12.6°, whereas the (004), (103), and (006) planes were clearly observed. The (002) peak of the CNTs was not clear until 15% CNT content was used in the samples. The WCNT03 sample showed a low-intensity peak at that position. Furthermore, the high crystallinity of W_2C and WS_2 and their large sizes also contributed to the high peak intensity, leading to decreased CNT peaks. In addition, the samples with CNTs showed significantly improved peaks for the stacking layer at $2\theta = 12.6°$. According to Bragg's law, $d = \lambda/2sin\ \theta$ (where d is the lattice spacing, λ is the incident X-ray wavelength, and θ is the diffraction angle); the average lattice spacing was ~7.2 Å. This lattice spacing is large compared to the ionic radius of Li^+, which is 0.76 Å, therefore, this spacing could provide a facile path for lithium ions to easily insert/desert into the material structures. The TEM images in Figure 2b–d also confirm the formation of W_2C/WS_2 on CNTs network, with the lattice spacing of the stacking layer in the range of 0.62–0.84 nm, which is consistent with the XRD results. Therefore, the presence of CNTs not only created a frame network but also facilitated the growth of W_2C/WS_2 alloys, forming average lattice spacing of ~7.2 Å, which is promising for metal-ion-storage applications. The TEM images with energy dispersive x-ray elemental mapping also confirmed the presence of W, S, C atoms on the W_2C/WS_2 and CNT structure in Figure S2. The TGA curves of bare W_2C/WS_2 and WCNT01 samples were presented in Figure S3. The mass of bare W_2C/WS_2 and WCNT01 sample reduced to ~90% and ~80% after the measurement. It is noted that both W_2C/WS_2 and CNTs were oxidized during the measurement. Therefore, the different mass percentage after the measurement is proportional to the mass change from CNTs. The amounts of W_2C/WS_2 and CNTs in WCNT01 were calculated to be 90 and 10 wt%, respectively. The increased quantity of CNTs could reduce the nanoflowers' size, leading to an increase in surface area and an improvement in the electrochemical performance of anode materials.

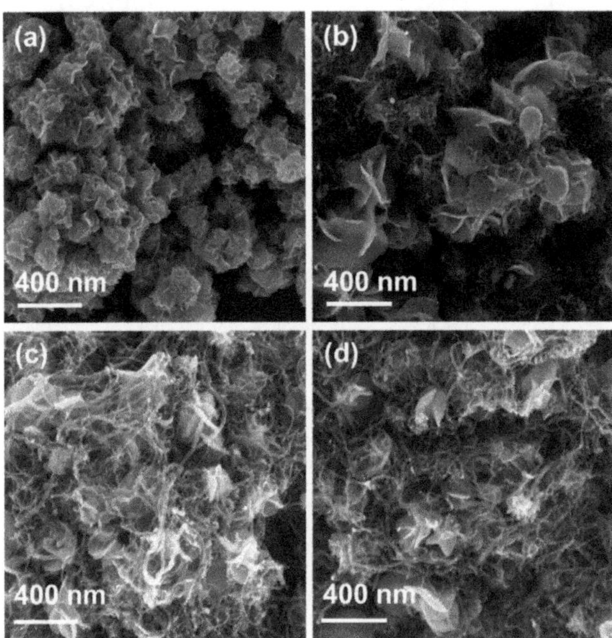

Figure 1. FESEM images of (**a**) W_2C/WS_2 (**b**) WCNT01, (**c**) WCNT02, and (**d**) WCNT03 samples.

To further confirm the structure of W_2C/WS_2 on CNTs, Figure 3a shows the Raman spectra of W_2C/WS_2 nanoflowers and WCNT01 samples. The optical phonon modes (E12g and A1g) of WS_2 were well-recorded at ~350 and 415 cm^{-1} [44]. The tungsten-carbide vibration modes were also detected at ~700 and ~800 cm^{-1} [10]. The W_2C/WS_2 alloys showed a low intensity of carbon sp3 and sp2 peaks, corresponding to the D and G band, respectively. These peaks are highly increased in the WCNT01 sample, indicating the presence of the CNT structure [45,46]. The XPS spectra of WCNT01 material are shown in Figure 3b–d. The W^{4+} peaks can be deconvoluted to the W–C binding and W–S binding, corresponding to the doublets with W 4f$_{7/2}$ at 32.2 and at 33.0 eV, respectively. Moreover, small W^{6+} peaks were observed, which could be due to the oxidation on the surface during sample preparation. Sulfur atoms show a doublet of S 2p$_{3/2}$ and 2p$_{1/2}$ peaks at 161.7 and 163.0 eV, respectively, indicating the S–W binding. A small peak at 169.5 eV was observed due to the surface oxidation of the material. The C 1s spectrum could be deconvoluted into six peaks at 284.1, 284.6, 285.5, 286.5, 287.2, and 290.6 eV, which correspond to the C–W, C=C, C–C, C–O, C=O, and O–C=O binding, respectively. These results are consistent with binding energy of CNT and carbide compounds, indicating the formation of W_2C/WS_2 on CNTs [30,45,47].

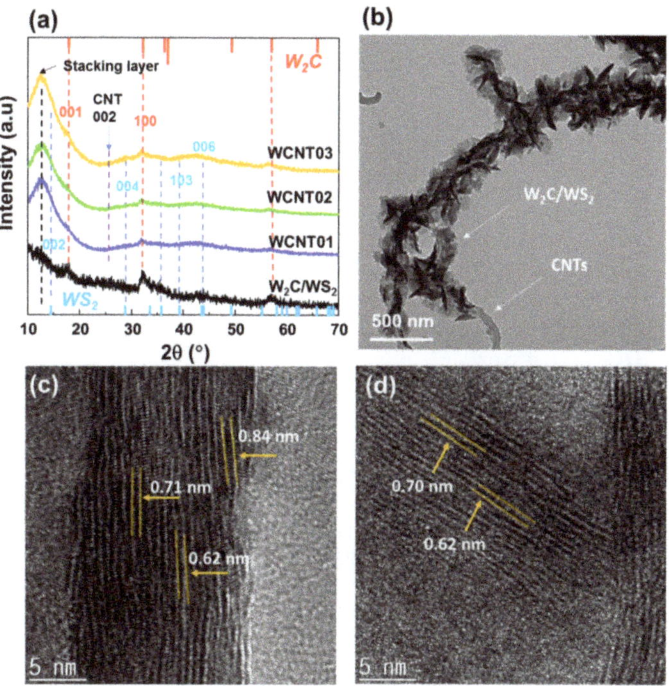

Figure 2. (**a**) XRD patterns of W_2C/WS_2 and WCNT01/02/03; (**b**) TEM images and (**c**,**d**) high-resolution TEM (HR-TEM) images of WCNT01.

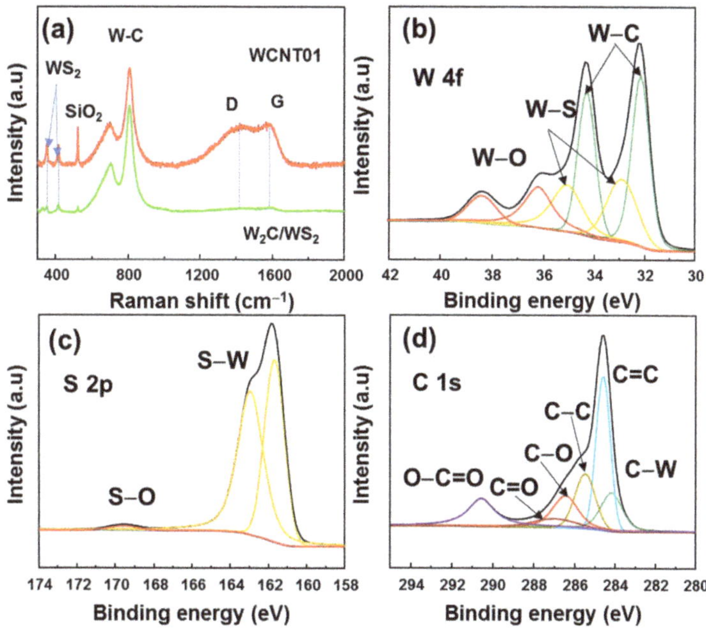

Figure 3. (**a**) Raman spectra of bare W_2C/WS_2 and WCNT01 samples. High-resolution XPS spectra of (**b**) W 4f, (**c**) S 2p, and (**d**) C 1s of WCNT01 sample.

To reveal the effectiveness of CNTs in W_2C/WS_2 materials, CV of bare and CNT-frame-networked samples was performed (Figure 4). The electrochemical process can be summarized in the following equations:

$$WS_2 + xLi^+ + xe^- \rightarrow Li_xWS_2 \tag{1}$$

$$W_2C + yLi^+ + ye^- \rightarrow Li_yW_2C \tag{2}$$

$$Li^+ + e^- + electrolyte \rightarrow SEI \tag{3}$$

$$Li_xWS_2 + (4-x)Li^+ + (4-x)e^- \rightarrow 2Li_2S + W \tag{4}$$

$$2Li_2S \rightarrow 4Li^+ + 4e^- + 2S \tag{5}$$

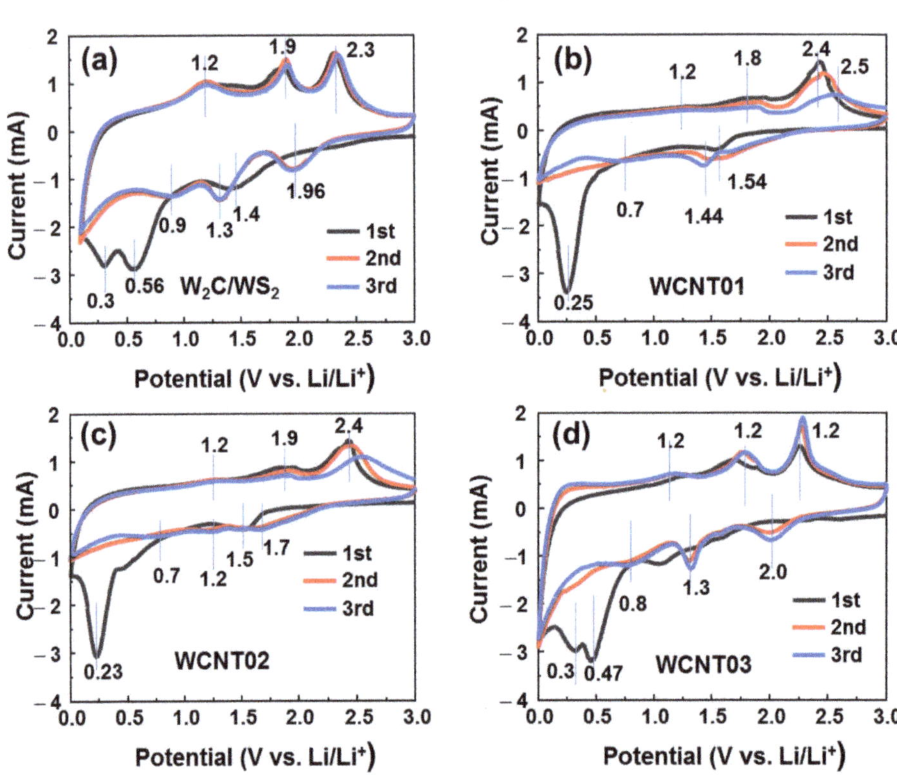

Figure 4. Cyclic voltammograms of the (**a**) W_2C/WS_2, (**b**) WCNT01, (**c**) WCNT02, and (**d**) WCNT03 anodes.

In the first cycle of the bare W_2C/WS_2 anode, the cathodic scan showed the insertion of lithium ions into the layered structure of W_2C and WS_2 at ~1.5 V (Equations (1) and (2)). The solid-electrolyte-interface (SEI) layer formed at ~0.6 V (Equation (3)) [48,49]. The peak at ~0.3 V is related to the deep insertion of Li^+ ions described by Equation (4) [50]. Meanwhile, from the second cycle onward, the cathodic scan demonstrated three major peaks at ~2.0, 1.3, and 0.9 V. As per the previous report, the dissolution of the S atom generated a gel-like SEI layer, which led to the shift of the cathodic peak to ~2.0 V [24]. The peaks at 1.3 and 0.9 V are related to the insertion of Li ions into WS_2 and W_2C [33]. In the anodic scan, Li_2S decomposes at ~ 2.3 V, as shown in Equation (5), and the oxidation of W to W^{4+} occurs at 1.9 V [23]. The anodic peak at 1.2 V may correspond to the desertion of Li^+ ions from Li_yW_2C as the reverse reaction of Equation (2). The presence of the

CNT network increased conductivity and prevented coverage of the gel-like SEI layer. Therefore, the WCNT01 and WCNT02 samples showed cathodic peaks of ~1.6 V and 0.25 V, which are lower than 2.0 V and 0.3 V, as demonstrated by the bare W_2C/WS_2 anode. The decomposition peak of Li_2S also shifted to ~2.5 V. Interestingly, the intensity of SEI formation peaks of the WCNT01/02 anodes at ~0.5–0.7 V dramatically decreased in comparison to that of the bare anode. At a CNT content above 15%, the WCNT03 anode showed a similar behavior to the bare W_2C/WS_2 anode, in which the cathodic peak at 2.0 V (Equation (1)) and the peak at ~0.5 V for SEI-layer formation (Equation (3)) appeared again with high intensity. This is attributed to the nonuniform W_2C/WS_2 on the CNTs, which originated from the aggregation of CNTs at high concentrations in the prepared mixture before the hydrothermal process. Thus, a low quantity of CNTs (below 10%) could enhance the electrochemical performance by preventing the coverage of the gel-like SEI layer.

The initial voltage profiles of the as-prepared anodes are shown in Figure 5. The WCNT01/02/03 samples showed a low open potential of ~1.1 V in comparison to the W_2C/WS_2 sample, which could be attributed to the contact of W_2C/WS_2 with CNTs. This behavior was also observed in MoS_2 and WS_2 grown with graphene or CNTs, as discussed in several reports [45,51–53]. The initial discharge capacity of the W_2C/WS_2 sample was in the range of 1000–1100 mAh g^{-1}. The voltage profiles of the bare W_2C/WS_2 alloy anode showed the discharge plateau at ~1.4 V and charge plateau at ~2.3 V. However, these plateaus gradually decreased after the first cycle. In contrast, the plateaus of the WCNT01/02/03 anodes were much more stable, demonstrating a similar flatform during the first three cycles. This indicates that the CNT network optimized the electrochemical reaction, resulting in a stable flatform of the voltage profiles.

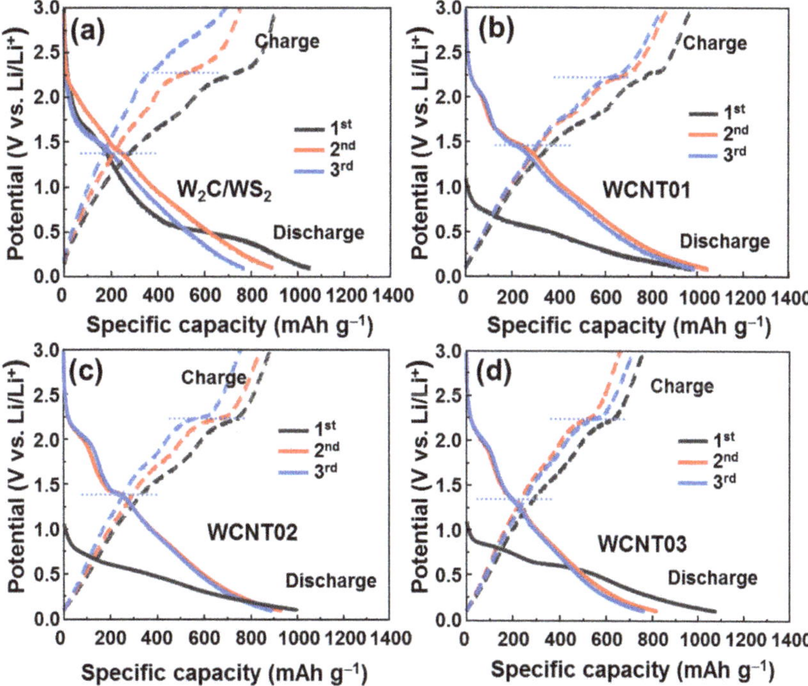

Figure 5. Initial voltage profiles of the (**a**) W_2C/WS_2 alloys, (**b**) WCNT01, (**c**) WCNT02, and (**d**) WCNT03 anodes.

The long-term cyclic stability of these four anodes was further evaluated, as illustrated in Figure 6a–d. The bare W_2C/WS_2 anode exhibited fast degradation for the first 20 cycles, and only approximately 40% of the initial capacity remained (~400 mAh g^{-1}); then, it gradually degraded to ~28% of the initial capacity (~300 mAh g^{-1}) after 100 cycles, as shown in Figure 6a. Meanwhile, the WCNT01 electrode underwent a fast degradation in only the first five cycles, and the capacity then slowly degraded to 650 mAh g^{-1} (67% of the initial capacity) after 100 cycles. Both the WCNT02 and WCNT03 anodes showed a fast reduction in capacity for the first ten cycles, followed by a slow reduction to 420 and 410 mAh g^{-1}, respectively. These results indicate that a high concentration of CNTs is not necessary and even reduces the overall capacity owing to the lower contribution of the lithium-ion host. Therefore, it was confirmed that only 5% CNTs in the alloys were sufficient to connect the network of W_2C/WS_2, prevent fast degradation, and stabilize the capacity. The rate performances of bare W_2C/WS_2 and WCNT01 anodes are shown in Figure S4. The bare W_2C/WS_2 anode shows a low performance at 1.0 A, which delivered a low capacity ~ 110 mAh g^{-1} and low recovered capacity ~79% when reducing the current rate from 1.0 to 0.1 A g^{-1}. on the other hand, the WCNT01 anode with CNT networks could remain at a capacity of ~250 mAh g^{-1} at 1.0 A g^{-1} and recovered ~92% capacity when reducing the current rate to 0.1 A. Moreover, the WCNT01 anode shows a trend to recover 100% capacity when increasing the number of cycles at 0.1 A g^{-1}. Furthermore, the capacity of the composite anode continuously degrades with the increase in cycles. This could be due to the main two reasons: First, the lithium-counter electrode could be degraded due to the imperfect recovery of *Li* and the SEI-layer formation on the surface [54–56]. Second, it could be due to the degradation of active materials. The CNT network can prevent the formation of a gel-like polymeric layer. However, the sulfur atom could be slowly dissolved into electrolyte during cycling, as discussed for bare WS_2 materials. Due to the conversion type of anode materials, the WS_2 will be converted to W-*Li* alloys and Li_2S when inserting the *Li* ions. Therefore, it is believed that the cycling process could slightly change the morphologies and material types. However, with the stability of CNT networks, it can prevent these changes with a slow rate.

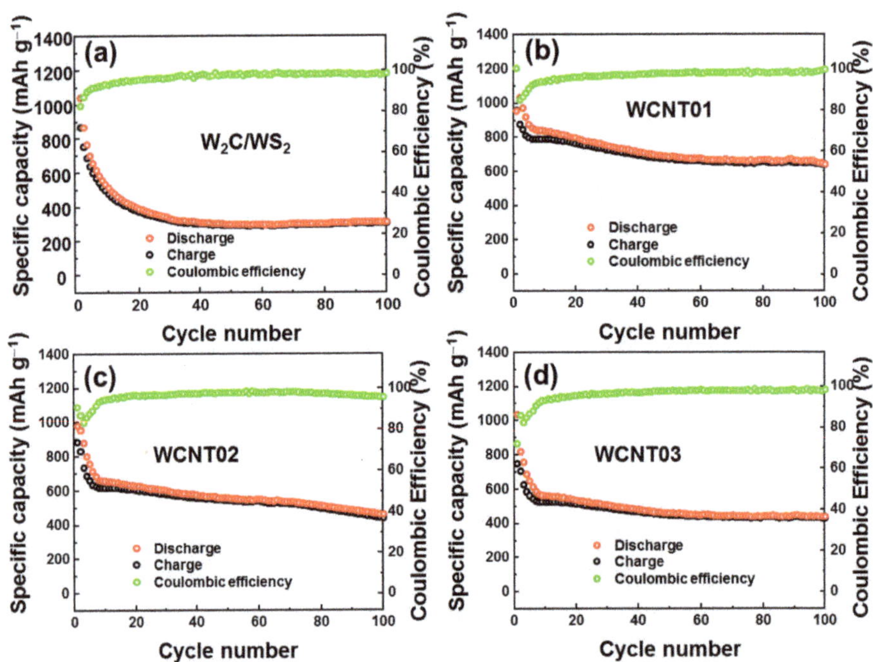

Figure 6. Cyclic performance of the (**a**) W_2C/WS_2 alloys, (**b**) WCNT01, (**c**) WCNT02, and (**d**) WCNT03 anodes under the current rate of 0.1 A g^{-1}.

Impedance measurements further confirmed the change in the electrical properties of the CNT networks in the W_2C/WS_2 materials, as shown in Figure 7a. The equivalent circuit was used with a modified Randle's model, which contains a series resistance R_s, charge-transfer resistance R_1, and SEI-layer resistance R_2 with a Warburg diffusion element and constant-phase elements Cpe1 and Cpe2. The extracted charge-transfer resistances of the bare W_2C/WS_2 and WCNT01/02/03 were 441.3, 125.9, 106.3, and 60.8 Ω, respectively. The enhancement in lithium diffusion can be estimated using the following equation [57–59]:

$$D = \frac{R^2T^2}{2A^2n^4F^4C^2\sigma^2}$$

(6)

where R and F are the gas constant and Faraday constant, respectively, T is the absolute temperature, A is the effective area of the working electrode, n is the electronic transport ratio during the redox process, C is the molar density of Li^+ in the electrode, and σ is the Warburg factor associated with the impedance of the cell, which can be obtained from the following equation [60]:

$$Z' = R_s + R_1 + R_2 + \sigma\omega^{-1/2}$$

(7)

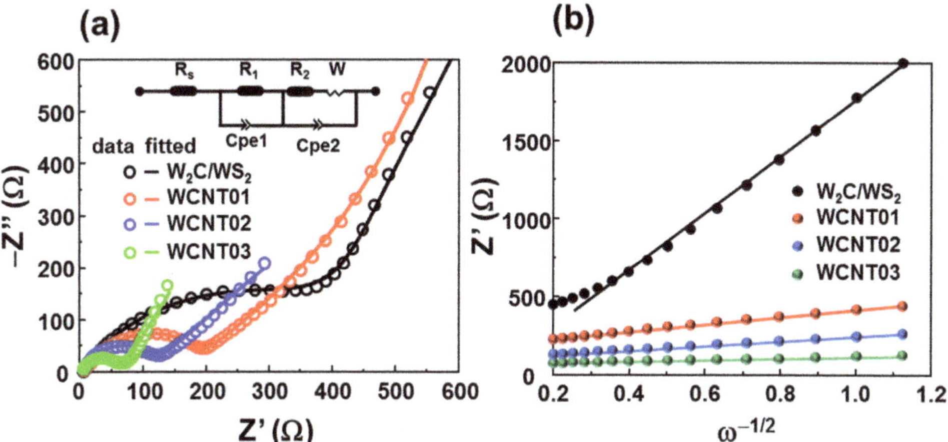

Figure 7. (a) Nyquist plots and (b) Z' vs. $\omega^{-1/2}$ plots of the bare W_2C/WS_2 alloys and WCNT01/02/03 anodes.

Figure 7b shows the fitting line of Z' vs. $\omega^{-1/2}$, in which the slopes of the lines of the bare W_2C/WS_2 and WCNT01/02/03 anodes are 1697.1, 389.9, 246.5, and 239.6, respectively. According to Equation (6), the diffusion coefficients of the lithium ions in WCNT01/02/03 are proportional to $\sigma^{-1/2}$ and were approximately 19, 49, and 50 times higher than those of the bare W_2C/WS_2 anode, respectively. Even though the WCNT02/03 samples showed a great improvement in the lithium diffusion coefficient, their stability in terms of structure and electrochemical properties was not as suitable as that of the WCNT01 electrode. Therefore, WCNT01 is recommended as the best combination of CNTs with W_2C/WS_2 alloys for high-performance anodes in lithium-ion cells.

In order to investigate the lithium-storage mechanism, the CV curves at scan rates from 10 to 100 mV s^{-1}, and the logarithm plot of peak currents with logarithm of scan rates are shown in Figure S5. The capacitive and diffusion contribution can be evaluated by the b factor in the following equation [61]:

$$i = kv^b$$

(8)

where i is the current density, v is the scan rate, k and b are adjustable factors. When $b = 1$, the storage mechanism is capacitive; when $b = 0.5$, the storage mechanism is diffusion. By the logarithm of Equation (8), the b factor can be obtained by plotting the fitting line of $\log(i)$ vs. $\log(v)$. The b values are 0.9 and 0.52 with cathodic peak and anodic peak, respectively. In the cathodic process, the cell behavior can be considered as a capacitor, while in the anodic process, the diffusion-controlled process is major contributor to the current. Therefore, it is noted that the high reversible capacity of the WCNT01 anode was based on capacitive behavior.

The comparison of the research on WS_2-based materials for the LIB anode is shown in Table 1. It clearly illustrates that the bare 2D or oxygen-modified WS_2 have low electrochemical performances, which reveal their storage capability only for 20 cycles. The optimized compositions of WS_2 with other stable materials are required to enhance their cyclability and rate performance. Most compositions of WS_2 with graphene, carbon, or CNTs can form with some modifications such as N-doping or three-dimensional morphologies, where stable capacity can go up to ~960 mAh g^{-1}. In our study, the ternary compound with W_2C, WS_2, and CNTs for the LIB anode is not the best material but shows a comparative result. Moreover, the simple preparation method could be an advantage for the application in lithium storage. Therefore, it is noted that control of the flowers' sizes and/or the compositions of functionalized CNTs and W_2C/WS_2 could be effective ways to further enhance their electrochemical performances for LIBs.

Table 1. Comparison of electrochemical performance of WS_2-based composite materials for lithium-ion batteries.

Anode Materials	Current Density (mA g^{-1})	Initial Discharge Capacity (mAh g^{-1})	Cycle Number	Specific Capacity (mAh g^{-1})	References
WS_2 nanoflakes	47.5	~1700	20	~700	[22]
Oxygen-functionalized WS_2	50	~920	20	~220	[62]
Hierarchical WS_2 on 3D graphene	100	~800	100	~740	[63]
Mesoporous WS_2	100	~1300	100	~800	[23]
N-graphene/WS_2	100	~1300	100	~800	[64]
N-carbon sphere/WS_2	100	~735	100	~630	[65]
N-graphene/WS_2	100	~950	100	~960	[66]
N-carbon/WS_2	100	~1000	100	~640	[67]
$W_2C/WS_2/CNTs$	100	~1000	100	~650	This work

4. Conclusions

In this study, W_2C/WS_2 was synthesized in situ using CNT networks via a hydrothermal method. The presence of CNTs led to a decrease in the number of multi-edge nanoflowers with a size range of 200–400 nm. The CNT networks enhanced the conductivity of anode materials, which in turn reduced the cathodic peak intensity from 2.0 to ~1.6 V. The impedance spectra also suggest that the lithium-ion diffusion in the WCNT01/02/03 samples was 19, 49, and 50 times higher than that of the W_2C/WS_2 sample, respectively. WCNT01 anodes with 5% CNTs showed the best performance, with a capacity of 650 mAh g^{-1} (67% of the initial value) remaining after 100 cycles. These results suggest that the utilization of CNT networks and a simple hydrothermal method can be appropriate for improving the overall stability of metal-sulfide anode materials.

Supplementary Materials: The following supporting information can be downloaded at: https://www.mdpi.com/article/10.3390/nano12061003/s1, Figure S1: SEM image of WCNT03 sample with and without CNT area; Figure S2: (a) TEM image, (b) Scanning TEM image and elemental mapped images (c) W-L, (d) S-K and (e) C-K of W_2C/WS_2 nanolowers on CNTs; Figure S3: TGA analysis of the synthesized W_2C/WS_2 and WCNT01 materials; Figure S4: Rate performance of (a) bare W_2C/WS_2 and (b)WCNT01 anodes at different current rate from 0.1 to 1.0 A g^{-1}; Figure S5: (a) CV curves of WCNT01 anode at different scan rate from 20–100 mV s^{-1} and (b) plots of log(current) with log(scan-rate).

Author Contributions: T.P.N.: Conceptualization, methodology, validation, visualization, writing, review, and editing; I.T.K.: project administration, funding acquisition, review, and editing. All authors have read and agreed to the published version of the manuscript.

Funding: This research was supported by the Basic Science Research Capacity Enhancement Project through a Korea Basic Science Institute (National Research Facilities and Equipment Center) grant funded by the Ministry of Education (2019R1A6C1010016). This research was supported by Korea Basic Institute (National Research facilities and Equipment Center) grant funded by the Ministry of Education (2020R1A6C103A050).

Institutional Review Board Statement: Not applicable.

Informed Consent Statement: Not applicable.

Data Availability Statement: The data presented in this study are available on request from the corresponding author.

Conflicts of Interest: The authors declare no conflict of interest.

References

1. Kuc, A.; Heine, T. The electronic structure calculations of two-dimensional transition-metal dichalcogenides in the presence of external electric and magnetic fields. *Chem. Soc. Rev.* **2015**, *44*, 2603–2614. [CrossRef] [PubMed]
2. Haque, F.; Daeneke, T.; Kalantar-zadeh, K.; Ou, J.Z. Two-Dimensional Transition Metal Oxide and Chalcogenide-Based Photocatalysts. *Nano-Micro Lett.* **2017**, *10*, 23. [CrossRef] [PubMed]
3. Wang, Q.H.; Kalantar-Zadeh, K.; Kis, A.; Coleman, J.N.; Strano, M.S. Electronics and optoelectronics of two-dimensional transition metal dichalcogenides. *Nat. Nanotechnol.* **2012**, *7*, 699–712. [CrossRef] [PubMed]
4. Chhowalla, M.; Shin, H.S.; Eda, G.; Li, L.J.; Loh, K.P.; Zhang, H. The chemistry of two-dimensional layered transition metal dichalcogenide nanosheets. *Nat. Chem.* **2013**, *5*, 263–275. [CrossRef] [PubMed]
5. Quesne, M.G.; Roldan, A.; de Leeuw, N.H.; Catlow, C.R.A. Bulk and surface properties of metal carbides: Implications for catalysis. *Phys. Chem. Chem. Phys.* **2018**, *20*, 6905–6916. [CrossRef]
6. Yu, Z.; Xinhui, X.; Fan, S.; Jiye, Z.; Jiangping, T.; Jin, F.H. Transition Metal Carbides and Nitrides in Energy Storage and Conversion. *Adv. Sci.* **2016**, *3*, 1500286.
7. Gao, G.; O'Mullane, A.P.; Du, A. 2D MXenes: A New Family of Promising Catalysts for the Hydrogen Evolution Reaction. *ACS Catal.* **2017**, *7*, 494–500. [CrossRef]
8. Pan, H. Ultra-high electrochemical catalytic activity of MXenes. *Sci. Rep.* **2016**, *6*, 32531. [CrossRef]
9. Nguyen, T.P.; Tuan Nguyen, D.M.; Tran, D.L.; Le, H.K.; Vo, D.-V.N.; Lam, S.S.; Varma, R.S.; Shokouhimehr, M.; Nguyen, C.C.; Le, Q.V. MXenes: Applications in electrocatalytic, photocatalytic hydrogen evolution reaction and CO$_2$ reduction. *Mol. Catal.* **2020**, *486*, 110850. [CrossRef]
10. Chen, Z.; Gong, W.; Cong, S.; Wang, Z.; Song, G.; Pan, T.; Tang, X.; Chen, J.; Lu, W.; Zhao, Z. Eutectoid-structured WC/W$_2$C heterostructures: A new platform for long-term alkaline hydrogen evolution reaction at low overpotentials. *Nano Energy* **2020**, *68*, 104335. [CrossRef]
11. Hussain, S.; Shaikh, S.F.; Vikraman, D.; Mane, R.S.; Joo, O.-S.; Naushad, M.; Jung, J. Sputtering and sulfurization-combined synthesis of a transparent WS$_2$ counter electrode and its application to dye-sensitized solar cells. *RSC Adv.* **2015**, *5*, 103567–103572. [CrossRef]
12. Gowtham, B.; Balasubramani, V.; Ramanathan, S.; Ubaidullah, M.; Shaikh, S.F.; Sreedevi, G. Dielectric relaxation, electrical conductivity measurements, electric modulus and impedance analysis of WO$_3$ nanostructures. *J. Alloys Compd.* **2021**, *888*, 161490. [CrossRef]
13. Al-Tahan, M.A.; Dong, Y.; Shrshr, A.E.; Liu, X.; Zhang, R.; Guan, H.; Kang, X.; Wei, R.; Zhang, J. Enormous-sulfur-content cathode and excellent electrochemical performance of Li-S battery accouched by surface engineering of Ni-doped WS^2@rGO nanohybrid as a modified separator. *J. Colloid Interface Sci.* **2022**, *609*, 235–248. [CrossRef] [PubMed]
14. Bai, Y.; Liu, C.; Chen, T.; Li, W.; Zheng, S.; Pi, Y.; Luo, Y.; Pang, H. MXene-Copper/Cobalt Hybrids via Lewis Acidic Molten Salts Etching for High Performance Symmetric Supercapacitors. *Angew. Chem. Int. Ed.* **2021**, *60*, 25318–25322. [CrossRef]

15. Li, X.; Yang, X.; Xue, H.; Pang, H.; Xu, Q. Metal–organic frameworks as a platform for clean energy applications. *EnergyChem* **2020**, *2*, 100027. [CrossRef]

16. Chen, J.; Chen, Y.; Feng, L.-W.; Gu, C.; Li, G.; Su, N.; Wang, G.; Swick, S.M.; Huang, W.; Guo, X.; et al. Hole (donor) and electron (acceptor) transporting organic semiconductors for bulk-heterojunction solar cells. *EnergyChem* **2020**, *2*, 100042. [CrossRef]

17. Kim, S.; Park, J.; Hwang, J.; Lee, J. Effects of functional supports on efficiency and stability of atomically dispersed noble-metal electrocatalysts. *EnergyChem* **2021**, *3*, 100054. [CrossRef]

18. Zheng, S.; Sun, Y.; Xue, H.; Braunstein, P.; Huang, W.; Pang, H. Dual-ligand and hard-soft-acid-base strategies to optimize metal-organic framework nanocrystals for stable electrochemical cycling performance. *Natl. Sci. Rev.* **2021**. [CrossRef]

19. Kim, H.; Choi, W.; Yoon, J.; Um, J.H.; Lee, W.; Kim, J.; Cabana, J.; Yoon, W.S. Exploring Anomalous Charge Storage in Anode Materials for Next-Generation Li Rechargeable Batteries. *Chem. Rev.* **2020**, *120*, 6934–6976. [CrossRef]

20. Zheng, Y.; Li, Y.; Yao, J.; Huang, Y.; Xiao, S. Facile synthesis of porous tubular NiO with considerable pseudocapacitance as high capacity and long life anode for lithium-ion batteries. *Ceram. Int.* **2018**, *44*, 2568–2577. [CrossRef]

21. Nguyen, T.P.; Giang, T.T.; Kim, I.T. Restructuring NiO to LiNiO$_2$: Ultrastable and reversible anodes for lithium-ion batteries. *Chem. Eng. J.* **2022**, *437*, 135292. [CrossRef]

22. Feng, C.; Huang, L.; Guo, Z.; Liu, H. Synthesis of tungsten disulfide (WS$_2$) nanoflakes for lithium-ion battery application. *Electrochem. Commun.* **2007**, *9*, 119–122. [CrossRef]

23. Liu, H.; Su, D.; Wang, G.; Qiao, S.Z. An ordered mesoporous WS$_2$ anode material with superior electrochemical performance for lithium-ion batteries. *J. Mater. Chem.* **2012**, *22*, 17437–17440. [CrossRef]

24. George, C.; Morris, A.J.; Modarres, M.H.; De Volder, M. Structural Evolution of Electrochemically Lithiated MoS$_2$ Nanosheets and the Role of Carbon Additive in Li-Ion Batteries. *Chem. Mater.* **2016**, *28*, 7304–7310. [CrossRef]

25. Jia, L.; Liu, B.; Zhao, Y.; Chen, W.; Mou, D.; Fu, J.; Wang, Y.; Xin, W.; Zhao, L. Structure design of MoS$_2$@Mo$_2$C on nitrogen-doped carbon for enhanced alkaline hydrogen evolution reaction. *J. Mater. Sci.* **2020**, *55*, 16197–16210. [CrossRef]

26. Liu, J.; Wang, P.; Gao, L.; Wang, X.; Yu, H. In situ sulfuration synthesis of heterostructure MoS$_2$–Mo$_2$C@C for boosting the photocatalytic H$_2$ production activity of TiO$_2$. *J. Mater. Chem. C* **2022**, *10*, 3121–3128. [CrossRef]

27. Ihsan, M.; Wang, H.; Majid, S.R.; Yang, J.; Kennedy, S.J.; Guo, Z.; Liu, H.K. MoO$_2$/Mo$_2$C/C spheres as anode materials for lithium-ion batteries. *Carbon* **2016**, *96*, 1200–1207. [CrossRef]

28. Li, Y.; Wu, X.; Zhang, H.; Zhang, J. Interface Designing over WS$_2$/W$_2$C for Enhanced Hydrogen Evolution Catalysis. *ACS Appl. Energy Mater.* **2018**, *1*, 3377–3384. [CrossRef]

29. Zhao, Z.; Qin, F.; Kasiraju, S.; Xie, L.; Alam, M.K.; Chen, S.; Wang, D.; Ren, Z.; Wang, Z.; Grabow, L.C.; et al. Vertically Aligned MoS$_2$/Mo$_2$C hybrid Nanosheets Grown on Carbon Paper for Efficient Electrocatalytic Hydrogen Evolution. *ACS Catal.* **2017**, *7*, 7312–7318. [CrossRef]

30. Cheng, Y.; Pang, K.; Wu, X.; Zhang, Z.; Xu, X.; Ren, J.; Huang, W.; Song, R. In Situ Hydrothermal Synthesis MoS$_2$/Guar Gum Carbon Nanoflowers as Advanced Electrocatalysts for Electrocatalytic Hydrogen Evolution. *ACS Sustain. Chem. Eng.* **2018**, *6*, 8688–8696. [CrossRef]

31. Faizan, M.; Hussain, S.; Vikraman, D.; Ali, B.; Kim, H.-S.; Jung, J.; Nam, K.-W. MoS$_2$@Mo$_2$C hybrid nanostructures formation as an efficient anode material for lithium-ion batteries. *J. Mater. Res. Technol.* **2021**, *14*, 2382–2393. [CrossRef]

32. Nguyen, T.P.; Choi, K.S.; Kim, S.Y.; Lee, T.H.; Jang, H.W.; Van Le, Q.; Kim, I.T. Strategy for controlling the morphology and work function of W$_2$C/WS$_2$ nanoflowers. *J. Alloys Compd.* **2020**, *829*, 154582. [CrossRef]

33. Nguyen, T.P.; Kim, I.T. W$_2$C/WS$_2$ Alloy Nanoflowers as Anode Materials for Lithium-Ion Storage. *Nanomaterials* **2020**, *10*, 1336. [CrossRef]

34. Sun, D.; Wang, M.; Li, Z.; Fan, G.; Fan, L.-Z.; Zhou, A. Two-dimensional Ti$_3$C$_2$ as anode material for Li-ion batteries. *Electrochem. Commun.* **2014**, *47*, 80–83. [CrossRef]

35. Wang, Y.; Li, Y.; Qiu, Z.; Wu, X.; Zhou, P.; Zhou, T.; Zhao, J.; Miao, Z.; Zhou, J.; Zhuo, S. Fe$_3$O$_4$@Ti$_3$C$_2$ MXene hybrids with ultrahigh volumetric capacity as an anode material for lithium-ion batteries. *J. Mater. Chem. A* **2018**, *6*, 11189–11197. [CrossRef]

36. Kong, F.; He, X.; Liu, Q.; Qi, X.; Sun, D.; Zheng, Y.; Wang, R.; Bai, Y. Enhanced reversible Li-ion storage in Si@Ti$_3$C$_2$ MXene nanocomposite. *Electrochem. Commun.* **2018**, *97*, 16–21. [CrossRef]

37. Zhang, Y. First principles prediction of two-dimensional tungsten carbide (W$_2$C) monolayer and its Li storage capability. *Comput. Condens. Matter* **2017**, *10*, 35–38. [CrossRef]

38. Bai, Y.-L.; Liu, Y.-S.; Ma, C.; Wang, K.-X.; Chen, J.-S. Neuron-Inspired Design of High-Performance Electrode Materials for Sodium-Ion Batteries. *ACS Nano* **2018**, *12*, 11503–11510. [CrossRef]

39. Landi, B.J.; Ganter, M.J.; Cress, C.D.; Di Leo, R.A.; Raffaelle, R.P. Carbon nanotubes for lithium-ion batteries. *Energy Environ. Sci.* **2009**, *2*, 638–654. [CrossRef]

40. Lu, C.; Liu, W.-W.; Li, H.; Tay, B.K. A binder-free CNT network–MoS$_2$ composite as a high performance anode material in lithium ion batteries. *Chem. Commun.* **2014**, *50*, 3338–3340. [CrossRef]

41. Chen, Y.; Hu, X.; Evanko, B.; Sun, X.; Li, X.; Hou, T.; Cai, S.; Zheng, C.; Hu, W.; Stucky, G.D. High-rate FeS$_2$/CNT neural network nanostructure composite anodes for stable, high-capacity sodium-ion batteries. *Nano Energy* **2018**, *46*, 117–127. [CrossRef]

42. Yan, G.; Wu, C.; Tan, H.; Feng, X.; Yan, L.; Zang, H.; Li, Y. N-Carbon coated P-W$_2$C composite as efficient electrocatalyst for hydrogen evolution reactions over the whole pH range. *J. Mater. Chem. A* **2017**, *5*, 765–772. [CrossRef]

43. Nguyen, T.P.; Kim, S.Y.; Lee, T.H.; Jang, H.W.; Le, Q.V.; Kim, I.T. Facile synthesis of $W_2C@WS_2$ alloy nanoflowers and their hydrogen generation performance. *Appl. Surf. Sci.* **2020**, *504*, 144389. [CrossRef]

44. Nguyen, T.P.; Sohn, W.; Oh, J.H.; Jang, H.W.; Kim, S.Y. Size-Dependent Properties of Two-Dimensional MoS_2 and WS_2. *J. Phys. Chem. C* **2016**, *120*, 10078–10085. [CrossRef]

45. Ren, J.; Ren, R.-P.; Lv, Y.-K. WS_2-decorated graphene foam@CNTs hybrid anode for enhanced lithium-ion storage. *J. Alloys Compd.* **2019**, *784*, 697–703. [CrossRef]

46. Vaziri, H.S.; Shokuhfar, A.; Afghahi, S.S.S. Synthesis of WS_2/CNT hybrid nanoparticles for fabrication of hybrid aluminum matrix nanocomposite. *Mater. Res. Express* **2020**, *7*, 025034. [CrossRef]

47. Zhang, L.-N.; Ma, Y.-Y.; Lang, Z.-L.; Wang, Y.-H.; Khan, S.U.; Yan, G.; Tan, H.-Q.; Zang, H.-Y.; Li, Y.-G. Ultrafine cable-like WC/W_2C heterojunction nanowires covered by graphitic carbon towards highly efficient electrocatalytic hydrogen evolution. *J. Mater. Chem. A* **2018**, *6*, 15395–15403. [CrossRef]

48. Chen, Y.; Song, B.; Tang, X.; Lu, L.; Xue, J. Ultrasmall Fe_3O_4 Nanoparticle/MoS_2 Nanosheet Composites with Superior Performances for Lithium-Ion Batteries. *Small* **2014**, *10*, 1536–1543. [CrossRef]

49. Jin, Y.; Li, S.; Kushima, A.; Zheng, X.; Sun, Y.; Xie, J.; Sun, J.; Xue, W.; Zhou, G.; Wu, J.; et al. Self-healing SEI enables full-cell cycling of a silicon-majority anode with a coulombic efficiency exceeding 99.9%. *Energy Environ. Sci.* **2017**, *10*, 580–592. [CrossRef]

50. Stephenson, T.; Li, Z.; Olsen, B.; Mitlin, D. Lithium-ion battery applications of molybdenum disulfide (MoS_2) nanocomposites. *Energy Environ. Sci.* **2014**, *7*, 209–231. [CrossRef]

51. Xiao, J.; Wang, X.J.; Yang, X.Q.; Xun, S.D.; Liu, G.; Koech, P.K.; Liu, J.; Lemmon, J.P. Electrochemically Induced High Capacity Displacement Reaction of PEO/MoS_2/Graphene Nanocomposites with Lithium. *Adv. Funct. Mater.* **2011**, *21*, 2840–2846. [CrossRef]

52. Ren, J.; Ren, R.-P.; Lv, Y.-K. A flexible 3D graphene@CNT@MoS_2 hybrid foam anode for high-performance lithium-ion battery. *Chem. Eng. J.* **2018**, *353*, 419–424. [CrossRef]

53. Wang, Y.; Chen, B.; Seo, D.H.; Han, Z.J.; Wong, J.I.; Ostrikov, K.; Zhang, H.; Yang, H.Y. MoS_2-coated vertical graphene nanosheet for high-performance rechargeable lithium-ion batteries and hydrogen production. *NPG Asia Mater.* **2016**, *8*, e268. [CrossRef]

54. Peled, E.; Menkin, S. Review—SEI: Past, Present and Future. *J. Electrochem. Soc.* **2017**, *164*, A1703–A1719. [CrossRef]

55. Lin, D.; Liu, Y.; Cui, Y. Reviving the lithium metal anode for high-energy batteries. *Nat. Nanotechnol.* **2017**, *12*, 194–206. [CrossRef] [PubMed]

56. Liu, J.; Bao, Z.; Cui, Y.; Dufek, E.J.; Goodenough, J.B.; Khalifah, P.; Li, Q.; Liaw, B.Y.; Liu, P.; Manthiram, A.; et al. Pathways for practical high-energy long-cycling lithium metal batteries. *Nat. Energy* **2019**, *4*, 180–186. [CrossRef]

57. Bisquert, J.; Garcia-Belmonte, G.; Bueno, P.; Longo, E.; Bulhões, L.O.S. Impedance of constant phase element (CPE)—Blocked diffusion in film electrodes. *J. Electroanal. Chem.* **1998**, *452*, 229–234. [CrossRef]

58. Piao, T.; Park, S.M.; Doh, C.H.; Moon, S.I. Intercalation of Lithium Ions into Graphite Electrodes Studied by AC Impedance Measurements. *J. Electrochem. Soc.* **1999**, *146*, 2794–2798. [CrossRef]

59. Ye, B.; Xu, L.; Wu, W.; Ye, Y.; Yang, Z.; Ai, J.; Qiu, Y.; Gong, Z.; Zhou, Y.; Huang, Q.; et al. Encapsulation of 2D MoS_2 nanosheets into 1D carbon nanobelts as anodes with enhanced lithium/sodium storage properties. *J. Mater. Chem. C* **2022**, *10*, 3329–3342. [CrossRef]

60. Rui, X.H.; Ding, N.; Liu, J.; Li, C.; Chen, C.H. Analysis of the chemical diffusion coefficient of lithium ions in $Li_3V_2(PO_4)_3$ cathode material. *Electrochim. Acta* **2010**, *55*, 2384–2390. [CrossRef]

61. Abdelaal, M.M.; Hung, T.-C.; Mohamed, S.G.; Yang, C.-C.; Huang, H.-P.; Hung, T.-F. A Comparative Study of the Influence of Nitrogen Content and Structural Characteristics of NiS/Nitrogen-Doped Carbon Nanocomposites on Capacitive Performances in Alkaline Medium. *Nanomaterials* **2021**, *11*, 1867. [CrossRef] [PubMed]

62. Bhandavat, R.; David, L.; Singh, G. Synthesis of Surface-Functionalized WS_2 Nanosheets and Performance as Li-Ion Battery Anodes. *J. Phys. Chem. Lett.* **2012**, *3*, 1523–1530. [CrossRef] [PubMed]

63. Huang, G.; Liu, H.; Wang, S.; Yang, X.; Liu, B.; Chen, H.; Xu, M. Hierarchical architecture of WS_2 nanosheets on graphene frameworks with enhanced electrochemical properties for lithium storage and hydrogen evolution. *J. Mater. Chem. A* **2015**, *3*, 24128–24138. [CrossRef]

64. Chen, D.; Ji, G.; Ding, B.; Ma, Y.; Qu, B.; Chen, W.; Lee, J.Y. In situ nitrogenated graphene—Few-layer WS_2 composites for fast and reversible Li^+ storage. *Nanoscale* **2013**, *5*, 7890–7896. [CrossRef] [PubMed]

65. Zeng, X.; Ding, Z.; Ma, C.; Wu, L.; Liu, J.; Chen, L.; Ivey, D.G.; Wei, W. Hierarchical Nanocomposite of Hollow N-Doped Carbon Spheres Decorated with Ultrathin WS_2 Nanosheets for High-Performance Lithium-Ion Battery Anode. *ACS Appl. Mater. Interfaces* **2016**, *8*, 18841–18848. [CrossRef] [PubMed]

66. Debela, T.T.; Lim, Y.R.; Seo, H.W.; Kwon, I.S.; Kwak, I.H.; Park, J.; Cho, W.I.; Kang, H.S. Two-Dimensional WS_2@Nitrogen-Doped Graphite for High-Performance Lithium-Ion Batteries: Experiments and Molecular Dynamics Simulations. *ACS Appl. Mater. Interfaces* **2018**, *10*, 37928–37936. [CrossRef]

67. Zhao, Z.; Wang, F.; Yuan, H.; Yang, Z.; Qin, Y.; Zheng, X.; Yang, Y. N-Doped Carbon—WS_2 Nanosheet Composites for Lithium-Ion Storage. *ACS Appl. Nano Mater.* **2021**, *4*, 7781–7787. [CrossRef]

 *nanomaterials*

Article

The Effects of the Binder and Buffering Matrix on InSb-Based Anodes for High-Performance Rechargeable Li-Ion Batteries

Vo Pham Hoang Huy, Il Tae Kim * and Jaehyun Hur *

Department of Chemical and Biological Engineering, Gachon University, Seongnam 13120, Gyeonggi, Korea;
vophamhoanghuy@yahoo.com.vn
* Correspondence: itkim@gachon.ac.kr (I.T.K.); jhhur@gachon.ac.kr (J.H.);
  Tel.: +82-31-750-8835 (I.T.K.); +82-31-750-5593 (J.H.)

Abstract: C-decorated intermetallic InSb (InSb–C) was developed as a novel high-performance anode material for lithium-ion batteries (LIBs). InSb nanoparticles synthesized via a mechanochemical reaction were characterized using X-ray diffraction (XRD), high-resolution transmission electron microscopy (HRTEM), scanning electron microscopy (SEM), X-ray photoelectron spectroscopy (XPS), and energy-dispersive X-ray spectroscopy (EDX). The effects of the binder and buffering matrix on the active InSb were investigated. Poly(acrylic acid) (PAA) was found to significantly improve the cycling stability owing to its strong hydrogen bonding. The addition of amorphous C to InSb further enhanced mechanical stability and electronic conductivity. As a result, InSb–C demonstrated good electrochemical Li-ion storage performance: a high reversible specific capacity (878 mAh·g^{-1} at 100 mA·g^{-1} after 140 cycles) and good rate capability (capacity retention of 98% at 10 A·g^{-1} as compared to 0.1 A·g^{-1}). The effects of PAA and C were comprehensively studied using cyclic voltammetry, differential capacity plots, ex-situ SEM, and electrochemical impedance spectroscopy (EIS). In addition, the electrochemical reaction mechanism of InSb was revealed using ex-situ XRD. InSb–C exhibited a better performance than many recently reported Sb-based electrodes; thus, it can be considered as a potential anode material in LIBs.

Keywords: InSb; InSb–C; PAA binder; anodes; Li-ion batteries

Citation: Hoang Huy, V.P.; Kim, I.T.; Hur, J. The Effects of the Binder and Buffering Matrix on InSb-Based Anodes for High-Performance Rechargeable Li-Ion Batteries. *Nanomaterials* **2021**, *11*, 3420. https://doi.org/10.3390/nano11123420

Academic Editor: Carlos Miguel Costa

Received: 25 November 2021
Accepted: 15 December 2021
Published: 17 December 2021

Publisher's Note: MDPI stays neutral with regard to jurisdictional claims in published maps and institutional affiliations.

1. Introduction

Lithium-ion batteries (LIBs) have been widely used in various portable devices and energy storage systems owing to their high energy density, high cell voltage, low self-discharge, and low memory effect [1–4]. Despite these beneficial features, current graphite anodes cannot satisfy the rapidly growing demands for their use in various applications, such as mobile devices, electrical vehicles, and large-scale grid storage systems. Therefore, the development of new anode materials with a high specific capacity, good rate capability, and long service life that can replace low theoretical capacity (372 mAh·g^{-1}) commercial graphitic anodes is required [4–17]. Li alloys with elements, such as Si, P, Sn, and Sb, are considered to be promising anode materials because of their higher theoretical capacities (Si: 4200, P: 2595, Sn: 993, and Sb: 660 mAh·g^{-1}). However, it is not straightforward to control the large volume change in these materials due to expansion/contraction during lithiation/delithiation, which leads to a deteriorated cell performance [18–23].

Recently, Sb-based materials have gained significant attention as promising anodes in LIBs owing to their low cost, high conductivity, high density, and high theoretical capacity [24–27]. Sb has higher conductivity and stability than P (the same element family) and Si (the material with the highest theoretical capacity), making it a suitable material for the development of high-performance anodes for LIBs. Because of these attractive characteristics, Sb has been intensively investigated for use in LIBs. However, satisfactory performance cannot be achieved using Sb alone because of its high-volume expansion

(135%) during the alloying reaction ($3Li^+ + Sb \rightarrow Li_3Sb$) [28]. Many strategies have been proposed to resolve this problem.

The formation of a nanoscale Sb-based intermetallic alloy is an effective approach that can improve the cycling stability of Sb-based electrodes. Nanoscale active materials reduce the Li-ion diffusion pathway and alleviate the stress and strain during the electrochemical reaction. In addition, the stepwise electrochemical reaction in bimetallic Sb-based alloy nanoparticles can mitigate a large volume change relative to a pure Sb electrode. Along this line, He et al. demonstrated monodisperse colloidal SnSb nanocrystals (approximately 20 nm) with a discharge capacity of 700 mAh·g^{-1} at 0.5 C after 100 cycles [29]. Yi et al. synthesized morphology-controllable Sn–Sb composites with micro- and nano-sized hollow, dendritic, or mixed-type structures; these designed composites also exhibited good cycling stability and rate performance in LIBs and sodium-ion batteries (SIBs) [29].

Another effective approach that can enhance the performance of Sb-based electrodes is to introduce various nanoscale conductive carbon materials to create nanostructured Sb/C composites (e.g., 1D carbon nanotubes, nanofibers, nanorods, 2D graphene, 3D graphite, and porous carbon) [30–33]. In this composite, carbon prevents the agglomeration of nanoparticles, increases the electrical conductivity, and reduces the volume change of the active Sb [34–36]. Therefore, the cycling stability is notably improved by adding carbon.

The binder is the crucial adhesive between the active material and conductive carbon on the current collector. The adhesion between the active component and the binder is very important during the electrochemical reaction because the stress on the active material caused by volume expansion can weaken the binding force. In a pioneering study on binder materials, Kim et al. studied the effect of a new binder material (a blend of poly(acrylic acid) (PAA) and poly(amide imide) (PAI)) on electrode adhesion and recovery characteristics. They demonstrated that the composite polymer binder exhibited superior properties compared to the individual polymers [37]. Similarly, Choi et al. developed a new polyrotaxane-based binder for active micro-silicon particle batteries in which they achieved a remarkably stable capacity of over 3000 mAh·g^{-1} after 150 cycles [38]. Wu et al. have shown that conductive binders based on polyfluorene (PF) exhibit superior performance owing to their electronic conductivity and mechanical strength [39]. Among the various binders studied, PAA has shown exceptionally good performance for the Si electrode because of (i) its abundant carboxylic acid functional groups (−COOH) that enable strong bonding to the native hydroxyl species on the Si particle surface [40,41]; (ii) good mechanical strength associated with low swelling in a liquid electrolyte [42]; and (iii) the formation of an artificial solid electrolyte interface (SEI) on the Si surface that stabilizes the electrode–electrolyte interface [43]. Accordingly, PAA is expected to be a promising binder for various anode materials with a large volume change in LIBs.

In this study, we demonstrate C-decorated InSb (InSb–C) as a novel Sb-based bimetallic high-performance anode for LIBs. InSb has been widely studied for use in transistors, magnetic sensors, and infrared photodetectors because of its semiconducting properties, which include a narrow band gap (0.17 eV), high electron mobility, and a high density of conduction states [44–46]. Although some In-based nanomaterials have been reported as good anode materials owing to the high theoretical capacity of In (1012 mAh·g^{-1}) [27,28], intermetallic InSb has rarely been investigated as an anode material for LIBs. To achieve a high-performance InSb electrode, we investigate the effects of the binder and buffering matrix on the performance of InSb. This study demonstrates that PAA is an effective binder that impedes volume expansion and limits the structural degradation of the electrode owing to its strong hydrogen bonding with the active InSb. The addition of amorphous C reduces the stresses in InSb during lithiation/delithiation and increases the electrical conductivity. Therefore, with an appropriate binder and matrix, InSb–C exhibits high performance in terms of specific capacity, cyclic stability, and rate performance. Various characterization techniques are used to elucidate the mechanism behind the improvement, including X-ray diffraction (XRD), scanning electron microscopy (SEM), high-resolution transmission electron microscopy (HRTEM), energy-dispersive X-ray spectroscopy (EDX),

Fourier-transform infrared spectroscopy (FTIR), X-ray photoelectron spectroscopy (XPS), and electrochemical impedance spectroscopy (EIS). Furthermore, the phase transformation mechanism of InSb during lithiation/delithiation is studied using ex-situ XRD.

2. Experimental Section

2.1. Synthesis of InSb and InSb–C

InSb was synthesized using high-energy mechanical milling (HEMM). In (99.99%, Sigma-Aldrich, St. Louis, MO, USA) and Sb (99.998%, Sigma-Aldrich, St. Louis, MO, USA) powders were mixed in a 1:1 molar ratio and then placed in an 80 cm^3 ZrO_2 bowl with hardened ZrO_2 balls in a 20:1 ball-to-powder ratio. The mixture was milled in an Ar atmosphere for 10 h at 300 rpm. InSb–C nanocomposites were prepared using HEMM, where a mixture of as-synthesized InSb and acetylene black powder (99.9%, 100% compressed, specific surface area of 75 $m^2 \cdot g^{-1}$, bulk density of 170–230 $g \cdot L^{-1}$, Alfa Aesar, Catalog No. 045527, Ward Hill, MA, USA) at a mass ratio of 9:1 was milled under the same conditions as the InSb synthesis. The mechanochemical synthesis reaction for InSb–C is described as follows:

$$In + Sb \rightarrow InSb + C \rightarrow InSb\text{-}C \tag{1}$$

2.2. Material Characterization

The crystal structures of the as-prepared InSb and InSb–C were measured using powder XRD (D/MAX–2200 Rigaku, Tokyo, Japan) with Cu Kα (λ = 1.54 Å) radiation. The microscopic morphology of the as-synthesized powder materials was observed using HRTEM (JEOL JEM-2100F) and SEM (Hitachi S4700, Tokyo, Japan). XPS (Kratos Axis Anova, Manchester, UK) was used to evaluate the chemical states of the synthesized materials. The elemental content and distribution of the as-prepared powder and electrode after electrochemical reactions were evaluated using EDX.

2.3. Electrochemical Measurements

All electrodes were prepared by casting a slurry containing 70% active material, 15% carbon (Super-P, 99.9%, Alfa Aesar), and 15% PAA (Mw 450000, Sigma Aldrich, St. Louis, MO, USA) or a poly(vinylidene fluoride) (PVDF, MW 534000, Sigma Aldrich, St. Louis, MO, USA) binder dissolved in N-Methyl-2-pyrrolidone. The cast electrodes were dried overnight in a vacuum oven at 70 °C and then transferred to an Ar glove box for cell assembly. A coin-type cell (CR2032) was used for half-cell testing. Li metal foil and polyethylene were used as the counter electrode and separating membrane, respectively. The electrolyte was 1 M $LiPF_6$ in ethylene carbonate/diethyl carbonate (EC/DEC, 1:1 v/v). The electrochemical performance of InSb and InSb–C was evaluated using a battery testing system (WBCS3000, WonATech, Seoul, South Korea). The galvanostatic charge–discharge (GCD) profile was measured from 0.01 to 2.5 V (vs. Li/Li$^+$). Cyclic voltammetry (CV) at a scanning rate of 0.1 $mV \cdot s^{-1}$ was used to characterize the electrochemical reactions of InSb with Li$^+$. The rate capability was measured using a battery cycler (WBCS3000, WonATech, Seoul, South Korea) at current densities of 0.1, 0.5, 1, 3, 5, and 10 $A \cdot g^{-1}$. EIS (ZIVE MP1, WonATech) was measured in the frequency range from 100 kHz to 100 mHz with an AC amplitude of 10 mV.

3. Results and Discussion

Figure 1a shows the XRD pattern of the as-prepared InSb powder obtained using the HEMM process. The XRD pattern coincided with the standard data of zinc blende InSb (JCPDS #06-0208) with no detected impurity phases. This indicated that a single phase of the zinc blende structure was successfully obtained, with a lattice constant of 0.646 Å and a space group of T_d^2-F43 m, as shown in the inset of Figure 1a. The average crystalline domain size of the as-prepared InSb was calculated to be 0.225 nm using the Scherrer formula (Table S1). The particle size of InSb ranged from hundreds of nanometers to a few micrometers (Figure 1b,c). One of the most important factors affecting the cell performance and safety of LIBs as well as reducing cell aging is the particle size of the active material. The particle size of the material affects the electrochemical performance of the battery [47–49]. In general, the small particles have short diffusion pathways (fast Li-ion diffusion), large surface area, and lower overpotential, thus allowing faster C-rate operation and high capacity. However, the beneficial effect of particle size reduction on cell performance is limited to certain particle sizes. The excessively large surface area can lead to large proportion of passivation layers, such as SEI, leading to an irreversible capacity loss [50–52]. Considering this, commercial batteries usually contain micrometer-sized particles for the electrode materials. However, the appropriate size of electrode material highly depends on the intrinsic properties of the electrode materials because they have different atomic structures that influence the electrochemical kinetics, Li-ion intercalation capacities, and structural stability. The size of InSb particles (mostly 200–400 nm in Figure 1c) is thought to be effective in terms of Li-ion diffusion kinetics and capacity while restraining the excessive surface passivation (e.g., SEI). EDX analysis of the InSb powder revealed that the elemental ratio of In and Sb was approximately 1:1 (Figure 1d). The presence of O in the InSb powder is due to the partially oxidized surface of the InSb particles. The composition and chemical state of InSb were examined using XPS (Figure 1e,f). The XPS signals observed at 452.1 and 444.5 eV (Figure 1e) can be ascribed to In $3d_{3/2}$ and In $3d_{5/2}$, respectively, while the peaks at 539.5 and 530.1 eV (Figure 1f) were indexed to Sb $3d_{3/2}$ and Sb $3d_{5/2}$, respectively, verifying the InSb alloy structure after the HEMM process. Meanwhile, the two small peaks at 536.9 and 527.2 eV (Figure 1f) are related to the surface oxidation of the InSb materials, consistent with the EDX analysis results (Figure 1d). The FTIR analysis of the InSb also confirmed the presence of hydroxide functional groups, as shown in Figure S1. The presence of hydroxyl groups on InSb should result in a high affinity for binders with polar functional groups (such as PAA), which can form strong hydrogen bonds. The binder can then serve as an elastic barrier that prevents InSb particles from aggregating while maintaining stable contact between the electrode and current collector during electrochemical reactions.

The half-cell performance of InSb was measured using two different binders (PAA and PVDF) to investigate its electrochemical behavior (Figure 2). The GCD voltage profiles of InSb_PAA and InSb_PVDF are shown in Figure 2a and Figure S2, respectively. The initial charge/discharge capacities of InSb_PAA and InSb_PVDF were 790/635 and 770/643 mAh·g^{-1}, respectively, corresponding to initial coulombic efficiencies (ICEs) of 80.9% and 83.5%. The irreversible capacity losses in the first cycle are associated with the formation of an SEI layer for both electrodes. Although the specific capacities of InSb were not significantly different for PAA and PVDF in the first cycle, a significant capacity reduction was observed for InSb_PVDF during the initial 10 cycles at both low (Figure 2b) and high current densities (Figure 2c). The specific capacities of InSb_PVDF were 203.3 mAh·g^{-1} after 140 cycles and 146.8 mAh·g^{-1} after 100 cycles at 100 and 500 mA·g^{-1}, respectively, corresponding to capacity retention values of 30.4% and 27.5%. Moreover, InSb_PAA displayed much better performance in terms of stability and capacity; it exhibited specific capacities of 639.5 mAh·g^{-1} after 140 cycles (93.2% capacity retention) and 558.3 mAh·g^{-1} after 100 cycles (92.3% capacity retention) at 100 and 500 mA·g^{-1}, respectively. Figure S3 displays the surface morphologies of pristine InSb_PAA and InSb_PVDF. InSb_PAA showed a more uniform surface with a lower roughness than InSb_PVDF owing to the strong hydrogen-bonding interaction between the hydroxyl groups on the InSb particles and the carboxylate groups in PAA, which is not present in InSb_PVDF. Figure 2d shows the first five CV cycles for InSb_PAA in the voltage range from 0.005 to 3.0 V vs. Li/Li$^+$. The initial CV curve was markedly different from those of the subsequent cycles due to the formation of an SEI layer on the electrode surface. In the first discharge step, a significant reduction peak emerged at 0.38 V, indicating the Li intercalation into InSb to form Li$_2$Sb and In. The peak emerging at 0.24 V can be due to the reaction between In and Li to form Li$_y$In. Thus, after completing the discharge step, Li$_2$Sb and Li$_2$In appear as final products. In the charge process, three oxidation peaks were observed at voltages of 0.70, 0.98, and 1.12 V.

Among them, the first peak at 0.70 V corresponds to the complete exclusion of Li, reverting Li_2In into In. When the anode was charged to 0.98 and 1.12 V, In began to intrude into Li_2Sb to form InSb. The detailed analysis of this phase transformation will be discussed in the ex-situ analyses. However, the curves nearly overlapped after the second cycle, demonstrating the high reversibility and stability of InSb_PAA. Compared to InSb_PAA, InSb_PVDF showed relatively unstable CV curves with polarized oxidation and reduction peaks even after the second cycle (Figure S4).

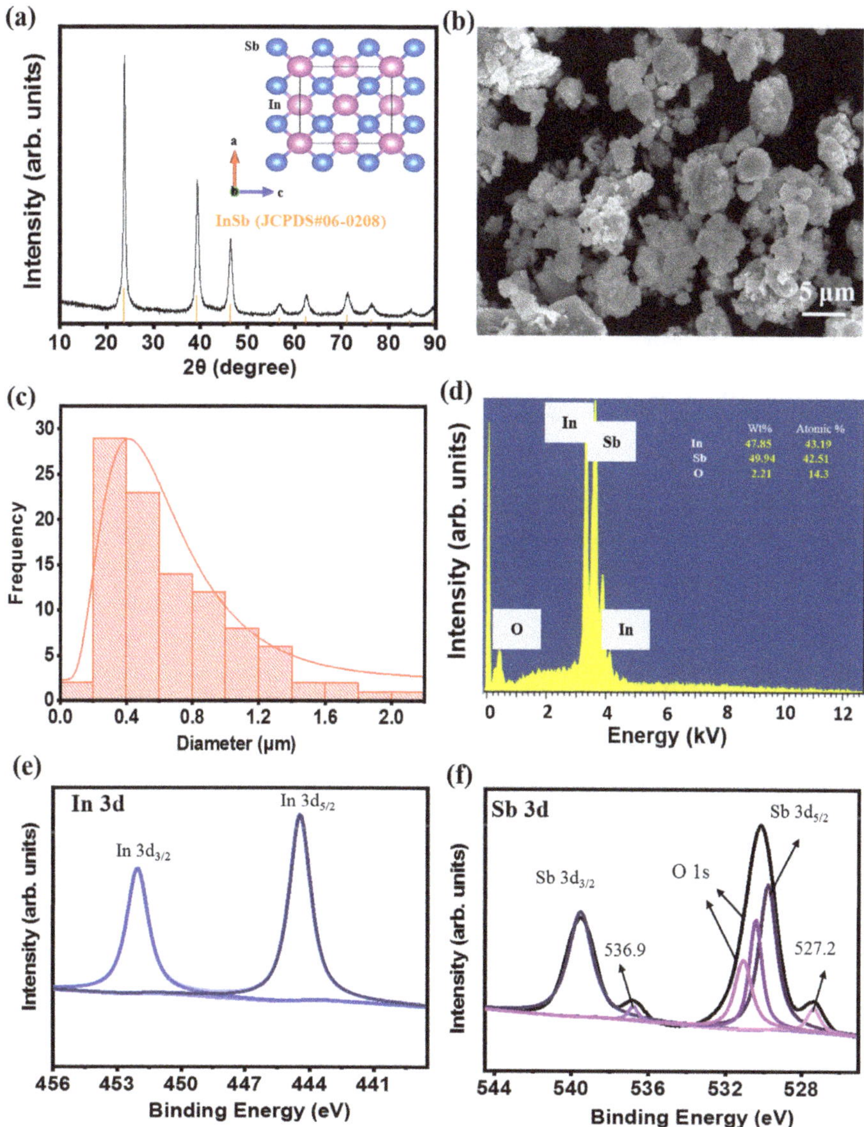

Figure 1. (**a**) XRD pattern (inset: crystalline structure), (**b**) SEM image, (**c**) particle size distribution, and (**d**) EDX spectrum of the as-synthesized InSb powder. XPS profiles of (**e**) In 3d, and (**f**) Sb 3d for the InSb powder.

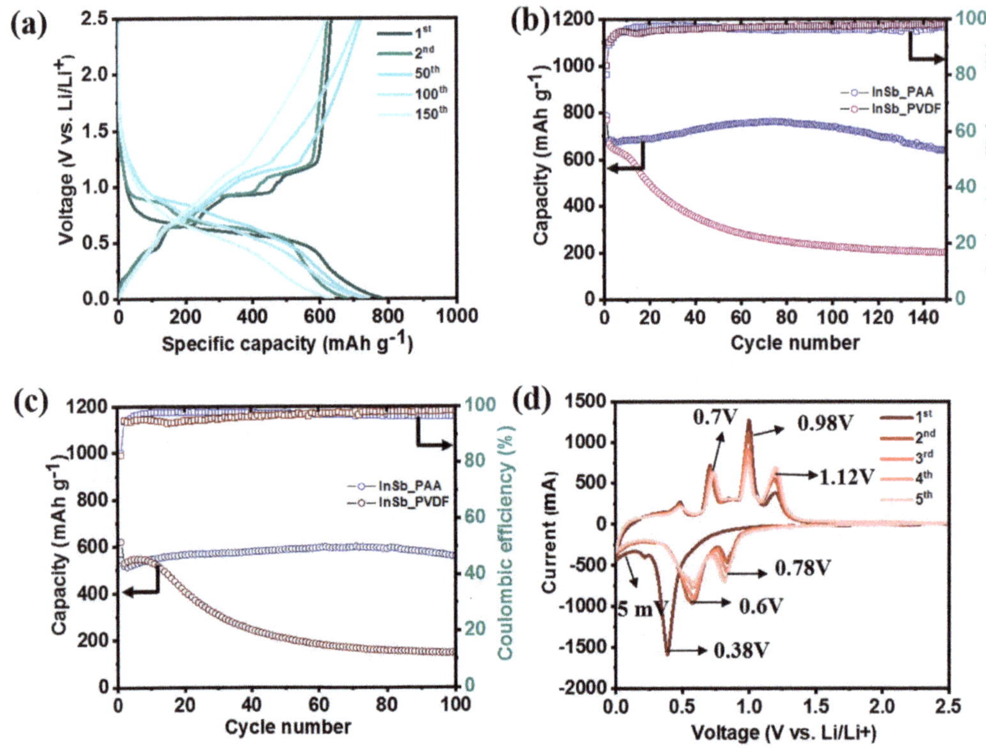

Figure 2. Electrochemical performance of the InSb electrode. (**a**) GCD voltage profiles of InSb_PAA at a current density of 100 mA·g^{-1}. Cyclic performance of the InSb_PAA and InSb_PVDF at a current density of (**b**) 100 and (**c**) 500 mA·g^{-1}. (**d**) CV curves of InSb_PAA.

Figure 3 compares the cross-sectional SEM images of InSb_PAA and InSb_PVDF in the pristine state and after 20 cycles. Although the thicknesses of InSb_PAA and InSb_PVDF were similar in the pristine states (10.2 μm in Figure 3a,d), InSb_PAA was thinner (10.8 μm in Figure 3b) than InSb_PVDF (12.4 μm in Figure 3e) after 20 cycles, indicating a smaller volume expansion of the InSb_PAA. In addition, the InSb_PAA maintained close contact between the electrode and current collector after 20 cycles (Figure 3c). However, the InSb_PVDF electrode partially delaminated from the current collector (Figure 3e) and aggregated (Figure 3f), because it failed to accommodate the large volume change of the InSb particles during repeated electrochemical reactions. These results justify the selection of PAA as an appropriate binder material for the InSb electrode.

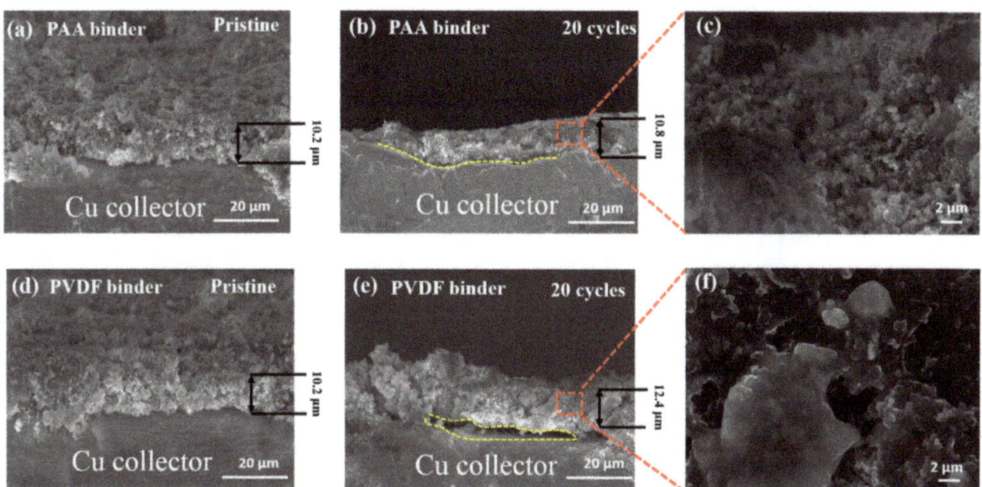

Figure 3. Comparison of the InSb_PAA and InSb_PVDF electrodes before and after 20 cycles. Cross-sectional images of (**a**) pristine InSb_PAA, (**b,c**) InSb_PAA after 20 cycles at different magnifications, (**d**) pristine InSb_PVDF, and (**e,f**) InSb_PVDF after 20 cycles at different magnifications. The dashed yellow lines in (**b,e**) indicate the boundary between electrode and Cu collector.

Ex-situ XRD was used to investigate the electrochemical reaction mechanism during the initial lithiation/delithiation process of the InSb electrode (Figure 4a). At a discharge voltage of 0.38 V (D-0.38 V), peaks corresponding to Li_2Sb and In emerged. When fully discharged (D-5 mV), Li_2In peaks appeared, while Li_2Sb and In peaks remained. Upon charging to 0.7 V (C-0.7 V), the Li_2In phase disappeared. At the charging states of 0.98 and 1.12 V the Li_2Sb phase vanished. When fully charged to 2.5 V (C-2.5 V), only the peaks matching with InSb re-emerged. The structural transformation of InSb during lithiation/delithiation is summarized as follows:

Discharging:

$$InSb \rightarrow Li_2Sb + In \rightarrow Li_2Sb + Li_2In + In \text{ (partly)} \tag{2}$$

Charging:

$$Li_2Sb + Li_2In + In \text{ (partly)} \rightarrow In + Li_2Sb \rightarrow InSb \tag{3}$$

Notably, the InSb phase (major peaks at 39.9° and 46.5°) fully recovered without any impurity peaks after the first cycle, indicating a highly reversible reaction of InSb with Li ions. This likely correlates with the robust binding between InSb and PAA, which effectively protects the active material from pulverization and delamination caused by volume changes. The ex-situ XRD results demonstrate the conversion and alloying/dealloying mechanism of the InSb electrode during discharge/charge, as schematically illustrated in Figure 4b.

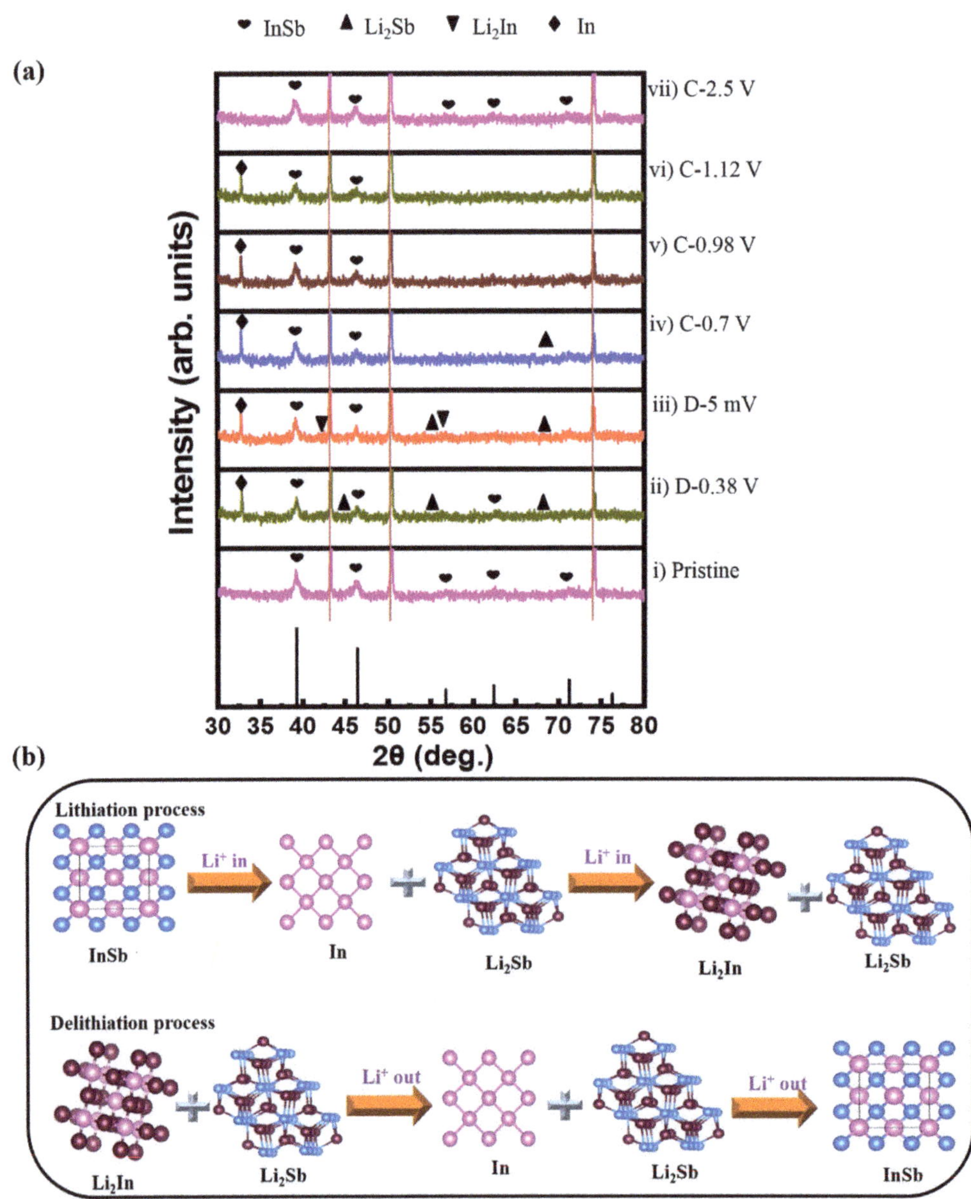

Figure 4. (**a**) XRD patterns collected at selected potential states in the initial lithiation/delithiation process. (**b**) Schematic of the electrochemical reaction mechanism of the InSb_PAA electrode during cycling.

Despite the better performance of InSb_PAA compared to that of InSb_PVDF, the InSb_PAA electrode still had a gradual decrease in capacity after ~80 cycles when measured at 100 mA·g^{-1} (Figure 2b). This behavior is also reflected in the coulombic efficiency (CE) variation (Table S2), where the CE steadily increased until ~60 cycles, then decreased afterwards. This might be associated with increasing side reactions between InSb_PAA and the electrolyte as the electrode was cycled. These side reactions can be further explained by a differential capacity plot (DCP) analysis of the initial 140 cycles (Figure S5). From this analysis, the main reduction (at ~0.86 and ~0.92 V) and oxidation (at ~0.59 and 0.81 V) peaks remained unchanged for 80 cycles, but then became broader and shifted after 80 cycles. This polarization leads to inefficient lithiation/delithiation and a progressive capacity drop after 80 cycles. A similar trend was observed at a high current density (Figure 2c). In this case, the capacity gradually increased until 80 cycles, followed by a slight decrease in subsequent cycles. This trend was also observed in the CE variation (Table S3) and DCP analysis (Figures S6 and S7), where the intensities of the oxidation (at ~0.59 and 0.81 V) and reduction (at ~0.86 and ~0.92 V) peaks generally increased for 60 cycles with a negligible polarization (Figure S6), then decreased in intensity after 60 cycles with a slight polarization (Figure S7). Therefore, the electrochemical performance of InSb_PAA at current densities of 100 and 500 mA·g^{-1} was not fully satisfactory, based on these results.

High-performance LIB anode materials frequently use C decoration to overcome the disadvantages of the active materials. Amorphous C provides improved electrical conductivity and acts as a buffer for withstanding the volume change of Li-active materials [53–56]. Therefore, InSb–C (or InSb–C_PAA) was prepared by two sequential steps of HEMM (adding acetylene black to the InSb electrode in the secondary HEMM). The XRD peaks of the as-prepared InSb–C matched well with those of InSb (Figure 5a). The size and shape of the InSb–C were almost unchanged compared to those of InSb (Figure 5b). The presence of InSb nanocrystallites was confirmed with HRTEM, with interplanar distances of 0.374 nm and 0.229 nm ((111) and (220) phases of InSb, respectively (Figure 5c)), which was consistent with the XRD analysis. The elemental mapping images (Sb, In, and C) revealed evenly distributed constituent elements (Figure 5d). In addition, the uniform distribution of O confirmed the oxidation of the functional groups on the InSb–C, as in the case of InSb.

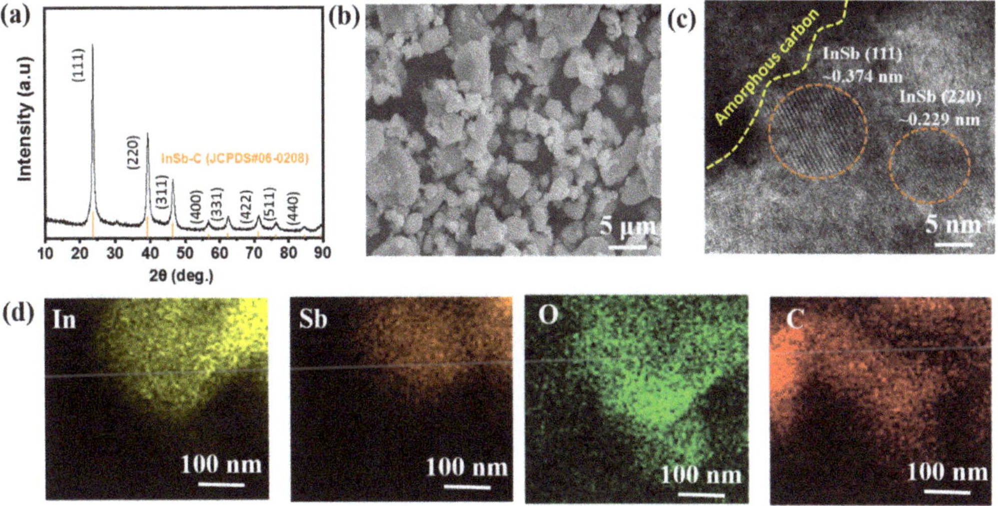

Figure 5. (**a**) XRD pattern, (**b**) SEM image, (**c**) HRTEM image, and (**d**) EDX elemental maps of In, Sb, O, and C of InSb–C.

The electrochemical performance of the InSb–C_PAA electrode is shown in Figure 6. The initial charge/discharge capacity of InSb–C_PAA was 669/540 mAh·g^{-1} with an ICE of 80.7% (Figure 6a). From the EDX analysis (Figure S8) and the calculated theoretical capacity of the individual components (Table S4), the capacity contribution from C in the InSb electrode was estimated to be ~10%. Therefore, the capacity of the electrode was mainly from the active InSb (90% of the total capacity) while C mainly functioned as a buffer matrix (10% capacity contribution) that mitigated the volume expansion of the electrode. Remarkably, the measured capacities of InSb_PAA and InSb–C_PAA were higher than their theoretical capacities (454 and 435.6 mAh·g^{-1}, respectively, as calculated in Table S5). This additional capacity is most likely due to electrolyte decomposition and interfacial Li-ion storage. Although the specific capacity of InSb–C_PAA was lower than that of InSb_PAA in the initial cycle, the long-term performance of InSb–C_PAA was superior to that of InSb_PAA. In particular, InSb–C_PAA delivered 878 and 634 mAh·g^{-1} at 100 mA·g^{-1} (Figure 6b) and 500 mAg^{-1} (Figure 6c) after 150 and 300 cycles, respectively. Notably, a steady capacity increase was observed for InSb–C_PAA during the repeated discharge/charge processes, which was attributed to the creation of a polymer-gelled film from electrolyte decomposition and interfacial Li-ion storage [57–59]. Furthermore, the variations in the DCP profiles as a function of cycle number were studied at current densities of 100 and 500 mA·g^{-1} to better understand the steady rise in capacity (Figure S9). In the DCP curves of the InSb–C_PAA electrodes, the overall intensity of the redox potentials increased with increasing cycle number. There was also a minor positive shift in the reduction peaks (at 0.86 and 0.92 V) and a slight negative shift in the oxidation peaks (at 0.59 and 0.81 V) in the capacity-increasing region. The degree of polarization in InSb–C_PAA was much lower than that of InSb–PAA (Figure S9). Figure S10 compares the CE variation in InSb–C_PAA and InSb_PAA at current densities of 100 and 500 mA·g^{-1}. The detailed CE values are summarized in Table S6 (at 100 mA·g^{-1}) and Table S7 (at 500 mA·g^{-1}) for the InSb_PAA, InSb_PVDF, and InSb–C_PAA electrodes during the first 10 cycles. As seen in Table S6, InSb–C_PAA had a slightly lower ICE (80.58%) than the InSb_PAA (ICE = 81.42%) and InSb_PVDF electrodes (ICE = 83.53%). However, the CE of the InSb–C_PAA electrode significantly increased after the first cycle, exhibiting the highest CE among the three different electrodes. This trend was also observed at high current densities (Table S7). The high CE of the InSb–C_PAA electrode after the first cycle indicated a high reversibility of lithiation/delithiation. Figure 6d shows the first five CV curves of InSb–C_PAA. In contrast to InSb_PAA and InSb_PVDF, the CV curves of InSb–C_PAA nearly overlapped after the second cycle, exhibiting exceptional cycling stability. The redox peak positions were exactly identical to those observed for InSb_PAA (Figure 2d), indicating that InSb is the main active material. The rate performance (Figure 6e) and normalized capacity retention (Figure 6f) of InSb–C_PAA were measured at various current densities. The average discharge capacities of InSb–C_PAA were 669, 660, 659, 645, 644, and 635 mAh·g^{-1} at current densities of 0.1, 0.5, 1.0, 3.0, 5.0, and 10.0 A·g^{-1}, respectively (Figure 6e), which were significantly greater than those of InSb_PAA and InSb_PVDF. Remarkably, even at a high current density of 10 A·g^{-1}, the capacity retention of InSb–C_PAA was as high as 98% of its initial capacity (Figure 6f). Even at the current densities higher than 10 A·g^{-1}, InSb–C_PAA still presented outstanding electrochemical performance with average specific capacities were 627 and 541 mAh·g^{-1} at 15 and 20 A·g^{-1}, respectively (Figure S11). In addition, a high-capacity retention (94.4%) was achieved when the discharge rate was returned to 0.1 A·g^{-1} from 10 A·g^{-1}, demonstrating the good rate performance of InSb–C_ PAA.

EIS profiles of the InSb_PAA, InSb_PVDF, and InSb–C_PAA electrodes were obtained at the 1st, 5th, and 20th cycles (Figure 7). The electrolyte resistance (R_b), SEI layer resistance (R_{SEI}), charge-transfer resistance (R_{ct}), interfacial double layer capacitance (C_{dl}), constant phase element (C_{PE}), and Warburg impedance (Z_w) are all included in the simplified equivalent circuit depicted in Figure 7a. The compressed semi-circles in the mid-frequency region of the Nyquist plots correspond to R_{ct} at the electrode–electrolyte interface. The cells containing the electrode with the PAA binder (InSb_PAA and InSb–C_PAA) showed decreasing semicircles in the low-frequency region with an increase in the cycle number (from 1 to 20 cycles), indicating a gradual decrease in R_{ct} and steady stabilization of the electrode (Figure 7b,d). The R_{ct} values of InSb_PAA and InSb–C_PAA (Figure 7b,d) were significantly lower than that of InSb_PVDF (Figure 7c). After 20 cycles, InSb–C_PAA exhibited the lowest R_{ct} value among the electrodes (Table S8). These results help to explain the gradual increase in capacity and performance of the InSb–C_PAA electrode during long-term cycling.

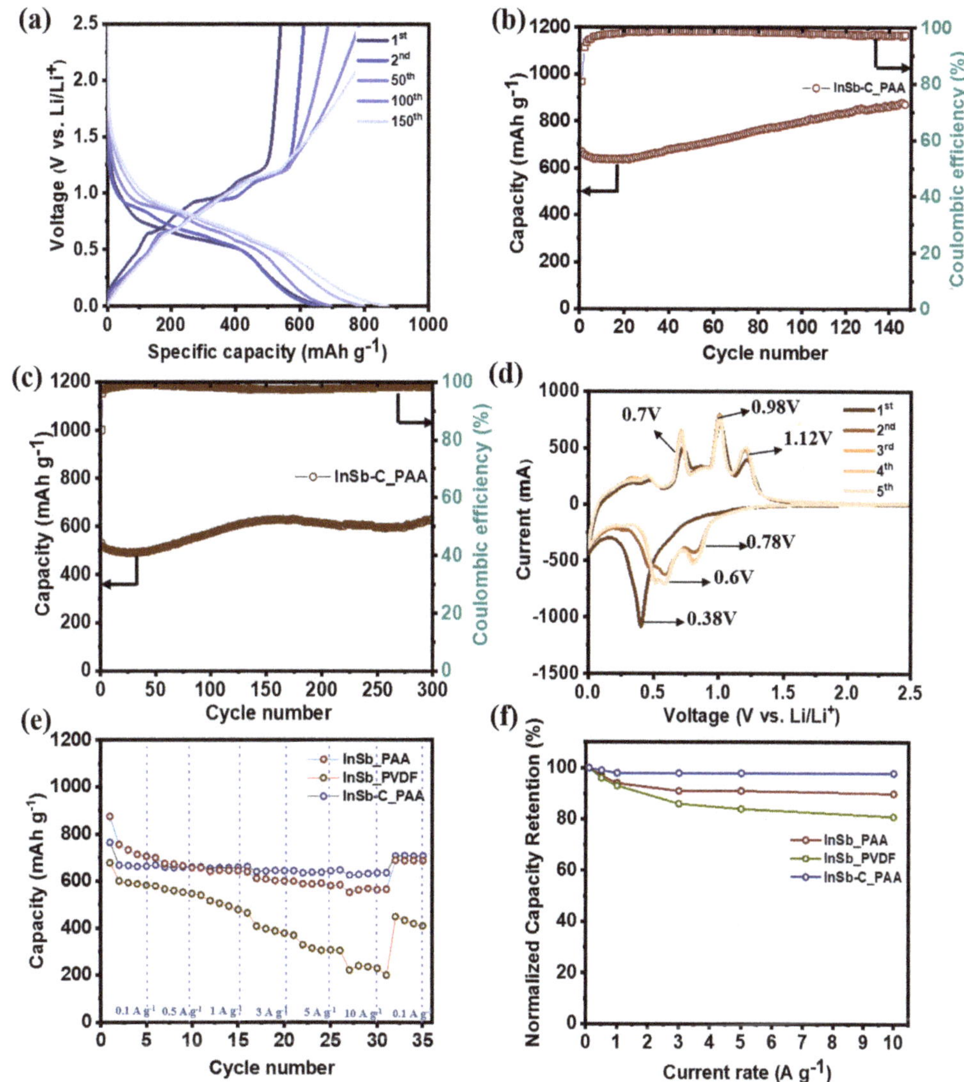

Figure 6. Electrochemical performance of the half-cells. (**a**) GCD profiles of InSb-C_PAA at a current density of 100 mA·g^{-1}, cyclic performance of InSb-C_PAA at (**b**) 100 mA·g^{-1} and (**c**) 500 mA·g^{-1}, (**d**) CV curves of InSb-C_PAA, (**e**) rate capabilities of the composites, and (**f**) capacity retention of the composites at different current densities.

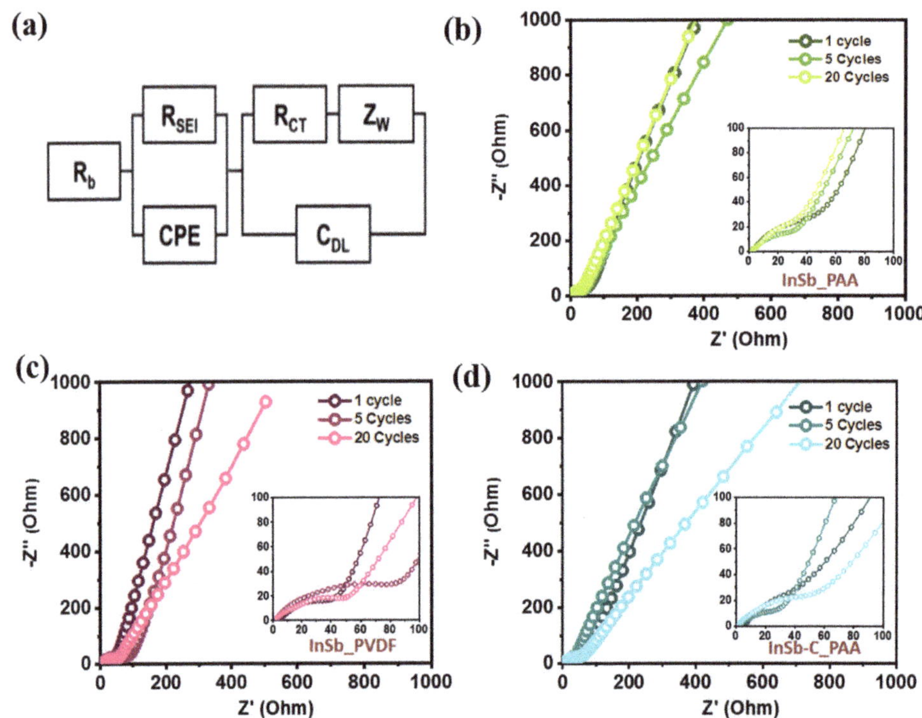

Figure 7. (**a**) The equivalent circuit. Nyquist plots after 1, 5, and 20 cycles for (**b**) InSb_PAA, (**c**) InSb_PVDF, and (**d**) InSb–C_PAA.

Considering all the results, the Li-ion storage mechanism of the InSb–C_PAA electrode is schematically presented in Figure 8. The overall electrochemical reaction is written as $InSb + 4Li^+ + 4e^- \rightleftarrows Li_2Sb + Li_2In$, neglecting the small capacity contribution from the C matrix. As the discharge proceeds, Li_2Sb and Li_2In are formed as products after the reaction with Li ions. During this reaction, a large volume expansion (Li_2Sb (~135%) and Li_2In (~297%)) causes mechanical stress on the active InSb. Under prolonged cycles, accumulated stress can result in particle agglomeration, pulverization, and delamination. These issues were effectively resolved by employing a PAA binder and a C buffering matrix. PAA is a binder with numerous COOH functional groups that can form hydrogen bonds with OH groups on the surfaces of active materials (as determined using FTIR (Figure S1) and XPS analyses (Figure 1e)), thereby stabilizing the electrode structure. The presence of amorphous C around InSb facilitates charge transport and provides a mechanical buffer for the active InSb. Therefore, the synergistic effect between the PAA binder and amorphous C contributes to a significant improvement in the electrochemical performance of InSb. Consequently, the performance of the InSb–C_PAA electrode is better than that of most previously reported Sb-based electrodes (Table 1).

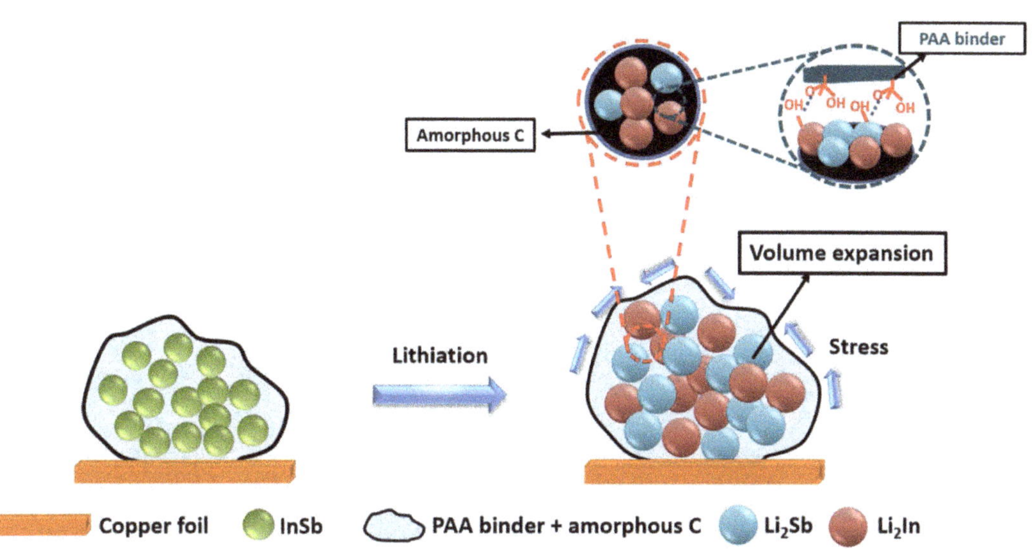

Figure 8. Illustration of the InSb–C_PAA reaction mechanism.

Table 1. Performance comparison of intermetallic Sb-based anodes for LIBs.

Anode	Cycling Performance	Rate Capability	Synthesis Method	Ref.
Cu$_2$Sb	290 mAh·g^{-1} after 25 cycles	-	Ball milling	[60]
Mo$_3$Sb$_7$	350 mAh·g^{-1} after 100 cycles at 0.12 C	300 mAh·g^{-1} at 100 C	Furnace	[61]
CoSb	448 mAh·g^{-1} after 1000 cycles at 0.66 A·g^{-1}	-	Facile colloidal synthesis	[62]
NiSb@C	405 mAh·g^{-1} after 1000 cycles at 0.1 A g^{-1}	393 mAh·g^{-1} at 2.0 A·g^{-1}	Freezing drying	[63]
NiSb hollow nanosphere	420 mAh·g^{-1} after 50 cycles at 0.1 A·g^{-1}	352 mAh·g^{-1} at 0.8 A·g^{-1}	Galvanic replacement reaction	[64]
NiSb/C nanosheet	393 mAh·g^{-1} after 1000 cycles at 2 C	325 mAh·g^{-1} at 10 C	Hydrothermal low-temperature carbothermic reduction	[65]
SnSb@Carbon fiber	674 mAh·g^{-1} after 100 cycles at 0.1 A·g^{-1}	163 mAh·g^{-1} at 1.6 A·g^{-1}	Electrospinning	[66]
ZnSb/C	481 mAh·g^{-1} after 240 cycles at 0.1 A·g^{-1}	426 mAh·g^{-1} at 0.5 A·g^{-1}	Annealing	[67]
TiSb$_2$	420 mAh·g^{-1} after 120 cycles at 1 C	300 mAh·g^{-1} at 12 C	Furnace	[68]
InSb_PAA InSb–C_PAA	640 mAh·g^{-1} after 140 cycles 846 mAh·g^{-1} after 150 cycles at 0.1 A·g^{-1}	594 mAh·g^{-1} at 10 A·g^{-1} 716 mAh·g^{-1} at 10 A·g^{-1}	Ball milling	This work

4. Conclusions

In summary, InSb and InSb–C were successfully synthesized via HEMM and studied as potential anodes for LIBs. The crystal structure, morphology, and chemical state of these materials were characterized using XRD, SEM, HRTEM, EDX, and XPS. Electrochemical measurements revealed that the PAA binder played a significant role in improving the performance of the InSb-based electrode over conventional PVDF owing to the formation of hydrogen bonds with InSb, which contributed to the strong adhesion between the active materials and current collectors. The addition of amorphous C to InSb improved the mechanical stability and electrical conductivity. As a result, InSb–C_PAA electrodes delivered a high reversible specific capacity (878 mAh·g^{-1} at 100 mA·g^{-1} after 140 cycles) and good rate capability (capacity retention of 98% at 10 A·g^{-1} as compared to 0.1 A·g^{-1}), which outperforms most of the Sb-based electrodes recently reported. The synergistic effect of the PAA binder and amorphous C is responsible for the improved electrochemical performance of InSb–C_PAA. Therefore, InSb–C_PAA can be considered as a potential anode material for next-generation LIBs.

Supplementary Materials: The following are available online at https://www.mdpi.com/article/10.3390/nano11123420/s1, Figure S1: FT-IR results of InSb powder; Figure S2: GCD curves of InSb_PVDF; Figure S3: SEM images of (a and b) InSb_PAA, (c and d) InSb_PVDF binder at different magnification; Figure S4: CV curves of InSb_PVDF from first to fifth cycle; Figure S5: DCP of InSb_PAA during 140 cycles measured at 100 mA·g^{-1}: (a) 1−60 cycle, (b) 80−140 cycle. Enlarged view of (c) reduction peak and (d) oxidation peak; Figure S6: (a) DCP of InSb_PAA during initial 60 cycles measured at 500 mA·g^{-1}. Enlarged view of (b) oxidation peak and (c) reduction peak; Figure S7: (a) DCP of InSb_PAA from 60 th to 100 th cycle measured at 500 mA·g^{-1}. Enlarged view of (b) oxidation peak and (c) reduction peak; Figure S8: EDX spectrum of synthesized InSb–C; Figure S9: DCP profiles of InSb–C_PAA electrodes at current density of (a) 100 mA·g^{-1} during 140 cycles and (b) 500 mA·g^{-1} during 300 cycles; Figure S10: Coulombic efficieny of InSb_PAA, InSb_PVDF, and InSb–C_PAA at current density of (a) 100 and (b) 500 mA·g^{-1}; Figure S11: Cyclic performance of InSb-C_PAA at 15 A·g^{-1} and 20 A·g^{-1}; Table S1: Crystallite size of InSb calculated using Scherrer equation; Table S2: Coulombic efficiency variation of InSb_PAA at various cycle numbers measured at 100 mA·g^{-1}; Table S3: Coulombic efficiency variation of InSb_PAA at various cycle numbers measured at 500 mA·g^{-1}; Table S4: Calculation of capacity contribution of InSb and C in the InSb-C composite; Table S5: Calculation of theoretical capacity of InSb and InSb-C; Table S6: Coulombic efficiency of InSb_PAA, InSb_PVDF, and InSb-C_PAA at current density of 100 mA·g^{-1} for initial 10 cycles; Table S7: Coulombic efficiency of InSb_PAA, InSb_PVDF, and InSb–C_PAA at current density of 500 mA·g^{-1} for initial 10 cycles; Table S8: The charge-transfer resistance (R_{ct}) of InSb_PAA, InSb_PVDF, InSb-C_PAA.

Author Contributions: Conceptualization, J.H. and I.T.K.; validation, V.P.H.H.; investigation, V.P.H.H.; writing—original draft preparation, V.P.H.H.; writing—review and editing, J.H. and I.T.K.; supervision, J.H. and I.T.K.; funding acquisition, J.H. All authors have read and agreed to the published version of the manuscript.

Funding: This work was supported by the Basic Science Research Capacity Enhancement Project through Korea Basic Science Institute (National Research Facilities and Equipment Center) grant funded by the Ministry of Education (2019R1A6C1010016) and the Korea Institute of Energy Technology Evaluation and Planning (KETEP) and the Ministry of Trade, Industry & Energy (MOTIE) of the Republic of Korea (No. 20194030202290).

Institutional Review Board Statement: Not applicable.

Informed Consent Statement: Not applicable.

Data Availability Statement: Not applicable.

Conflicts of Interest: The authors declare no competing interest.

References

1. Chen, T.; Jin, Y.; Lv, H.; Yang, A.; Liu, M.; Chen, B.; Xie, Y.; Chen, Q. Applications of lithium-ion batteries in grid-scale energy storage systems. *Trans. Tianjin Univ.* **2020**, *26*, 208–217. [CrossRef]
2. Armand, M.; Tarascon, J.-M. Building better batteries. *Nature* **2008**, *451*, 652–657. [CrossRef]
3. Xu, J.; Wang, D.; Kong, S.; Li, R.; Hong, Z.; Huang, F.Q. Pyrochlore phase Ce$_2$Sn$_2$O$_7$via an atom-confining strategy for reversible lithium storage. *J. Mater. Chem. A* **2020**, *8*, 5744–5749. [CrossRef]
4. Kim, T.; Song, W.; Son, D.-Y.; Ono, L.K.; Qi, Y. Lithium-ion batteries: Outlook on present, future, and hybridized technologies. *J. Mater. Chem. A* **2019**, *7*, 2942–2964. [CrossRef]
5. Meng, J.; Guo, H.; Niu, C.; Zhao, Y.; Xu, L.; Li, Q.; Mai, L. Advances in structure and property optimizations of battery electrode materials. *Joule* **2017**, *1*, 522–547. [CrossRef]
6. Goriparti, S.; Miele, E.; De Angelis, F.; Di Fabrizio, E.; Zaccaria, R.P.; Capiglia, C. Review on recent progress of nanostructured anode materials for Li-ion batteries. *J. Power Sources* **2014**, *257*, 421–443. [CrossRef]
7. Jiao, X.; Liu, Y.; Li, B.; Zhang, W.; He, C.; Zhang, C.; Yu, Z.; Gao, T.; Song, J. Amorphous phosphorus-carbon nanotube hybrid anode with ultralong cycle life and high-rate capability for lithium-ion batteries. *Carbon* **2019**, *148*, 518–524. [CrossRef]

8. Mun, Y.S.; Pham, T.N.; Bui, V.K.H.; Tanaji, S.T.; Lee, H.U.; Lee, G.-W.; Choi, J.S.; Kim, I.T.; Lee, Y.-C. Tin oxide evolution by heat-treatment with tin-aminoclay (SnAC) under argon condition for lithium-ion battery (LIB) anode applications. *J. Power Sources* **2019**, *437*, 226946. [CrossRef]

9. Nguyen, T.L.; Park, D.; Kim, I.T. $Fe_xSn_yO_z$ composites as anode materials for lithium-ion storage. *J. Nanosci. Nanotechnol.* **2019**, *19*, 6636–6640. [CrossRef] [PubMed]

10. Nguyen, T.L.; Kim, J.H.; Kim, I.T. Electrochemical performance of $Sn/SnO/Ni_3Sn$ composite anodes for lithium-ion batteries. *J. Nanosci. Nanotechnol.* **2019**, *19*, 1001–1005. [CrossRef]

11. Pham, T.N.; Tanaji, S.T.; Choi, J.S.; Lee, H.U.; Kim, I.T.; Lee, Y.C. Preparation of Sn-aminoclay (SnAC)-templated Fe_3O_4 nanoparticles as an anode material for lithium-ion batteries. *RSC Adv.* **2019**, *9*, 10536–10545. [CrossRef]

12. Nguyen, T.; Kim, I. Ag nanoparticle-decorated MoS_2 nanosheets for enhancing electrochemical performance in lithium storage. *Nanomaterials* **2021**, *11*, 626. [CrossRef] [PubMed]

13. Nguyen, T.P.; Kim, I.T. Self-assembled few-layered MoS_2 on SnO_2 anode for enhancing lithium-ion storage. *Nanomaterials* **2020**, *10*, 2558. [CrossRef]

14. Preman, A.N.; Lee, H.; Yoo, J.; Kim, I.T.; Saito, T.; Ahn, S.-K. Progress of 3D network binders in silicon anodes for lithium ion batteries. *J. Mater. Chem. A* **2020**, *8*, 25548–25570. [CrossRef]

15. Nguyen, T.P.; Kim, I.T. W_2C/WS_2 alloy nanoflowers as anode materials for lithium-ion storage. *Nanomaterials* **2020**, *10*, 1336. [CrossRef] [PubMed]

16. Kim, W.S.; Vo, T.N.; Kim, I.T. GeTe-TiC-C composite anodes for li-ion storage. *Materials* **2020**, *13*, 4222. [CrossRef]

17. Vo, T.N.; Kim, D.S.; Mun, Y.S.; Lee, H.J.; Ahn, S.-K.; Kim, I.T. Fast charging sodium-ion batteries based on Te-P-C composites and insights to low-frequency limits of four common equivalent impedance circuits. *Chem. Eng. J.* **2020**, *398*, 125703. [CrossRef]

18. Park, C.-M.; Kim, J.-H.; Kim, H.; Sohn, H.-J. Li-alloy based anode materials for Li secondary batteries. *Chem. Soc. Rev.* **2010**, *39*, 3115–3141. [CrossRef]

19. Obrovac, M.N.; Chevrier, V.L. Alloy negative electrodes for li-ion batteries. *Chem. Rev.* **2014**, *114*, 11444–11502. [CrossRef]

20. Lai, S.Y.; Knudsen, K.D.; Sejersted, B.T.; Ulvestad, A.; Mæhlen, J.P.; Koposov, A.Y. Silicon nanoparticle ensembles for lithium-ion batteries elucidated by small-angle neutron scattering. *ACS Appl. Energy Mater.* **2019**, *2*, 3220–3227. [CrossRef]

21. Sun, Y.; Wang, L.; Li, Y.; Li, Y.; Lee, H.R.; Pei, A.; He, X.; Cui, Y. Design of red phosphorus nanostructured electrode for fast-charging lithium-ion batteries with high energy density. *Joule* **2019**, *3*, 1080–1093. [CrossRef]

22. Liu, S.; Feng, J.; Bian, X.; Qian, Y.; Liu, J.; Xu, H. Nanoporous germanium as high-capacity lithium-ion battery anode. *Nano Energy* **2015**, *13*, 651–657. [CrossRef]

23. Wang, L.; Wang, Y.; Xia, Y. A high performance lithium-ion sulfur battery based on a Li_2S cathode using a dual-phase electrolyte. *Energy Environ. Sci.* **2015**, *8*, 1551–1558. [CrossRef]

24. Nam, K.-H.; Park, C.-M. Layered Sb_2Te_3 and its nanocomposite: A new and outstanding electrode material for superior rechargeable Li-ion batteries. *J. Mater. Chem. A* **2016**, *4*, 8562–8565. [CrossRef]

25. Hou, H.; Jing, M.; Yang, Y.; Zhu, Y.; Fang, L.; Song, W.; Pan, C.; Yang, X.; Ji, X. Sodium/lithium storage behavior of antimony hollow nanospheres for rechargeable batteries. *ACS Appl. Mater. Interfaces* **2014**, *6*, 16189–16196. [CrossRef] [PubMed]

26. Luo, W.; Li, F.; Gaumet, J.-J.; Magri, P.; Diliberto, S.; Zhou, L.; Mai, L. Bottom-up confined synthesis of nanorod-in-nanotube structured Sb–N–C for durable lithium and sodium storage. *Adv. Energy Mater.* **2018**, *8*, 1703237. [CrossRef]

27. Nam, K.-H.; Park, C.-M. 2D layered Sb_2Se_3-based amorphous composite for high-performance Li- and Na-ion battery anodes. *J. Power Sources* **2019**, *433*, 126639. [CrossRef]

28. Darwiche, A.; Marino, C.; Sougrati, M.T.; Fraisse, B.; Stievano, L.; Monconduit, L. Better cycling performances of bulk Sb in Na-Ion batteries compared to Li-Ion Systems: An unexpected electrochemical mechanism. *J. Am. Chem. Soc.* **2012**, *134*, 20805–20811. [CrossRef]

29. Yi, Z.; Han, Q.; Geng, D.; Wu, Y.; Cheng, Y.; Wang, L. One-pot chemical route for morphology-controllable fabrication of Sn-Sb micro/nano-structures: Advanced anode materials for lithium and sodium storage. *J. Power Sources* **2017**, *342*, 861–871. [CrossRef]

30. Hou, H.; Jing, M.; Yang, Y.; Zhang, Y.; Song, W.; Yang, X.; Chen, J.; Chen, Q.; Ji, X. Antimony nanoparticles anchored on interconnected carbon nanofibers networks as advanced anode material for sodium-ion batteries. *J. Power Sources* **2015**, *284*, 227–235. [CrossRef]

31. Liu, Z.; Yu, X.-Y.; Lou, X.W.; Paik, U. Sb–C coaxial nanotubes as a superior long-life and high-rate anode for sodium ion batteries. *Energy Environ. Sci.* **2016**, *9*, 2314–2318. [CrossRef]

32. Zhao, X.; Vail, S.A.; Lu, Y.; Song, J.; Pan, W.; Evans, D.R.; Lee, J.-J. Antimony/graphitic carbon composite anode for high-performance sodium-ion batteries. *ACS Appl. Mater. Interfaces* **2016**, *8*, 13871–13878. [CrossRef]

33. Yi, Z.; Han, Q.; Zan, P.; Wu, Y.; Cheng, Y.; Wang, L. Sb nanoparticles encapsulated into porous carbon matrixes for high-performance lithium-ion battery anodes. *J. Power Sources* **2016**, *331*, 16–21. [CrossRef]

34. Yang, Q.; Zhou, J.; Zhang, G.; Guo, C.; Li, M.; Zhu, Y.; Qian, Y. Sb nanoparticles uniformly dispersed in 1-D N-doped porous carbon as anodes for Li-ion and Na-ion batteries. *J. Mater. Chem. A* **2017**, *5*, 12144–12148. [CrossRef]

35. Wan, F.; Guo, J.-Z.; Zhang, X.-H.; Zhang, J.-P.; Sun, H.-Z.; Yan, Q.; Han, D.-X.; Niu, L.; Wu, X.-L. In situ binding Sb nanospheres on graphene via oxygen bonds as superior anode for ultrafast sodium-ion batteries. *ACS Appl. Mater. Interfaces* **2016**, *8*, 7790–7799. [CrossRef] [PubMed]

36. Yang, C.; Li, W.; Yang, Z.; Gu, L.; Yu, Y. Nanoconfined antimony in sulfur and nitrogen co-doped three-dimensionally (3D) interconnected macroporous carbon for high-performance sodium-ion batteries. *Nano Energy* **2015**, *18*, 12–19. [CrossRef]
37. Yim, T.; Choi, S.J.; Jo, Y.N.; Kim, T.-H.; Kim, K.J.; Jeong, G.; Kim, Y.-J. Effect of binder properties on electrochemical performance for silicon-graphite anode: Method and application of binder screening. *Electrochim. Acta* **2014**, *136*, 112–120. [CrossRef]
38. Choi, S.; Kwon, T.-W.; Coskun, A.; Choi, J.W. Highly elastic binders integrating polyrotaxanes for silicon microparticle anodes in lithium ion batteries. *Science* **2017**, *357*, 279–283. [CrossRef] [PubMed]
39. Wu, M.; Xiao, X.; Vukmirovic, N.; Xun, S.; Das, P.K.; Song, X.; Olalde-Velasco, P.; Wang, D.; Weber, A.Z.; Wang, L.-W.; et al. Toward an ideal polymer binder design for high-capacity battery anodes. *J. Am. Chem. Soc.* **2013**, *135*, 12048–12056. [CrossRef] [PubMed]
40. Koo, B.; Kim, H.; Cho, Y.; Lee, K.T.; Choi, N.-S.; Cho, J. A highly cross-linked polymeric binder for high-performance silicon negative electrodes in lithium ion batteries. *Angew. Chem. Int. Ed.* **2012**, *51*, 8762–8767. [CrossRef]
41. Mazouzi, D.; Karkar, Z.; Hernandez, C.R.; Manero, P.J.; Guyomard, D.; Roué, L.; Lestriez, B. Critical roles of binders and formulation at multiscales of silicon-based composite electrodes. *J. Power Sources* **2015**, *280*, 533–549. [CrossRef]
42. Magasinski, A.; Zdyrko, B.; Kovalenko, I.; Hertzberg, B.; Burtovyy, R.; Huebner, C.F.; Fuller, T.F.; Luzinov, I.; Yushin, G. Toward efficient binders for Li-Ion battery Si-Based anodes: Polyacrylic acid. *ACS Appl. Mater. Interfaces* **2010**, *2*, 3004–3010. [CrossRef] [PubMed]
43. Key, B.; Bhattacharyya, R.; Morcrette, M.; Seznéc, V.; Tarascon, J.-M.; Grey, C. Real-time NMR investigations of structural changes in silicon electrodes for Lithium-Ion batteries. *J. Am. Chem. Soc.* **2009**, *131*, 9239–9249. [CrossRef]
44. Madelung, O.; Rössler, U.; Schulz, M. Indium antimonide (InSb), carrier concentrations: Datasheet from Landolt-Börnstein— Group III Condensed Matter Volume 41A1β. In *Group IV Elements, IV-IV and III-V Compounds. Part b-Electronic, Transport, Optical and Other Properties*; Springer: Berlin, Germany, 2002; pp. 1–6. [CrossRef]
45. Isaacson, R.A. Electron spin resonance in n-Type InSb. *Phys. Rev.* **1968**, *169*, 312–314. [CrossRef]
46. Orlov, V.G.; Sergeev, G. Numerical simulation of the transport properties of indium antimonide. *Phys. Solid State* **2013**, *55*, 2215–2222. [CrossRef]
47. Agubra, V.; Fergus, J. Lithium ion battery anode aging mechanisms. *Materials* **2013**, *6*, 1310–1325. [CrossRef]
48. Joho, F.; Rykart, B.; Blome, A.; Novák, P.; Wilhelm, H.; Spahr, M.E. Relation between surface properties, pore structure and first-cycle charge loss of graphite as negative electrode in lithium-ion batteries. *J. Power Sources* **2001**, *97*, 78–82. [CrossRef]
49. Tran, T.D.; Feikert, J.H.; Pekala, R.W.; Kinoshita, K. Rate effect on lithium-ion graphite electrode performance. *J. Appl. Electrochem.* **1996**, *26*, 1161–1167. [CrossRef]
50. Bläubaum, L.; Röder, F.; Nowak, C.; Chan, H.S.; Kwade, A.; Krewer, U. Impact of particle size distribution on performance of Lithium-Ion batteries. *ChemElectroChem* **2020**, *7*, 4755–4766. [CrossRef]
51. Roselin, L.S.; Juang, R.-S.; Hsieh, C.-T.; Sagadevan, S.; Umar, A.; Selvin, R.; Hegazy, H.H. Recent advances and perspectives of carbon-based nanostructures as anode materials for Li-ion batteries. *Materials* **2019**, *12*, 1229. [CrossRef]
52. Röder, F.; Braatz, R.D.; Krewer, U. Multi-scale simulation of heterogeneous surface film growth mechanisms in Lithium-Ion batteries. *J. Electrochem. Soc.* **2017**, *164*, E3335–E3344. [CrossRef]
53. Dong, Y.; Yang, S.; Zhang, Z.; Lee, J.-M.; Zapien, J.A. Enhanced electrochemical performance of lithium ion batteries using Sb_2S_3 nanorods wrapped in graphene nanosheets as anode materials. *Nanoscale* **2018**, *10*, 3159–3165. [CrossRef]
54. Wang, X.; Hwang, J.-Y.; Myung, S.-T.; Hassoun, J.; Sun, Y.-K. Graphene decorated by indium sulfide nanoparticles as high-performance anode for sodium-ion batteries. *ACS Appl. Mater. Interfaces* **2017**, *9*, 23723–23730. [CrossRef] [PubMed]
55. Ganesan, V.; Nam, K.-H.; Park, C.-M. Robust polyhedral $CoTe_2$–C nanocomposites as high-performance Li- and Na-Ion battery anodes. *ACS Appl. Energy Mater.* **2020**, *3*, 4877–4887. [CrossRef]
56. Zhang, W.; Zhang, Q.; Shi, Q.; Xin, S.; Wu, J.; Zhang, C.-L.; Qiu, L.; Zhang, C. Facile synthesis of carbon-coated porous Sb_2Te_3 nanoplates with high alkali metal ion storage. *ACS Appl. Mater. Interfaces* **2019**, *11*, 29934–29940. [CrossRef] [PubMed]
57. Kang, W.; Tang, Y.; Li, W.; Yang, X.; Xue, H.; Yang, Q.; Lee, C.S. High interfacial storage capability of porous $NiMn_2O_4$/C hierarchical tremella-like nanostructures as the Lithium-Ion battery anode. *Nanoscale* **2015**, *7*, 225–231. [CrossRef]
58. Zeng, Z.; Zhao, H.; Lv, P.; Zhang, Z.; Wang, J.; Xia, Q. Electrochemical properties of iron oxides/carbon nanotubes as anode material for lithium ion batteries. *J. Power Sources* **2015**, *274*, 1091–1099. [CrossRef]
59. Zhu, X.J.; Guo, Z.P.; Zhang, P.; Du, G.D.; Zeng, R.; Chen, Z.X.; Li, S.; Liu, H.K. Highly porous reticular tin–cobalt oxide composite thin film anodes for lithium ion batteries. *J. Mater. Chem.* **2009**, *19*, 8360–8365. [CrossRef]
60. Fransson, L.; Vaughey, J.; Benedek, R.; Edström, K.; Thomas, J.; Thackeray, M. Phase transitions in lithiated Cu_2Sb anodes for lithium batteries: An in situ X-ray diffraction study. *Electrochem. Commun.* **2001**, *3*, 317–323. [CrossRef]
61. Baggetto, L.; Allcorn, E.; Unocic, R.R.; Manthiram, A.; Veith, G.M. Mo_3Sb_7 as a very fast anode material for lithium-ion and sodium-ion batteries. *J. Mater. Chem. A* **2013**, *1*, 11163–11169. [CrossRef]
62. Wang, S.; He, M.; Walter, M.; Kravchyk, K.V.; Kovalenko, M.V. Monodisperse CoSb nanocrystals as high-performance anode material for Li-ion batteries. *Chem. Commun.* **2020**, *56*, 13872–13875. [CrossRef]
63. Pan, Q.; Wu, Y.; Zhong, W.; Zheng, F.; Li, Y.; Liu, Y.; Hu, J.; Yang, C. Carbon nanosheets encapsulated NiSb nanoparticles as advanced anode materials for Lithium-Ion batteries. *Energy Environ. Mater.* **2020**, *3*, 186–191. [CrossRef]
64. Hou, H.; Cao, X.; Yang, Y.; Fang, L.; Pan, C.; Yang, X.; Song, W.; Ji, X. NiSb alloy hollow nanospheres as anode materials for rechargeable lithium ion batteries. *Chem. Commun.* **2014**, *50*, 8201–8203. [CrossRef] [PubMed]

65. Song, G.; Choi, S.; Hwang, C.; Ryu, J.; Song, W.-J.; Song, H.-K.; Park, S. Rational structure design of fast-charging NiSb Bimetal nanosheet anode for Lithium-ion batteries. *Energy Fuels* **2020**, *34*, 10211–10217. [CrossRef]

66. Ning, X.; Li, Z. Centrifugally spun SnSb nanoparticle/porous carbon fiber composite as high-performance lithium-ion battery anode. *Mater. Lett.* **2021**, *287*, 129298. [CrossRef]

67. Fan, L.; Liu, Y.; Tamirat, A.G.; Wang, Y.; Xia, Y. Synthesis of ZnSb@C microflower composites and their enhanced electrochemical performance for lithium-ion and sodium-ion batteries. *New J. Chem.* **2017**, *41*, 13060–13066. [CrossRef]

68. Gómez-Cámer, J.L.; Novák, P. Polyacrylate bound $TiSb_2$ electrodes for Li-ion batteries. *J. Power Sources* **2015**, *273*, 174–179. [CrossRef]

nanomaterials

MDPI

Review

Zn Metal Anodes for Zn-Ion Batteries in Mild Aqueous Electrolytes: Challenges and Strategies

Vo Pham Hoang Huy [†], Luong Trung Hieu [†] and Jaehyun Hur *

Department of Chemical and Biological Engineering, Gachon University, Seongnam 13120, Gyeonggi, Korea; vophamhoanghuy@yahoo.com.vn (V.P.H.H.); LuongTrungHieu290694@gmail.com (L.T.H.)
* Correspondence: jhhur@gachon.ac.kr; Tel.: +82-31-750-5593; Fax: +82-31-750-8839
† These authors contributed equally to this study.

Abstract: Over the past few years, rechargeable aqueous Zn-ion batteries have garnered significant interest as potential alternatives for lithium-ion batteries because of their low cost, high theoretical capacity, low redox potential, and environmentally friendliness. However, several constraints associated with Zn metal anodes, such as the growth of Zn dendrites, occurrence of side reactions, and hydrogen evolution during repeated stripping/plating processes result in poor cycling life and low Coulombic efficiency, which severely impede further advancements in this technology. Despite recent efforts and impressive breakthroughs, the origin of these fundamental obstacles remains unclear and no successful strategy that can address these issues has been developed yet to realize the practical applications of rechargeable aqueous Zn-ion batteries. In this review, we have discussed various issues associated with the use of Zn metal anodes in mildly acidic aqueous electrolytes. Various strategies, including the shielding of the Zn surface, regulating the Zn deposition behavior, creating a uniform electric field, and controlling the surface energy of Zn metal anodes to repress the growth of Zn dendrites and the occurrence of side reactions, proposed to overcome the limitations of Zn metal anodes have also been discussed. Finally, the future perspectives of Zn anodes and possible design strategies for developing highly stable Zn anodes in mildly acidic aqueous environments have been discussed.

Keywords: Zn metal anode; aqueous Zn ion batteries; mildly acidic electrolyte; dendrite-free; hydrogen evolution reaction suppression

Citation: Hoang Huy, V.P.; Hieu, L.T.; Hur, J. Zn Metal Anodes for Zn-Ion Batteries in Mild Aqueous Electrolytes: Challenges and Strategies. *Nanomaterials* **2021**, *11*, 2746. https://doi.org/10.3390/nano11102746

Academic Editor: Christophe Detavernier

Received: 28 September 2021
Accepted: 14 October 2021
Published: 17 October 2021

Publisher's Note: MDPI stays neutral with regard to jurisdictional claims in published maps and institutional affiliations.

1. Introduction

Renewable energy supplies have drawn worldwide attention because of the scarcity of fossil fuels and the increasing global warming [1–3]. Nevertheless, renewable energy resources, such as solar, wind, and geothermal are interrupted by various climatic and natural factors; thus, grid-scale energy is a vital underpinning for the continued development of large-scale energy storage techniques, and secondary batteries are an indispensable choice for achieving this [4–6]. Since their successful commercialization in the 1980s, lithium-ion batteries (LIBs) have dominated the energy market and have been employed in various applications, from portable electronics to grid-scale energy storage systems [7–22]. Nevertheless, the growth of the LIB technology has been limited owing to its safety issues, limited Li supplies, and high intrinsic prices [23–26]. Sodium-ion batteries (SIBs) and potassium-ion batteries (PIBs) have been intensively studied as alternatives to LIBs. However, SIBs and PIBs utilize volatile, flammable, and toxic organic electrolytes, which lead to safety and environmental issues [27–33]. Therefore, various efforts have been made to develop potential alternatives to these batteries that are suitable for grid-scale applications [34–37]

Aqueous batteries are very promising alternatives for LIBs, SIBs, and PIBs because aqueous electrolytes are inexpensive and environmentally friendly, and hence can alleviate the risk of fire hazards and explosions [38–42]. In addition, fast charging and high energy density can be achieved with aqueous batteries because water has considerably higher (by two to three orders of magnitude) ionic conductance (i.e., ~1 S cm^{-1}) than most organic solvents [43–45].

Among the various aqueous batteries investigated to date, aqueous Zn ion batteries (AZIBs) have recently emerged as a promising technology capable of satisfying today's stringent battery requirements. The use of Zn metal as an anode in aqueous media is a unique feature of these batteries that contributes to their excellent performance. First, as compared to monovalent metals, such as Li or Na, Zn is not only readily available and cheap but is also known for its non-toxicity and chemical durability in aqueous systems [46]. Second, in mildly acidic electrolytes (pH = 4–6), Zn oxidizes to Zn^{2+} without producing intermediate products and possesses a high overpotential for the hydrogen evolution reaction (HER) [47]. Third, given the relatively narrow operating window in which gas (H$_2$ and O$_2$) formation can be prevented in water, Zn has a redox potential of -0.76 V (vs. the typical hydrogen electrode), which is suitable for battery applications. Finally, in its metallic state, Zn has a high theoretical capacity (820 mAh g^{-1}, 5854 mAh L^{-1}) [45,48,49]. However, the practical applications of AZIBs are significantly hindered by the formation of Zn dendrites and the occurrence of side reactions (HER and corrosion) on the Zn anode. Zn dendrites can transpierce the battery separator, causing a short circuit, while the corrosion of the Zn metal anode results in irreversible electrolyte consumption and the generation of insoluble byproducts, increasing the electrode polarity and degrading the battery performance. On the other hand, the H$_2$ gas produced from the electrochemical reaction elevates the internal pressure of the battery, causing safety hazards, as illustrated in Figure 1. As a result, in recent years, various efforts have been made to develop appropriate strategies to overcome the limitations associated with Zn metal anodes. Figure 2a highlights some representative reviews on Zn metal anodes published over the last three years [45,50–64]. Currently, AZIBs can be largely categorized into two types: those with alkaline electrolytes, normally Zn-Ag, Zn-air, and Zn-Ni batteries [65–67]. and those with mildly acidic electrolytes, such as Zn-MnO$_2$ and Zn-V$_2$O$_5$ batteries (as shown in Figure 2b) [68,69]. Although Zn-based batteries with alkaline electrolytes generally have higher energy densities than those with non-alkaline electrolytes, mildly acidic aqueous electrolytes can effectively inhibit the growth of Zn dendrites, thus improving the cyclic performance of the battery through highly reversible electrochemical plating/stripping of Zn^{2+} on the Zn anode. As shown in Figure 2c, the number of publications on aqueous electrolytes with mildly low pH has increased significantly year by year. However, comprehensive and systematic reviews on the challenges and strategies for the development of next-generation Zn metal anodes suitable for application in mildly acidic electrolytes are scarce. Therefore, it will be beneficial to review the recent studies on Zn metal anodes in mildly acidic electrolytes to provide guidelines for the development of high-performance AZIBs.

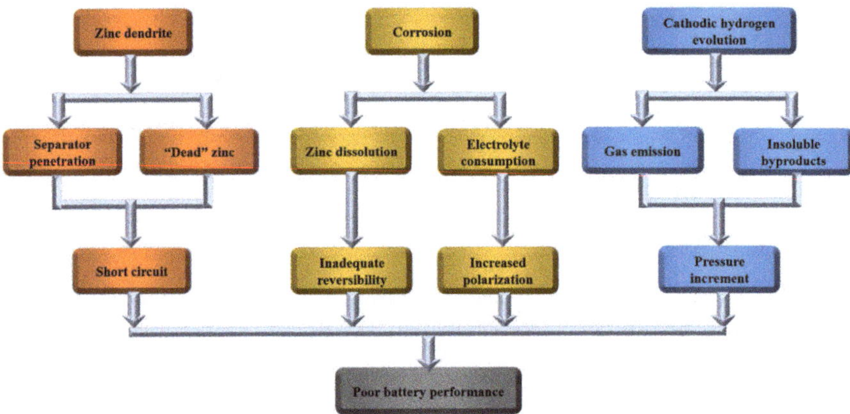

Figure 1. Major limitations of Zn metal anodes for battery applications.

(a)

2019

> ➤ Challenges in Zn electrodes for alkaline zinc–air batteries commercialization [63]
> ➤ Nanoscale design of Zn anodes for high–energy aqueous rechargeable batteries [64]

2020

> ➤ Aqueous Zn ion batteries: focus on zinc metal anodes [45]
> ➤ Issues and future perspective on Zn metal anode for rechargeable AZIBs [50]
> ➤ Recent advances in Zn anodes for high–performance aqueous Zn–ion batteries [51]
> ➤ Principals and strategies for a highly reversible Zn metal anode in aqueous batteries [52]
> ➤ Issues and solution toward zinc anode in aqueous zinc–ion batteries: A mini review [53]
> ➤ Inhibition of zinc dendrite in Zn–based flow batteries [54]
> ➤ Challenges in the material and structural design of Zn anode towards AZIBs [55]
> ➤ Dendrite issues and advances in Zn anode for aqueous rechargeable Zn–based batteries [56]
> ➤ Understanding the Zn anode behavior and improvement strategies in different AZIBs [57]

2021

> ➤ Zn–ion batteries: strategies for the stabilization of Zn metal anodes [58]
> ➤ The strategies of boosting the performance of highly reversible zinc anode in ZIBs [59]
> ➤ Strategies towards the challenges of Zn metal anode in rechargeable AZIBs [60]
> ➤ Controlling electrochemical growth of metallic Zn electrodes [61]
> ➤ The rising Zn anode for high–energy aqueous batteries [62]
> ➤ **This work**: Challenges and strategies towards Zn metal anode in mildly acidic aqueous electrolytes for Zn–ion batteries

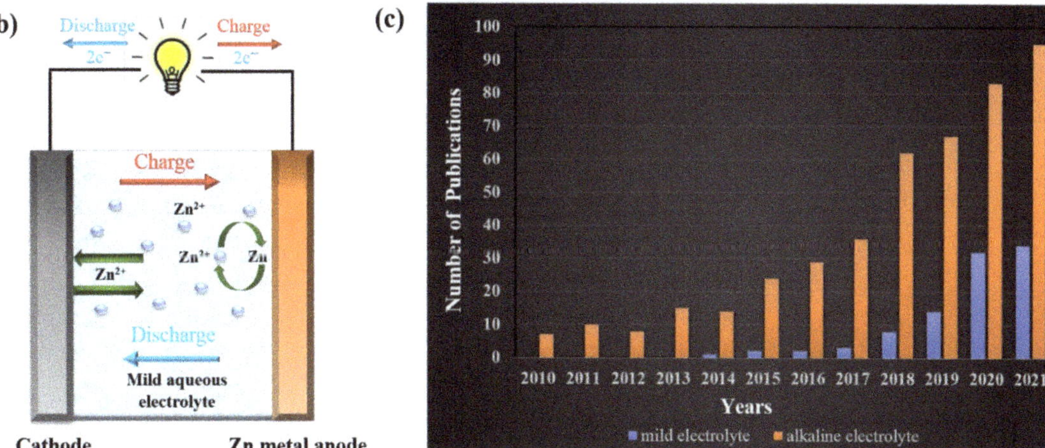

Figure 2. (**a**) Brief summary of recent reviews on Zn metal anodes for AZIBs [45,50–64], (**b**) a typical full cell battery configuration of AZIBs in a mildly acidic aqueous electrolyte. (**c**) The number of publications on AZIBs from 2010 to 2021 (search from Google Scholar; search time: 7 September 2021).

This review summarizes the strategies for the development of high-performance AZIB anode materials. First, the major limitations of zinc metal anodes are discussed followed by the approaches developed to overcome them. Finally, the future perspectives for next-generation AZIB research are discussed. This review will be beneficial for the rational design of Zn metal anodes in mildly acidic aqueous systems.

2. Challenges in the Commercialization of Zn Metal Anodes

Zn Anode Reactions

To date, various types of metal-ion aqueous rechargeable batteries (M = Li, K, Na, Ca, Zn, Mg, Al) operable in the pH range of 3–11 have been extensively investigated [70–72]. While most metals do not function properly in aqueous media because of their limited redox voltages, beyond which water becomes unstable, Zn exhibits advantages, such as a low redox potential (-0.76 V vs. the standard H_2 evolution), high overpotential for the HER, and high theoretical capacity (820 mAh g^{-1}, 5855 mAh cm^3), which make it suitable for application as a metal anode in aqueous systems [57,73,74] (Figure 3a–c).

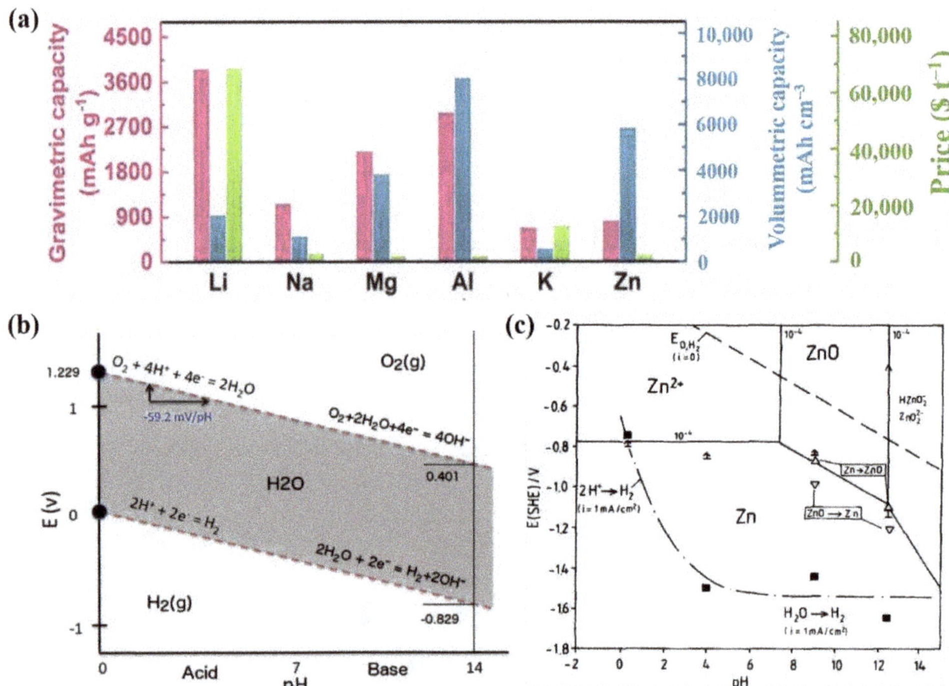

Figure 3. (**a**) Gravimetric capacity, volumetric capacity, and price of typical metal anodes. Reprinted with permission from Liang et al. [58]. Copyright 2021 Wiley-VCH GmbH. (**b**) Pourbaix diagram of water. Reprinted with permission from Zeng et al. [75]. Copyright 2019 Elsevier B.V. (**c**) Pourbaix diagram of the Zn/H_2O system with HER overpotential considerations. Reprinted with permission from Wippermann et al. [76]. Copyright 1990 Elsevier Ltd.

In AZIBs, the cathode reactions depend on the cathode material, whereas the Zn anode reactions are highly influenced by various electrolyte conditions. In alkaline media, Zn metal anodes oxidize to the zincate ion complex (Zn(OH)$_4^{2-}$) because of the presence of numerous OH$^-$ ions in the surroundings [77]. When the concentration gradient changes, these zincate ions disperse away from the electrode surface, causing the loss of the active material. In addition, when the solubility of the zincate ions decreases locally, a passive ZnO layer is formed on the electrode surface [78]. This precipitate results in dendritic growth and/or passivation, which reduces the rechargeable capacity of the alkaline aqueous electrolyte battery.

In mildly acidic aqueous media (virtually neutral media), the charge carrier is mainly Zn^{2+} owing to the lack of OH^-. Unlike the case in alkaline systems, only Zn^{2+} ions are reversibly stripped/plated on the Zn metal anode surface during the battery operation in mildly acidic aqueous media, which is similar to the Li metal anode mechanism in LIBs. Despite this, Zn electrodes are highly reactive and induce many side reactions with the electrolyte, which results in low Coulombic efficiency (CE) and poor cycle life. Typically, highly reversible Zn plating/stripping is possible in mildly acidic aqueous electrolytes, such as $ZnSO_4$ or $Zn(CF_3SO_3)_2$. In these media, the reactions of the Zn metal anode occur as follows:

Discharge process:

$$Zn \rightarrow Zn^{2+} + 2e^- \text{ (Zn stripping)} \tag{1}$$

Charge process:

$$Zn^{2+} + 2e^- \rightarrow Zn \text{ (Zn plating)} \tag{2}$$

Figure 4 summarizes the common problems encountered in alkaline (Figure 4a) and mild-pH media (Figure 4b). Although some typical problems (e.g., shape change and ZnO passivation) encountered in alkaline electrolytes are not serious in mildly acidic electrolytes because of their different Zn electrode reaction mechanisms, the most common problems, such as dendrite growth and H_2 evolution are the main reasons for the occurrence of irreversible reactions in ZIBs, which reduces the CE of Zn electrodes and deteriorates their performance. In the following section, we will focus on these issues of Zn metal anodes in mildly acidic electrolyte systems.

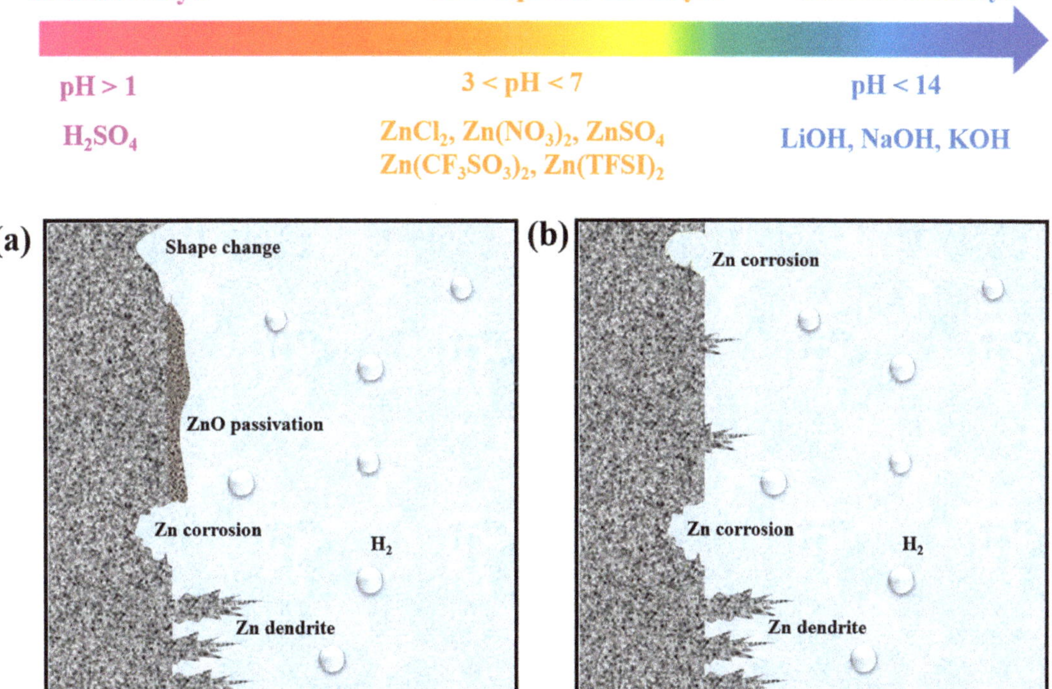

Figure 4. Schematic illustration of the phenomena observed on Zn electrodes in (**a**) alkaline and (**b**) mild aqueous electrolytes.

3. Assembly and Test Technology of Zn-Ion Batteries with Zn Metal Anodes

3.1. Cell Assembly

A typical AZIB consists of the following components: Zn metal (anode), an aqueous electrolyte, a separator, and cathode material. Figure 5 and Table 1 show the list of components and typical materials used in the AZIB coin-cell. Because of its tunnel or layered structure which allows reversible insertion/extraction of Zn^{2+} ions, MnO_2 has been extensively used as a cathode material in the early stages of mild aqueous ZIBs. Furthermore, manganese (Mn)-based oxides have been considered as promising energy storage materials due to their low cost, abundance, environmental friendliness, low toxicity, and numerous valence states (Mn^0, Mn^{2+}, Mn^{3+}, Mn^{4+}, and Mn^{7+}) [79,80]. The cathode material is prepared as a slurry by mixing it with conductive carbon and polymer binder and dispersing it in the organic solvent. The slurry is cast on the current collector (typically, stainless steel) with a well-defined thickness using the doctor blade technique. The electrolyte, as a component in direct contact with the Zn anode and directing the plating/stripping process of Zn, is critical to the electrochemical reversibility and stability of the Zn metal anode in AZIB systems. Currently, the $ZnSO_4$ or $Zn(CF_3SO_3)_2$ salt-based electrolytes are considered to be promising electrolytes in mild AZIBs [57]. Although a commercial Zn foil has been widely adopted as an anode and directly used as a current collector in most ZIBs, electrodeposited Zn electrodes on appropriate current collectors can also be the Zn metal anodes. The selection of the current collector is of great importance for the deposited Zn electrode; carbon-based, copper-based, and MOF-based current collectors have been widely used to support Zn, owing to their great chemical and electrochemical stability in various electrolytes, robust mechanical strength to accommodate deposition, high electrical conductivity, and close affinity for Zn [81].

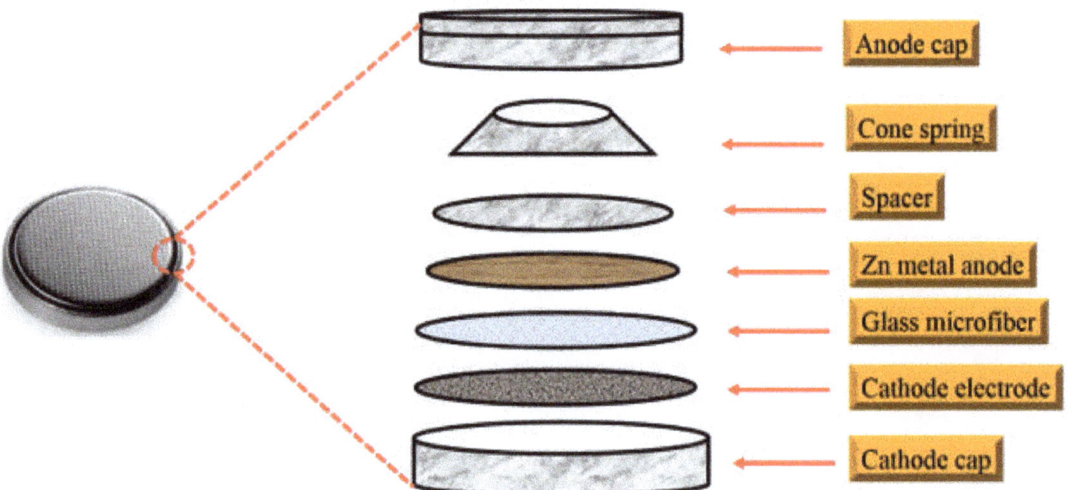

Figure 5. Components of the AZIB in coin cell assembly.

Table 1. List of components used in the typical coin-cell assembly of the ZIB with Zn metal anode [82].

Components	Representative Material
Anode material	Zn foil (80.0 mm in diameter, 0.25 mm in thickness)
Cathode material	α-MnO_2, β-MnO_2, V-based materials, Prussian blue analogues
Cathode current collector	Stainless steel spring (15.4 mm in diameter and 1.1 mm in thickness)
Separator	Whatman glass fiber filter
Electrolyte	2 M $ZnSO_4$ with 0.1 M $MnSO_4$

3.2. Cell Test

In AZIBs, the cell tests are typically performed with symmetric and full cell configuration. In evaluating the symmetrical cell performance, two identical electrodes (Zn/Zn) are used to investigate the coulombic efficiency (CE) and degree of polarization as a function of cycle number. CE is one of the important parameters to quantify the reversibility of an electrochemical system. The tendency of CE can be a useful tool in accurately predicting cycle life and the reversibility of Zn deposition/stripping, which is calculated based on the capacity ratio of stripping to plating. In Zn/Zn symmetric cells, there are three key parameters: cycle life, current density, and cycling capacity, which can provide a comprehensive picture of the electrochemical performance of symmetric batteries. Figure 6 is an example of evaluating a material's effectiveness through the symmetric cell test [83]. The stripping/plating stability and polarization of electrodes were evaluated at various current densities and cycling capacities with the symmetric cell test. For example, Yu et al. demonstrated the enhanced cycling stability of a Zn metal anode by using a Sn-coated separator. It was shown that a Zn/Zn symmetric cell with Sn-coated separator exhibits a dramatically improved cycle life of 3800 h (current density: $2\,mA\,cm^{-2}$, cycling capacity: $2\,mAh\,cm^{-2}$), 1000 h (current density: $5\,mA\,cm^{-2}$, cycling capacity: $5\,mAh\,cm^{-2}$) (Figure 6a,b). Furthermore, a highly reversible stripping/plating reversibility was achieved with a CE of ~99% at $0.3\,A\,g^{-1}$ after 600 cycles for the Sn-coated separator (Figure 6c).

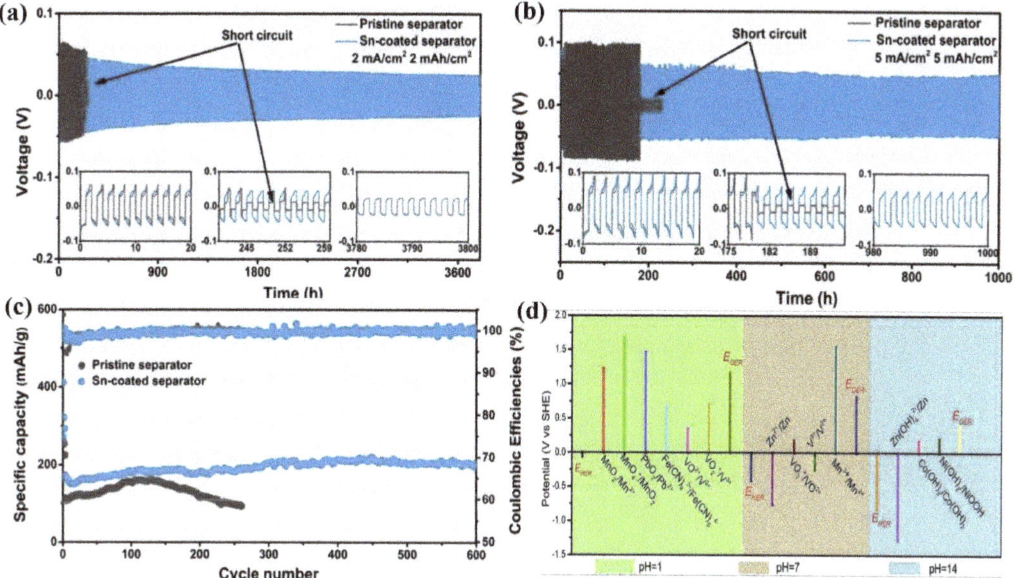

Figure 6. Cycling performance of Zn/Zn symmetric cells tested at (**a**) $2\,mA\,cm^{-2}$ and $2\,mAh\,cm^{-2}$, (**b**) $5\,mA\,cm^{-2}$ and $5\,mAh\,cm^{-2}$, (**c**) cycling performance at $0.3\,A\,g^{-1}$. Reprinted with permission from Zhen et al. [83]. (**d**) The standard potentials of redox couple in some reported cathode materials in ZIB systems. Reprinted with permission from [84]. Copyright 2021 Wiley-VCH GmbH.

In a full cell measurement, two different materials are used for cathode and anode which determines the voltage range that can be applied. In most cases, the practical output voltage of AZIBs cannot be fully in agreement with the theoretical voltage range. Theoretically, the maximum output voltage can be easily obtained by choosing a strong oxidizer for the cathode and a strong reducing agent for the anode, respectively. However, for AZIBs with Zn metal anode, the electrode potential is fixed (-0.76 V vs. SHE in a neutral or acidic solution, 1.23 V vs. SHE in an alkaline solution). Thus, the key to constructing high-voltage AZBs is to choose a suitable cathode material that has a lower electrode potential than Zn metal [84]. Figure 6d shows the standard potentials of redox couples in some reported cathode materials. For instance, Zn//MnO$_2$ batteries in a mild electrolyte deliver an output voltage that is below 1.5 V (vs. Zn/Zn^{2+}). The practical energy density, power density, and energy efficiency of the cell can be estimated from the full cell configuration.

4. Drawbacks of Zn Metal Anodes in Mildly Acidic Electrolytes
4.1. Zn Dendrite Growth

In AZIBs with mild aqueous electrolytes, the formation and growth of Zn dendrites within the battery is a major issue. The term "Zn dendrite" refers to a wide range of Zn morphologies with sharp ends or edges that pierce the separator and eventually lead to a short circuit and the breakdown of the cell. When Zn dendrites detach from an electrode, "dead" or "orphaned" Zn is easily generated, leading to a quick drop in the CE of the battery and an irreversible drop in its capacity. The Zn^{2+} concentration gradient in the proximity of the electrode surface influences the formation of Zn dendrites in a mildly acidic electrolyte. Zn ions diffuse on the electrode surface and accumulate easily on the nucleation sites, forming an initial protrusion. This further exacerbates the unequal electric field distribution, causing more Zn ions to be collected and the growth of dendrites over time via repeated plating/stripping at the anode. This inhomogeneous Zn^{2+} deposition directly affects the development of Zn dendrites [53,85]. Figure 7a shows the free energy diagram of the Zn reduction process, where Zn ions must overcome a nucleation energy barrier to reach a new solid stage. After the nucleation, Zn tends to be plated and accumulated in the pre-deposited region owing to the combined actions of the electric field and concentration gradient [86,87]. Figure 7b shows the voltage profile during the Zn deposition process. The difference between the potential minima and the subsequent stable potential, which indicates the thermodynamic cost of establishing a crucial atom cluster, is known as the nucleation overpotential, whereas the plateau overpotential is related to the Zn growth after the initial nucleation [88,89]. The lower the value of these two parameters, the better is the progress of the nucleation and growth processes (with low energy consumption). In addition, the electric field, ion concentration, and surface energy all influence the Zn nucleation. The "tip effect" causes Zn^{2+} to be concentrated in the protruded regions with high surface energy because the electric field at these points is substantially higher than those in the remaining areas. As a result, the Zn nucleation and growth are inclined to occur at such points. The "tip effect" leads to an uneven distribution of the electric field intensity at the surface of the electrode, a non-uniform ion concentration distribution, and the preferential deposition of Zn at locations with higher Zn^{2+} concentrations (Figure 7c,d). It is obvious that the Zn^{2+} concentration is high in a region with a high electric field where dendrites are easily formed [90]. While Zn dendrites are homogeneously formed at surfaces with high surface energies, which can minimize the nucleation barriers by generating a large number of nucleation sites, irregular Zn dendrites are formed at the surfaces with a large number of nucleation barriers because of the unequal distribution of the electric field or a small number of nucleation sites [53,58,91].

Kang et al. proposed an intriguing concept concerning the dimensions and distribution of pores in the coating material, which influences the evolution of Zn protrusions/dendrites, to understand the formation of Zn dendrites [92]. They proposed that the dimensions and dispensation of pores in a filter paper resemble those of the immense protuberances on cycled Zn foils (Figure 8a,b vs. Figure 8c,d). Therefore, a study with a filter paper can provide an understanding of the stripping/plating behavior of Zn^{2+} on Zn metal anodes. During cycling, the pores in a filter paper (an example of a porous coating material) behave similarly to the holes in water-permeable bricks (Figure 8e), forming a region highly active for Zn stripping/plating. As the number of cycles increases, additional Zn accumulates on the Zn metal surface, which makes contact with the porous regions in the coating material and generates a large number of protrusions, affecting the efficiency of the Zn^{2+}

stripping/plating processes in diverse ways. First, the separator with uneven voids (such as filter paper) kinetically deteriorates the electrochemical performance of Zn anodes by limiting the local electrolyte transport. Second, the effect of the composition of the filter paper pores on the Zn dendrite evolution is pore dimension dependent; that is, a tiny pore promotes the development of small protrusions/dendrites. As a result, the dimensional characteristics of the pores in the covering material are important for regulating the Zn dendrite formation.

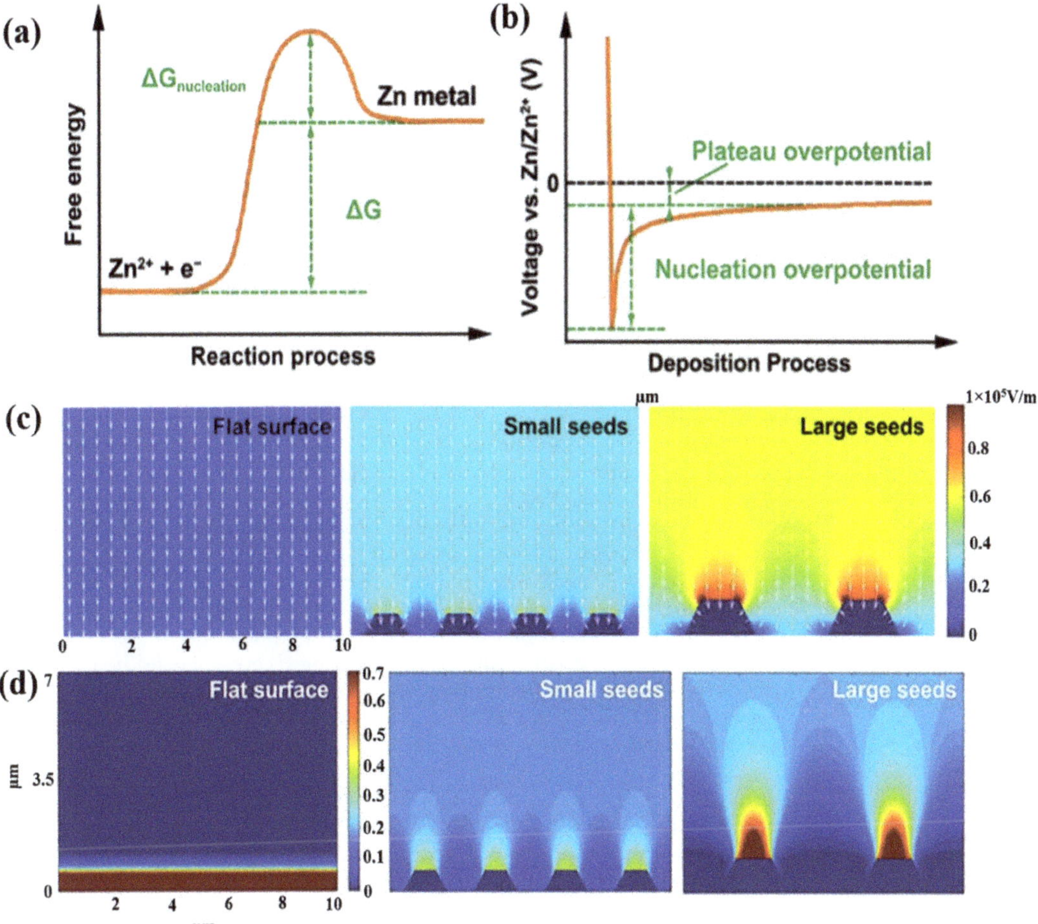

Figure 7. (a) Energy barrier for the Zn nucleation process, (b) voltage profile during Zn deposition. Reprinted with permission from Pei et al. [87]. Copyright 2017 American Chemical Society. Simulation of (c) electric field and (d) ion distribution on the Zn anode surface under different dendrite formation conditions: flat surface, small dendritic seeds, and large dendritic seeds. Reprinted with permission from [90]. Copyright 2019 WILEY-VCH Verlag GmbH&Co. KgaA, Weinheim.

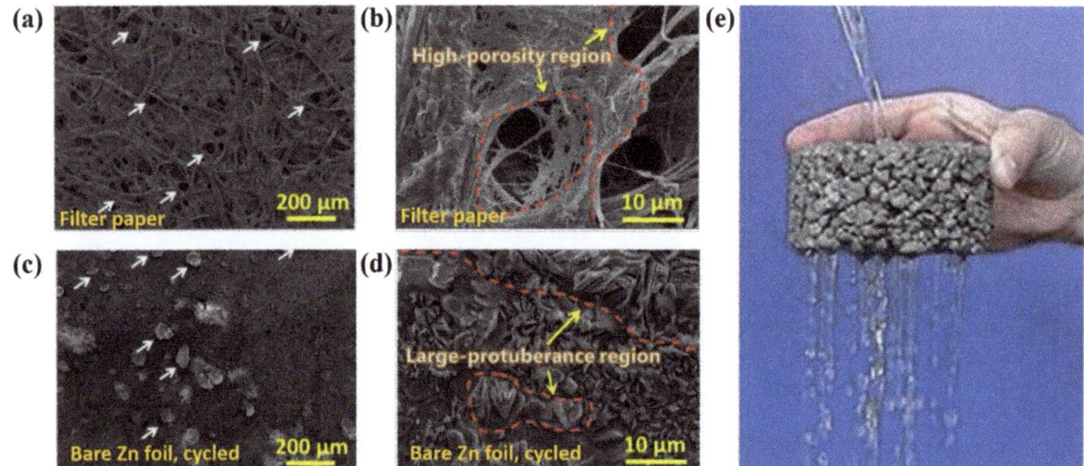

Figure 8. SEM images of (**a**,**b**) a fresh filter paper, and (**c**,**d**) a cycled bare Zn foil. The pores in filter papers act as "highways" for electrolyte transport, like the (**e**) pores in water-permeable bricks. Reprinted with permission from [92]. Copyright 2018 WILEY-VCH Verlag GmbH&Co. KgaA, Weinheim.

4.2. Zn Electrode Corrosion

Owing to its amphoteric nature, Zn can react with OH^- and H^+ ions. The presence of a considerable number of OH^- species in alkaline electrolytes deteriorates the performance of Zn anodes by forming $Zn(OH)_4^{2-}$ or ZnO by-products. On the other hand, electrolytes with mildly low pH can limit the occurrence of such unfavorable reactions because of the presence of modest amounts of H^+ species [93]. According to the Pourbaix diagram, although Zn metal is thermodynamically stable over a wide pH range in mild aqueous media, the corrosion caused by free water should still be considered [94]. Unlike the case of Li metal batteries, in which the solid electrolyte interface (SEI) prevents the excessive consumption of Li metal, in AZIBs, Zn deposition is hampered by H_2 evolution, leading to severe Zn metal corrosion. When the corrosion process continues, the Zn anode surface becomes more uneven and unsafe because of the increased pressure inside the cell. A recent study demonstrated that some Zn salts (e.g., $Zn(NO_3)_2$ and $Zn(ClO_4)_2$) may play a negative role in the Zn corrosion reaction [95]. For example, in the early stages of corrosion in a neutral electrolyte, a strong oxidizing agent with NO_3^- anions leads to the corrosion of the Zn foil and cathode materials. Although $Zn(ClO_4)_2$ with ClO_4^{2-} having four O atoms and one Cl atom located at the corners and center of the tetrahedral structure, respectively, shows low reactivity, it requires an excessive operational voltage, which leads to slow reaction kinetics because of the formation of by-products [96]. Zinc halides (e.g., $ZnCl_2$ and ZnF_2) have been widely studied as Zn-based electrolytes owing to their poor oxidative properties [97]. However, there are certain disadvantages of these electrolytes: (i) the utilization of the ZnF_2 electrolyte is limited by its low water solubility; (ii) although $ZnCl_2$ is highly soluble in water, it has a narrow stable potential window for the occurrence of anodic electrochemical reactions without any side reactions, which limits its application in AZIBs. On the other hand, $ZnSO_4$ has been extensively employed as an electrolyte for AZIBs because of its low cost, good solubility, wide potential window, and mild pH in water. Furthermore, in such mild $ZnSO_4$ electrolytes, Zn anodes exhibit excellent dissolution/deposition reaction dynamics, little dendritic development, and moderate corrosion [98–100]. Nevertheless, the exact electrochemical reaction mechanism of Zn metal anodes in $ZnSO_4$ electrolytes is not known yet, partly because of the formation of $Zn_4(OH)_6SO_4nH_2O$ (ZHS). It is believed that the pH fluctuations of the electrolytes throughout the discharge process are normally regulated by ZHS according to the following reaction.

$$4\,Zn^{2+} + SO_4{}^{2-} + 6\,OH^- + 5\,H_2O \rightarrow Zn_4(OH)_6(SO_4)\cdot 5\,H_2O \tag{3}$$

These by-products are formed by prolonged electrolysis and Zn ion consumption, which generally reduces the plating/stripping CE of the Zn anode to some extent [13]. As a result, extra Zn is required to ensure continuous cycling to prevent the Zn anode from reaching its maximum theoretical capability. Furthermore, the inert by-products deposited on the Zn surface obstruct the ion transfer and reduce the reversibility of the Zn anode.

Recently, Cai et al. studied the corrosion process of Zn metal anodes in $ZnSO_4$ electrolysis [101]. They found that the corrosion of the Zn metal anode started from the surface layer, leading to the formation of $Zn_4(OH)_6SO_4$, followed by H_2 evolution. Subsequently, $Zn_4(OH)_6SO_4$ is further hydrated and converted into $Zn_4(OH)_6SO_4\cdot 5H_2O$, leading to the considerable corrosion of the Zn metal anode (Figure 9a–c). In this way, the Zn metal surface could not be passivated, and the corrosion proceeded until the liquid electrolyte or active Zn metal was completely consumed. Notably, the corrosion process continued until an uneven corrosion depth of 132.2 µm was achieved (Figure 9d), which could be clearly detected by the O atom signal in the energy-dispersive X-ray spectroscopy (EDS) elemental mapping image.

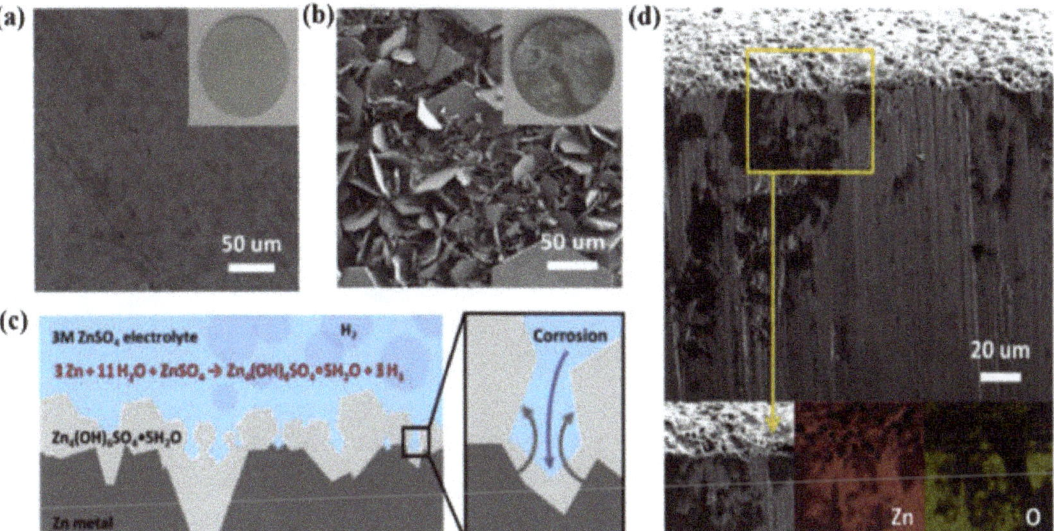

Figure 9. SEM images of a Zn electrode (**a**) before and (**b**) after 30 days of operation, (**c**) Chemical corrosion of the Zn metal electrode in a $ZnSO_4$ electrolyte. (**d**) cross-section SEM and EDS elemental mapping results of the electrode operated for 30 days. Reprinted with permission from Cai et al. [101]. Copyright 2020 Elsevier B.V.

4.3. Hydrogen Evolution

Metal corrosion is a common side effect of the HER, which is another key issue limiting the applications of ZIBs. Owing to its lower electronegativity than that of H, Zn prefers to react with water in neutral or slightly acidic electrolytes [102]. In addition, the Zn^{2+} deposited on the surface of highly reactive Zn metal anodes enhances the occurrence of side reactions. The hydrogen evolution that occurs over time not only dries out the electrolyte but also speeds up the hydration of the Zn cations in the aqueous electrolyte, resulting in suboptimal Zn usage as compared to the theoretical capacity. This generates H_2 gas, which also corrodes the Zn metal surface, deteriorates the battery performance, and poses a safety risk. Furthermore, the consumption of H^+ in water leads to an increase in the OH^- ion concentration at the interface of the Zn anode/electrolyte, which produces insulating by-products and causes non-uniform Zn plating [103]. In mildly acidic electrolysis, the occurrence of the HER is inevitable because the standard reduction potential of Zn/Zn^{2+} lags behind the H_2 evolution potential (0 V vs. SHE) [104]. as follows [105]:

$$Zn^{2+} + 2e^- \leftrightarrow Zn \ (-0.76 \ V) \tag{4}$$

$$2 \, H^+ + 2e^- \leftrightarrow H_2 \ (0 \ V) \tag{5}$$

Because Zn has a large overpotential for H_2 evolution in aqueous electrolytes, the HER is not as problematic as the values indicate [106], which could be described by the Tafel equation as follows [107].

$$\eta = b \, \log i + a \tag{6}$$

where η is the H_2 evolution overpotential, i is the current density, b is the Tafel slope constant, and a is the overpotential when the current density i is equal to the unit current density. Because b is roughly the same for all metals (0.112 V), the value of the overpotential for H_2 evolution is mostly determined by the value of a [108]. Zn has a strong H_2 overpotential as well as a high a value. According to the Pourbaix diagram of Zn in aqueous media, the high overpotential of H_2 evolution on the Zn metal surface suppresses the evolution of H_2 [46,109]. Nevertheless, in reality, the H_2 overpotential is also influenced by various other factors, including the roughness of the Zn surface [110], operating temperature [111], and Zn concentration [112]. As a result, under certain conditions, H_2 evolution can be observed even in mildly acidic electrolytes. Because the HER is affected by the Zn surface conditions and the interaction of the Zn anode with the electrolyte, the HER is related to the formation of dendrites. On the one hand, the dendritic growth leads to a porous structure of the Zn anode with a larger specific surface area, which provides more reaction sites for the HER. On the other hand, it can be thought that an increase in the specific surface area lowers the current density, which suppresses the HER and corrosion on the Zn surface by increasing the overpotential and forming nonconductive by-products capable of obstructing the electron transfer and Zn deposition.

Overall, the three drawbacks of Zn dendrite growth, Zn electrode corrosion, and hydrogen evolution have been discussed in the Zn metal anode. These drawbacks are not simple independent problems but are closely correlated with one another. The formation of dendrite increases the surface area of the Zn metal anode, which contributes to the accelerated hydrogen evolution. Hydrogen evolution causes a change in the local pH due to an increase in the OH^- concentration; however, it simultaneously accelerates the electrochemical corrosion reaction and changes the anode surface. In addition, the inert byproducts from the corrosion on the anodic surface can lead to the non-uniform surface and increased electrode polarization, which in turn facilitates the dendrite formation. Considering this complex phenomenon, it is necessary to tackle all these entangled problems with a comprehensive viewpoint rather than addressing each one of these problems separately.

5. Common Strategies for Modifying the Surface of Zn Metal Anodes

The electrochemical behavior of AZIBs is highly dependent on the structure of the Zn metal electrode surface. Therefore, various methods have been proposed to modify the surface of Zn metal electrodes. These methods can be categorized into several main approaches, including shielding the Zn metal to prevent side reactions, regulating the Zn deposition behavior, and creating a uniform electric field, as illustrated in Figure 10.

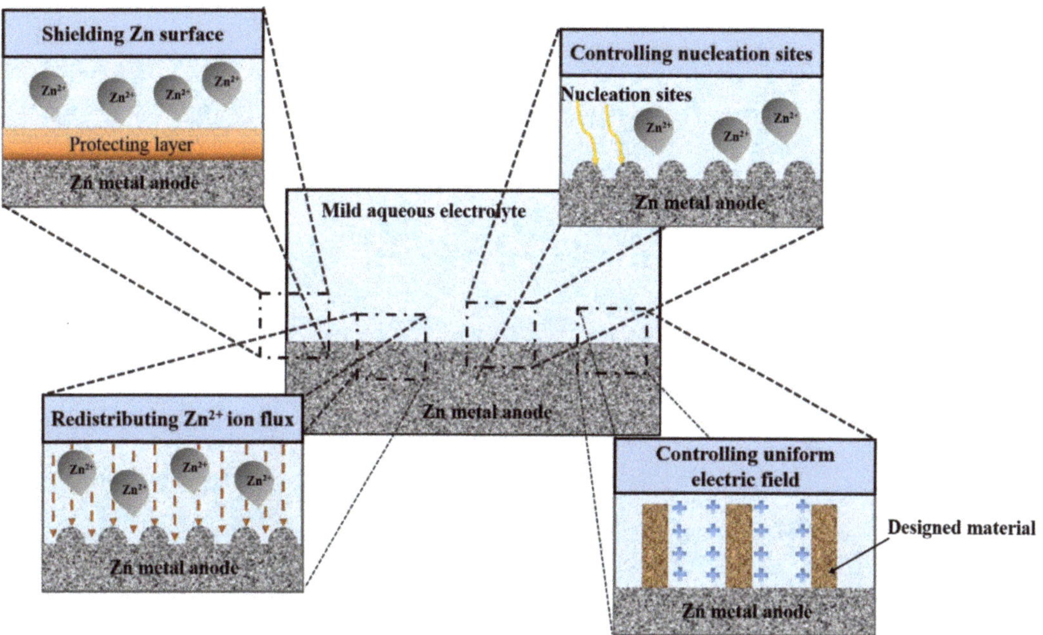

Figure 10. Modification strategies for enhancing the electrochemical performance of Zn metal anodes.

5.1. Shielding the Zn Surface

In LIBs and other battery systems with alkali metal anodes (e.g., Na and K), the SEI is automatically formed, which acts as a protective barrier to prevent undesired reactions [45]. However, in AZIBs, an SEI is indispensable. Because most of the electrolytes used in AZIBs are highly corrosive ($ZnSO_4$, $ZnCl_2$, etc.) and Zn metal is quite steady in both aqueous and non-aqueous conditions, the natural formation of the SEI layer becomes insignificant. As a result, covering a Zn metal electrode with an artificial protective coating is one of the simplest yet efficient ways to enhance its stability. Preventing immediate contact between the electrode and electrolyte is the key to enhancing the electrochemical behavior of AZIBs. To overcome the harsh operational conditions of AZIBs, which are more complex than those of LIBs, the shielding material on the Zn metal anode must possess electrochemical and chemical stability. Atomic layer deposition (ALD) is an efficient technique for coating Zn metal electrodes as it offers advantages such as large coverage, conformal deposition, and precisely controllable film thickness at the nanoscale. The basic working principle of ALD relies on the chemical reactions between two or more precursors pumped alternately into a chamber containing a substrate at a specific temperature and pressure, allowing materials to be deposited in a layer-by-layer fashion on the substrate surface [113,114]. Unlike chemical vapor deposition (CVD) and other related deposition processes, ALD pumps the precursors progressively rather than simultaneously. Although ALD and CVD have certain similarities, they are different in terms of the self-limiting properties for precursor adsorption, as well as the alternate and sequential entry of the precursors and reactants [115]. A general ALD process is illustrated in Figure 11.

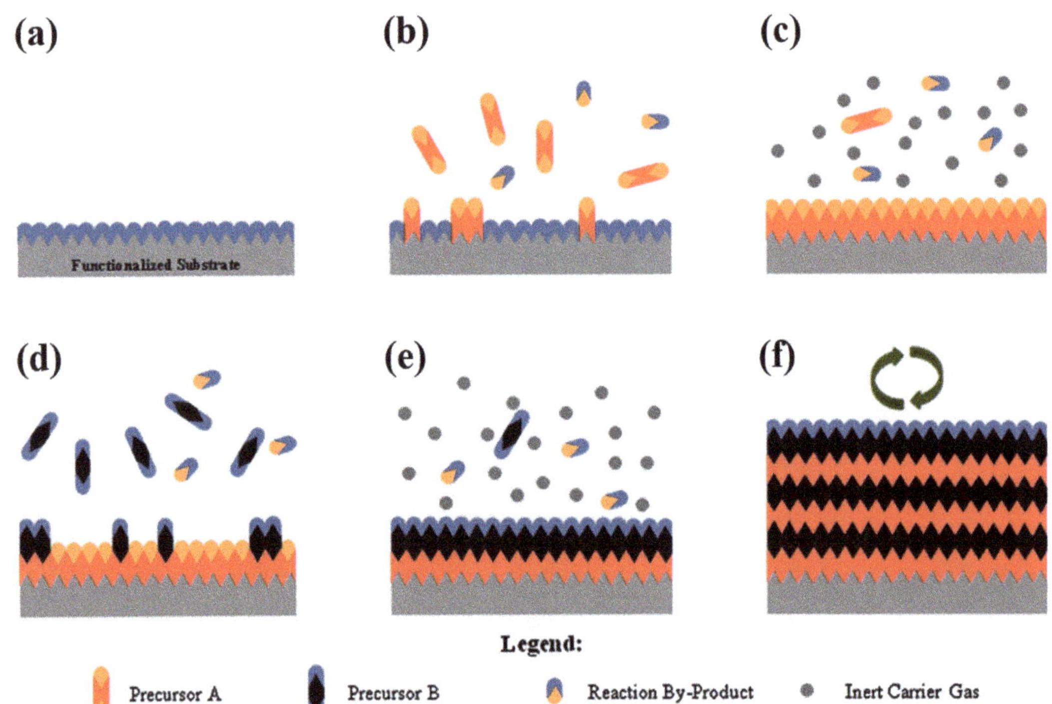

Figure 11. Schematic illustration of the ALD process; (**a**) the substrate surface is naturally functionalized or treated to be functionalized; (**b**) precursor A reacts with the surface after being pulsed; (**c**) an inert carrier gas is used to remove the excess precursor and by-products; (**d**) the surface reacts with the pulsed precursor B; (**e**) the inert carrier gas is used to remove excess precursor and byproducts; (**f**) repeat steps 2–5 until the desired material thickness is achieved. Reprinted with permission from Johnson et al. [114]. Copyright 2014 Elsevier Ltd.

Inspired by these advantages of ALD, Zhao et al. fabricated an ultrathin TiO_2 coating using the ALD method for the first time [116]. This passivation layer was chemically stable enough to resist the mildly acidic state and prevented the electrolyte and Zn plate electrode from coming into direct contact (as shown in Figure 12a,b). The undesirable HER was efficiently restricted under the protection of amorphous TiO_2 (8 nm in thickness), resulting in decreased gas generation, thus reducing the risk of cell breakage by the increased internal pressure. As a result, the symmetric cell with the ALD TiO_2@Zn electrodes had a low overpotential of 72.5 mV at 1 mA cm^{-2} and could hold it for 150 h without fluctuation, whereas the cell with the bare Zn electrodes only managed to cycle for 10 h (Figure 12c). The scanning electron microscopy (SEM) images of the electrodes cycled for 150 h revealed the presence of a large number of flakes on the surface of the Zn plate without the TiO_2 coating (Figure 12e); however, in the case of the Zn plate coated with ALD TiO_2, only a few flakes were visible, and their dimensions were substantially smaller than those of the flakes observed on the uncoated plate (Figure 12d). After several cycles, the hydrolysis of Zn^{2+} caused the loss of the solvent (water), leading to the formation of $Zn(OH)_2$. The presence of $Zn(OH)_2$ rather than Zn dendrites on the electrode surface is another issue that needs to be resolved because it is thermodynamically unfavorable in slightly acidic solutions. In this study, the full cell performance was tested using an ALD TiO_2@Zn anode and a MnO_2 cathode (ALD TiO_2@Zn-MnO_2). With TiO_2 protection, the ALD TiO_2@Zn-MnO_2 full cell exhibited a discharge capacity of 235 mAh g^{-1} after 60 cycles. In contrast, the Zn–MnO_2 cell showed a rapid capacity decay (155 mAh g^{-1} after 60 cycles) (Figure 12f). Additionally, the ALD TiO_2 coating also enhanced the CE of the Zn plate (Figure 12g).

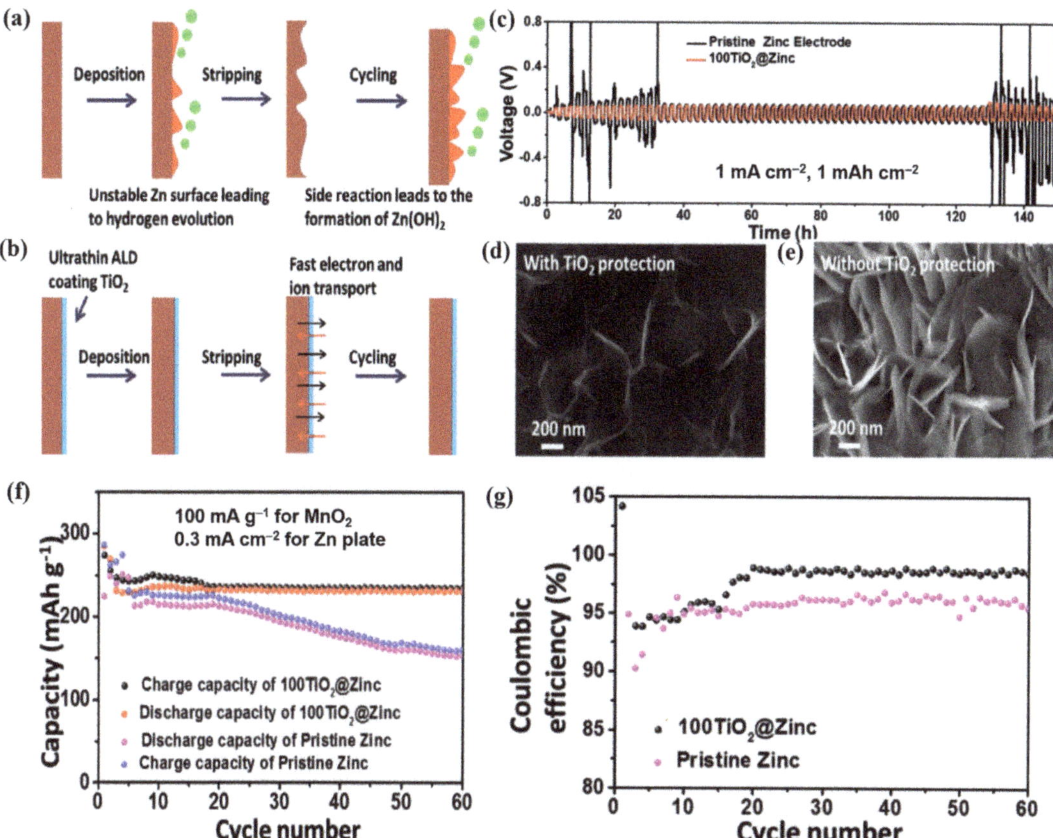

Figure 12. (**a**) Schematic illustration of Zn corrosion and H_2 evolution under repeated plating/stripping cycles, (**b**) stable deposition/stripping process with a thin layer of TiO_2 coated on the Zn anode, (**c**) symmetric cell performances of pristine Zn and TiO_2@Zn, (**d**,**e**) ex-situ SEM images of the (**d**) TiO_2@Zn and (**e**) pristine Zn anodes, (**f**) full cell performances of ALD TiO_2@Zn-MnO$_2$ and Zn-MnO$_2$ at 100 mA g^{-1}, (**g**) CEs of the ALD TiO_2@Zn-MnO$_2$ and Zn-MnO$_2$ full cells at 100 mA g^{-1}. Reprinted with permission from [116]. Copyright 2018 WILEY-VCH Verlag GmbH&Co. KgaA, Weinheim.

Inspired by Zhao's work, He et al. [117]. coated an ultrathin Al_2O_3 film on a Zn metal foil via ALD. Unlike sol-gel Al_2O_3 coatings, the ALD Al_2O_3 layer was homogeneous and thin enough (10 nm) to act as a corrosion inhibitor. Because the ALD Al_2O_3 coating provided effective surface wetting, less electrolyte was used during the repeated Zn stripping/plating while maintaining high efficiency. The parasitic processes were successfully suppressed by shielding the Zn surface, resulting in a significant reduction of inactive by-products such as $Zn(OH)_2$. Furthermore, the hydrophilic Al_2O_3 coating significantly improved the wettability of the Zn anode surface. It is believed that excess electrolyte plays a key role in improving the half-cell performance of Zn metal anodes. He et al. evaluated the charge-discharge profiles of symmetric cells with different electrolyte contents and found that the cells could show competitive results even at low electrolyte concentrations because of the presence of the ALD Al_2O_3 thin coatings. The improved wettability of the Zn foil not only reduced the charge-transfer barrier but also facilitated additional Zn^{2+} ion flux through the surface. As a result, the Al_2O_3@Zn symmetric cells could cycle for up to 500 h while maintaining a minimal overpotential (36.5 mV) at 1 mA cm^{-2}. The Zn nucleation overpotential was related not only to the dynamics of the Zn plating/stripping reaction but also to the transfer of Zn^{2+} and electrons. Zn^{2+} and electron transport are normally the intrinsic features of the electrolyte and electrode employed; however, the Zn stripping/plating kinetics are affected by a variety of variables, including the Zn nucleus dimension and surface tension, size distribution, and nucleation substrate shape. Furthermore, the ALD Al_2O_3 showed a high surface wetting ability. As a result, the cells showed excellent electrochemical performance even at low electrolyte contents, which made them suitable for practical applications.

Meanwhile, Cai et al. [101]. further explored this strategy by decorating a Zn anode with an inert metal (Cu) via a facile replacement reaction (Figure 13a). Cu shows excellent chemical stability and high conductivity in aqueous electrolytes; thus, the deposition of a Cu-rich composite surface efficiently prevents the corrosion of Zn metal. The corrosion potential of the Cu-Zn electrode (-0.964 V) was higher than that of the pristine Zn electrode (-0.976 V), which indicates that the deposition of Cu improved the chemical stability of the Zn electrode, as illustrated in Figure 13b. As compared to the pristine Zn anode, the Cu-Zn anode showed significantly high resistance. The Cu-Zn alloy (primarily Cu_5Zn_8) was produced in situ on the Cu/Zn anode during the electrochemical cycling, leading to the formation of a dense and Cu-rich surface layer. This alloy acted as a corrosion inhibitor and stabilizer for the electrode. Furthermore, its shape remained compact and smooth over time and prevented the deep penetration of bulk zinc metal by the electrolyte, which resulted in the formation of a significant amount of "dead Zn" and accelerated the corrosion. Another interesting approach was reported by Xie et al., who reformed a Zn plate with three-dimensional (3D)-nanoporous ZnO (3D-ZnO@Zn) [118]. The activation of the thin surface layer in the electric double layers (EDLs) by alien molecules on the inner Helmholtz plane improved the stability and long-term cycling performance of the electrode. Because ZnO had a 3D architecture with uniformly dispersed O on the surface of Zn, it not only reduced the current density by limiting the "tip effect," but also suppressed the side-reactions at the interface and the formation of H_2 by creating a tight exterior solvate sheath. In contrast, in the case of the bare Zn anode, Zn^{2+} showed sluggish transport kinetics into the host, resulting in strong polarization, high nucleation potential, and low stripping/plating efficiency. Another objective of this research was to gain a better understanding of the dynamic tuning of the Zn^{2+} transport toward the anode. In contrast to TiO_2 or Al_2O_3, which completely shielded the Zn surface from the corrosive electrolyte, the porous ZnO layer showed strong electrostatic attraction for Zn^{2+} ions (preferably the solvated ones) in the EDL. H_2 evolution was suppressed because of this propensity. Furthermore, because of its conductive and well-connected framework, the 3D architecture permitted quick Zn^{2+} transport, resulting in dramatically enhanced deposition kinetics. For a more precise assessment of the deposition dynamics, the exchange current density in the Zn electrodeposition process was calculated using the following equation:

$$i \sim i_0 \frac{F}{RT} \frac{\eta}{2} \tag{7}$$

where i is the exchange current density, i_0 is the reference exchange current density, F is the Faraday constant, R is the gas constant, T is the absolute temperature, and η is the total overpotential. The exchange current density represents the redox reaction rate of the electrode at the equilibrium potential. The value was determined from the temperature and H_2 concentration as well as the surface modification and cycling parameters. In addition to improving the deposition kinetics, the Arrhenius equation was used to examine the activation energy (E_a), which can provide information on the transfer and desolvation of Zn^{2+}.

$$\frac{1}{R_{ct}} = A exp\left(-\frac{E_a}{RT}\right) \tag{8}$$

where R_{ct} is the charge transfer resistance, A is the pre-exponential factor, E_a is the activation energy, R is the gas constant, and T is the absolute temperature. Meanwhile, Zn deposition unavoidably competed with H_2 evolution. The HER performance of the electrode was evaluated using linear sweep voltammetry (LSV). This novel structure and artificial surface improved the Zn^{2+} deposition kinetics (nucleation potential of only 42.4 mV for 3D-ZnO@Zn vs. 66.9 mV for bare Zn; charge transfer resistance of 292.7 Ω for 3D-ZnO@Zn vs. 1240 Ω for bare Zn), reduced the de-solvation energy consumption (51.0 kJ mol^{-1} for 3D-ZnO@Zn vs. 77.2 kJ mol^{-1} for bare Zn) in the EDLs, and suppressed the HER. Besides this, 3D-ZnO@Zn showed a low current density of 7.938 mA cm^{-2}, whereas pristine Zn exhibited a value of 19.68 mA cm^2, indicating the paradoxically sluggish deposition dynamics of 3D-ZnO@Zn.

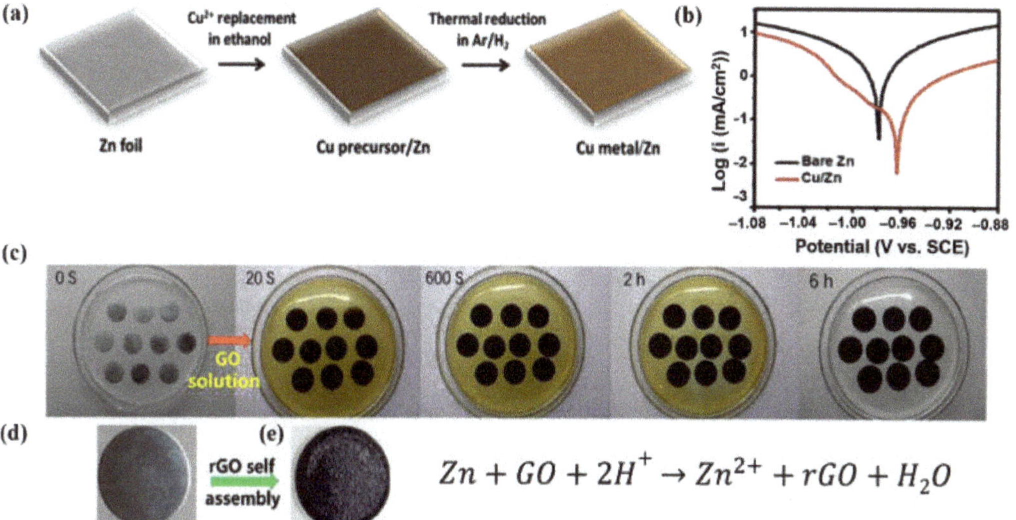

Figure 13. (a) Schematic illustration of the fabrication of the Cu-Zn electrode. (b) linear polarization curve of the Cu/Zn electrode in a 3 M ZnSO$_4$ electrolyte. Reprinted with permission from Cai et al. [101]. Copyright 2020 Elsevier B.V. (c) Photographs depicting the preparation of the Zn/rGO. Zn plate (d) before and (e) after coating the rGO film. Reprinted with permission from Xia et al. [119]. Copyright 2019 Elsevier B.V.

The idea of shielding the Zn surface to prevent the formation of undesired by-products is not limited to coating with an inert metal or oxide passivation layer. Xia et al. employed a casting method to modify a Zn mesh anode with reduced graphene oxide (rGO) for application in ZIBs [119]. As shown in Figure 13c, when the pretreated Zn foil was dipped into a brown aqueous solution with a dispersion of GO, the Zn surface quickly became dark, and with an increase in the incubation time, the brown GO aqueous solution turned colorless and transparent, indicating that the GO was completely reduced to rGO. Finally, the color of the Zn plate turned black (Figure 13d), indicating that a uniform black rGO layer was coated onto the Zn foil (Figure 13e). The rGO coating deposited on Zn plates offered many advantages. First, the low density of rGO increased the energy density of the full cell. Second, the layered rGO provided a flexible framework, which significantly reduced the volume change during Zn stripping/plating while improving the cyclic stability of the cell. Finally, owing to its large specific surface area, rGO promoted uniform Zn deposition during cycling and inhibited the formation of Zn dendrites on the Zn plate. In this study, the porous rGO foam on the Zn anode scaffold successfully prevented the production of Zn dendrites.

5.2. Regulating the Zn Deposition Behavior

5.2.1. Controlling the Nucleation Sites

In traditional planar metal electrodes, uncontrollable dendrite formation has been an unresolved issue. Large protuberances on the Zn surface can pierce the separator, causing a short circuit and cell failure. The basic strategy of controlling the nucleation sites is to create additional Zn^{2+} nucleation sites, regulate the Zn plating/stripping behavior by Zn^{2+} nucleation and deposition homogeneity, restrict the dendrite growth, and prevent the Zn corrosion. As a result, several studies focusing on regulating this protrusion-forming tendency aimed to mitigate the uncontrollable dendrite formation in Zn electrodes. For instance, Cui et al. proposed the novel concept of covering Zn electrodes with Au nanoparticles (Au NPs) [120]. They decorated a Zn anode with quasi-isolated nano-Au particles to manage its Zn striping behavior, promoting the nucleation/deposition on Zn, reducing the "tip effect", and therefore eliminating the dendritic/protuberance growth. During the stripping stage, the exposed Zn metal between the Au-NPs tended to strip faster than the Zn metal covered by the Au-NPs, leading to the formation of artificial nano-Zn tips. Because of their high Zn^{2+} affinity, the exposed Zn areas could serve as nucleation sites for directing the homogenous Zn deposition process in consecutive cycles. Because of the presence of the Au-NPs, the protrusions could be managed well and replaced by an ordered Zn plating layer even in the first few cycles. As a result, the plating layer aided the homogenous nucleation of Zn^{2+} ions, resulting in the formation of well-organized Zn flake arrays. The cell performance improved significantly because of the nucleation control mechanism, which allowed the cell to function for up to 2000 cycles with a capacity of 67 mAh g^{-1} (500 mA g^{-1}).

Following this approach, Liang et al. exploited the Maxwell–Wagner–Sillars polarization phenomenon as the working principle for their ZrO_2@Zn anode, which was prepared using the sol-gel method (Figure 14a,b) [121]. The Maxwell–Wagner–Sillars polarization (or more commonly, Maxwell–Wagner (MW) polarization) occurs at the interface of two structures with different relative permittivities (ε) and electrical conductivities (δ). As a result, the charges are separated over a significant distance. This phenomenon was intentionally implemented in Liang's research by coating ZrO_2 nanoparticles (a typical ceramic material with low electrical conductivity) onto a highly conductive Zn metal anode. Although the insulating nature of ZrO_2 necessitated an activation step for the movement of Zn^{2+} ions through the coating layer, resulting in a high interface resistance in the first plating stage, the impedance then reduced in the subsequent cycles by the MW polarization phenomenon. The CV curves of the symmetric and complete cells with the ZrO_2@Zn anode revealed the unexpected effects of ZrO_2 on the stripping/plating reactions of the Zn anode. At 0.125 mAh cm^{-2}, the ZrO_2@Zn symmetric cell showed a substantially lower initial overpotential (38 mV vs. 74 mV for the bare Zn anode) and a much longer lifespan (3800 h) than the cell with the bare Zn electrode (Figure 14c). In addition, as shown in Figure 14d, the ZrO_2-coated Zn anode showed a longer cycle lifespan (up to 2100 h) and lower polarization (32 mV) than the bare Zn anode at 5 mA cm^{-2}. This can be attributed to the increase in the number of nucleation sites and the ion diffusion rate of the ZrO_2-coated Zn anode due to the MW polarization effect. Furthermore, because ZrO_2 is a chemically inert oxide, it can serve as a protective barrier, preventing Zn from contacting mildly acidic electrolytes. This dual-functional coating effectively reduced the uncontrollable dendrite growth and its potentially dangerous side effects. As a result, even after cycling at 5 mA cm^{-2}, the ZrO_2-coated Zn anode retained its flat and dense surface.

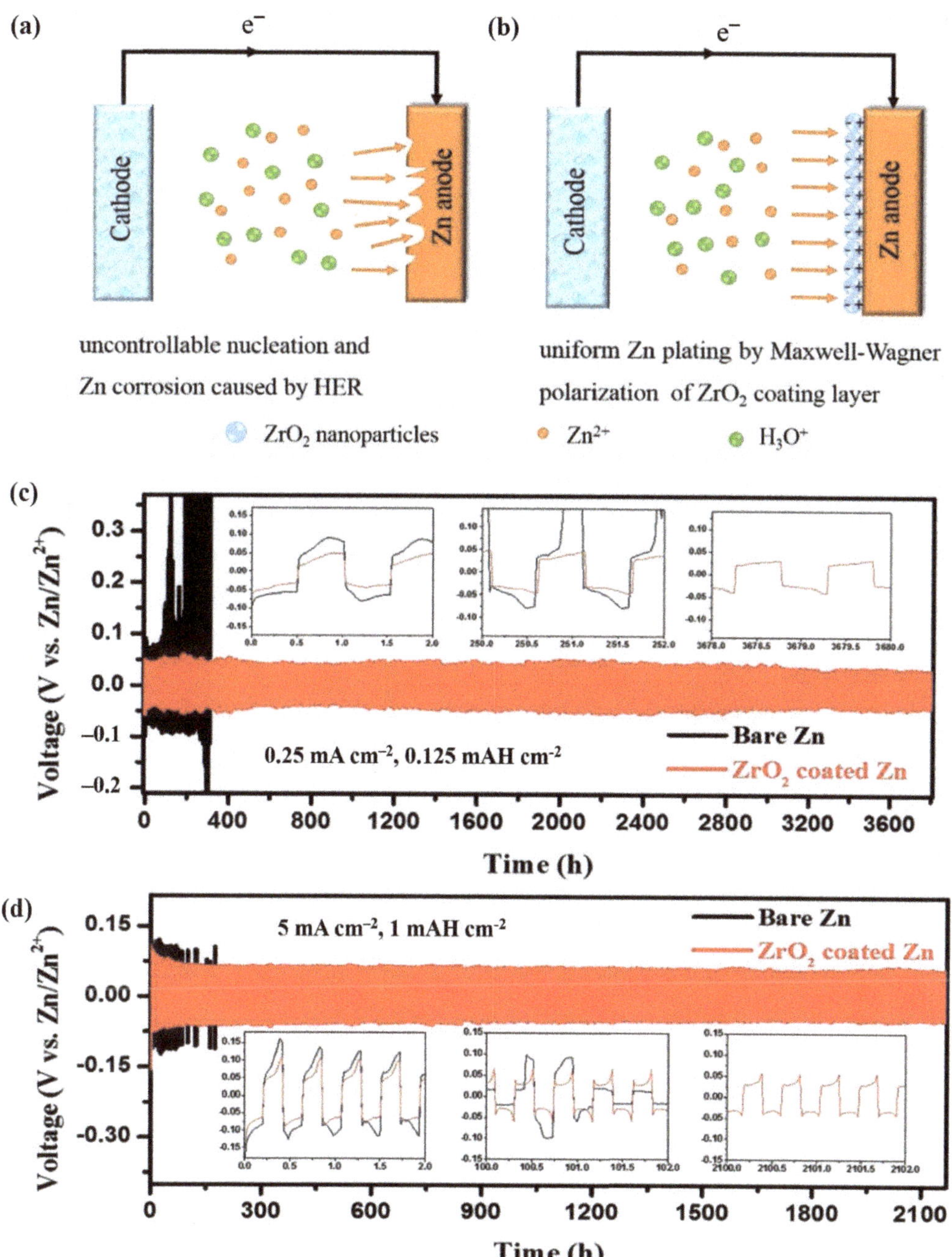

Figure 14. Schematics for the stripping/plating processes of (**a**) bare Zn and (**b**) ZrO₂-coated Zn. Voltage profiles of bare Zn and ZrO₂-coated Zn in a symmetric cell at (**c**) 0.25 mA cm⁻² for 0.125 mAh cm⁻² and (**d**) 5 mA cm⁻² for 1 mAh cm⁻². Reprinted with permission from Liang et al. [121]. Copyright 2020 WILEY-VCH Verlag GmbH&Co. KgaA, Weinheim.

Along this line, Kang et al. used a nanoporous $CaCO_3$ coating to control the Zn deposition (Figure 15a) [92]. The holes in the $CaCO_3$ coating on the Zn metal anode were crucial for controlling the formation of Zn dendrites. The electrolyte quickly penetrated the nano $CaCO_3$ coating owing to its high porosity, resulting in a consistent electrolysis flux and plating of Zn throughout the Zn metal surface. After continuous running, large protuberances and detached zinc flakes were formed as a large amount of Zn was deposited in these areas. The merits of nanoporous $CaCO_3$ were utilized based on this assumption. Owing to its high porosity, the nanoporous $CaCO_3$ buffer layer facilitated the electrolyte flux and prevented the "local bias" behavior. In addition, the fine pores and holes at the nanoscale enclosed the Zn nuclei, thus reducing the polarization. In addition, the potential variation induced by the electrically insulating characteristics of $CaCO_3$ was favorable for countering the "tip effect" as Zn^{2+} ions could be converted into Zn only in the area with enough negative potential, namely the area near the anode surface. Consequently, the battery with the nano-$CaCO_3$-coated Zn anode showed a capacity of 206 mAh g^{-1} at 1 A g^{-1} and a CE of 84.7%. These values are higher than those of the pristine Zn anode (188 mAh g^{-1} and 77.5%, respectively) (Figure 15b). The battery with the nano-$CaCO_3$-coated Zn anode exhibited substantially better cycling steadiness than that with the pristine Zn anode at 1 A g^{-1} (Figure 15c). Its capacity gradually increased from 206 to a maximum of 236 mAh g^{-1} in the first 500 cycles, and then reached 177 mAh g^{-1} after the 1000th cycle (capacity retention of 86%). On the other hand, the battery with the pristine Zn anode had a lower initial capacity of 188 mAh g^{-1}, which remained only 124 mAh g^{-1} after 1000 cycles. Later, Zeng et al. proposed a similar strategy using conductive CNT scaffolds [122]. The highly porous CNT skeleton helped to ensure uniform seeding sites across the electrode (similar to $CaCO_3$). Moreover, this flexible sheet also served as a "supplemental host" because of the occurrence of Zn stripping/plating in it, thus providing additional space for Zn deposition and alleviating the aggressive dendrite growth. More impressively, no coating process was employed; instead, the CNT scaffolds only needed to be placed between the Zn foil anode and separator, which indeed set this work apart from others because of its simplicity and scalability. In addition to applying CNTs in AZIBs, Dong et al. further applied this Zn@CNT electrode for hybrid capacitors. The Zn ion capacitor delivered remarkable stability with an average capacity of 47 mAh g^{-1} at 2 A g^{-1} (CE ~100%) for 7000 cycles.

5.2.2. Redistributing Zn^{2+} Ion Flux

Another efficient strategy to improve the Zn deposition behavior of Zn metal anodes is to redistribute the Zn^{2+} ion flux. As a pioneer in this field, Zhao et al. proposed a solid-state interphase comprising of polyamide (PA) and zinc trifluoromethane sulfonate $(Zn(TfO)_2)$ [94]. They demonstrated that ions tended to move horizontally across the electrode surface and aggregated at already existing nucleation sites to optimize the surface energy. However, this inclination was intentionally restrained by introducing the PA chains. Because of their higher energy barrier to diffuse laterally, the Zn^{2+} ions were obligated to deposit at the premier position where they were initially adsorbed through the interface, resulting in an increase in the number of nucleation sites. In addition, the synergistic interaction between the Zn^{2+} ions and polar groups (C=O) in the PA backbones improved the nucleation overpotential, which favored the formation of smaller nuclei. These dual effects ultimately led to the formation of a dense and smooth Zn layer when the cells were continuously cycled. It is important to note that the PA coating did not allow electrons to cross over it, preventing the external reduction of Zn^{2+}. Therefore, the electrolyte could not contact the newly formed Zn without shielding. The water-resistant characteristics of the coating also suppressed the adverse HERs. The carbonyl-rich networks offered plentiful hydrogen bonds to bind and impair the Zn^{2+} ion solvation-sheath, preventing H_2O and O_2 from participating in such reactions. As a result, the discharge capacity of the Zn/MnO_2 battery increased with an increase in the number of cycles from 450 (below 30 mAh g^{-1}) to 1000 (155.4 mAh g^{-1}, 88% capacity retention) at 2C.

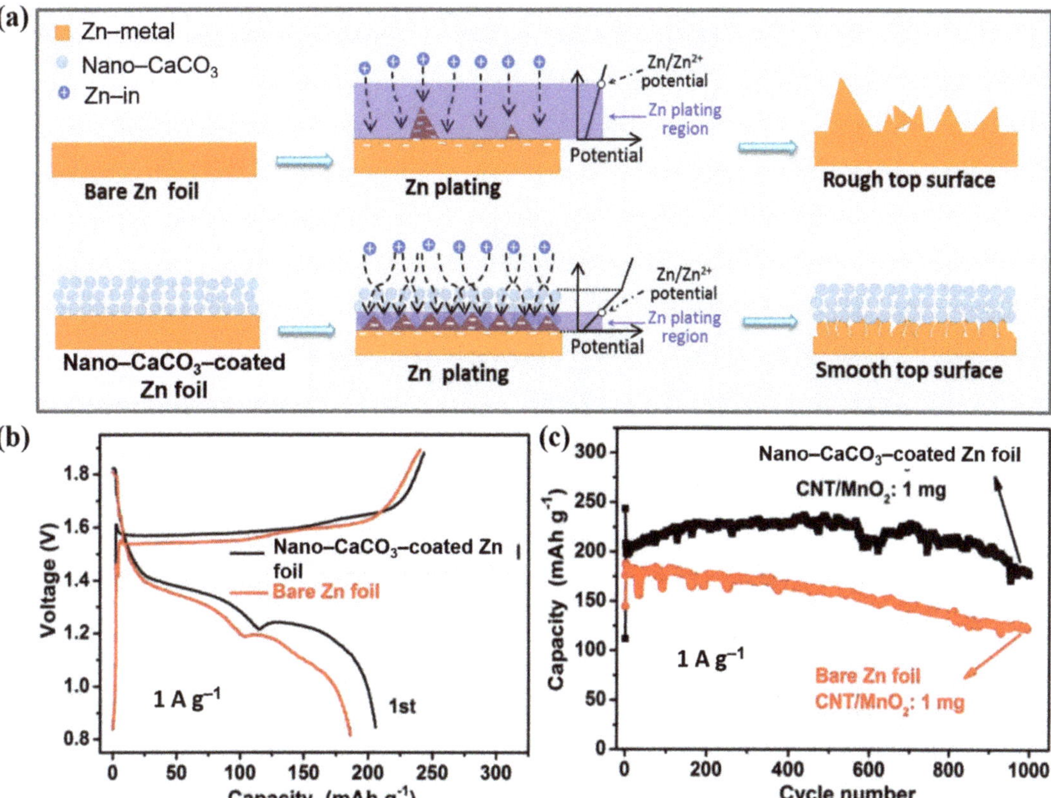

Figure 15. (a) Schematic illustration of morphology evolution for bare and nano-CaCO$_3$-coated Zn foils during Zn stripping/plating cycling, (b) charge-discharge profiles and (c) full cell performance of nano-CaCO$_3$-coated Zn foil and bare Zn foil. Reprinted with permission from [92]. Copyright 2018 WILEY-VCH Verlag GmbH&Co. KgaA, Weinheim.

Although the coating of metal anodes with polymers has been widely practiced with positive results, an appropriate polymer should be chosen based on some fundamental principles. To withstand the volume changes during repeated cycling, a metal anode must be mechanically strong and flexible. Chemical stability, water insolubility, and hydrophilicity are also essential features of an ideal metal anode. Finally, the polymer network used for coating Zn anodes must possess abundant polar groups to interact with the metal ions. Considering these requirements, polyvinylidene fluoride (PVDF) has been used to modify the surface of Zn anodes via a facile spin-coating process (Figure 16a) [123]. Under slow evaporation conditions, the F atoms in the PVDF side chains rearrange to form an all-trans conformation known as the β-phase, in which all the dipoles align on the same side. Owing to its high polarity, β-PVDF shows excellent ferroelectric, piezoelectric, and pyroelectric properties. When symmetric β-PVDF@Zn | ZnSO$_4$ | β-PVDF@Zn was cycled at 0.25 mA cm^{-2} and an area capacity of 0.05 mAh cm^{-2}, a small overpotential was achieved even after 2000 h of operation. This performance is far superior to that shown by Zn | ZnSO$_4$ | Zn, which withered after 200 h (Figure 16b). This superior performance of β-PVDF@Zn | ZnSO$_4$ | β-PVDF@Zn can be attributed to the presence of the multifunctional β-PVDF membrane in it. On the one hand, the highly electronegative C-F alignments acted as preferable diffusion paths for Zn^{2+} and distributed these ions uniformly onto the entire anode surface. This efficient channel network helped to abate the local current density and developed a homogeneous layer of Zn plating. On the other hand, owing to its resilience, the β-PVDF polymer could withstand dendrite growth and hindered the formation of inactive side-products (ZnO) in acidic electrolytes. Protected by the versatile β-PVDF cover, the Zn anode retained its morphology without serious damage and, as a result, the full cell delivered an appealing cycling performance (discharge capacity of 60 mAh g^{-1} after 4000 cycles).

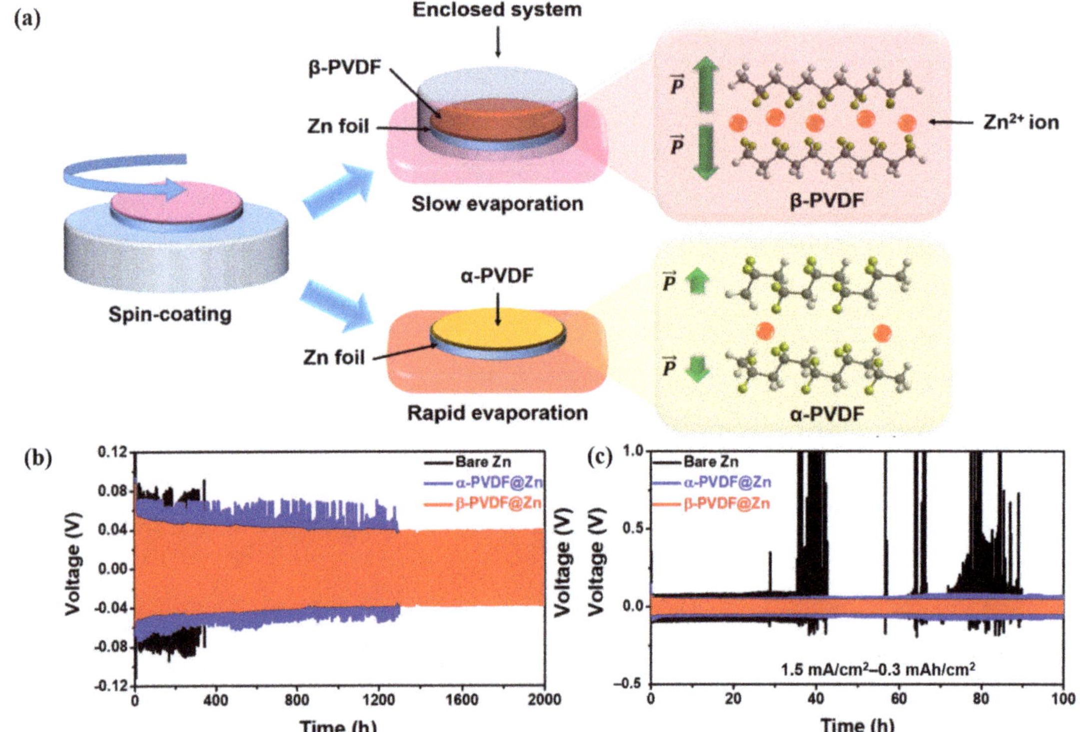

Figure 16. (a) Schematic illustration of the β- and α-PVDF coating processes, (b) long-term profiles of β-PVDF@Zn (red), α-PVDF@Zn (blue), and bare Zn (black) with symmetrical cells at a current density of (b) 0.25–0.05 mAh cm^{-2}, (c) 1.5–0.3 mAh cm^{-2}. Reprinted with permission from Hieu et al. [123]. Copyright 2021 Elsevier B.V.

With regard to PVDF, Liu et al. synthesized a MOF-PVDF composite coating layer consisting of PVDF and hydrophilic microporous metal-organic framework (MOF) particles [124]. The metal ions could evenly access the anode surface because of their low wettability with common aqueous electrolytes, resulting in local ion deposition and troublesome protrusion growth. The microporous architecture of the MOF induced a wetting effect on the Zn electrode at the nanoscale, generating a hydrophilic surface. Each MOF nanoparticle made intimate contact with Zn and served as a nanoscale electrolyte reservoir. As a result, the ion flux cooperated well with the Zn plate to the greatest extent, promoting a uniform Zn stripping/plating process. Furthermore, the compact MOF particles enabled fast ion diffusion through the fully wetted surface, dramatically lowering the charge-transfer resistance.

5.3. Creating Uniform Electric Field

One of the most important issues affecting the stability of Zn anodes is the formation of Zn dendrites. Engineering a synthetic SEI on the surface of Zn is an effective method for limiting the Zn dendrite formation. The fundamental goal of this strategy is to develop materials that can direct the orderly migration of Zn ions while providing a homogenous electric field at the electrode-electrolyte interface, preventing the formation of Zn dendrites. Recently, it has been demonstrated that an artificial SEI based on $BaTiO_3$ (BTO) can effectively restrain the Zn dendrite growth (Figure 17a,b) [125]. The polarized BTO layer played a significant role in controlling the orderly migration of Zn ions because of the uniform electric field through it. The polarization of BTO was induced by the Ti ions that deviated from the center of the symmetrical site in BTO under an external electric field. In this situation, the two-dimensional diffusion of Zn ions was highly restricted, and the vertical movement of Zn ions through the BTO layer was thermodynamically preferable. Moreover, according to the density functional theory calculation results, the formation energy of Zn deposition was much lower for BTO@Zn than that for bare Zn (-2.28 vs. -0.35 eV), thus BTO promoted dense Zn plating without dendrites. In addition, the water molecules in the solvated Zn ions ($Zn(H_2O)_6{}^{2+}$) were easily attracted by the O atom in BTO via hydrogen bonding, which facilitated the diffusion of Zn ions and restricted the Zn dendrite formation [91]. In contrast, in the absence of the BTO coating, the Zn dendrite formation could not be regulated because of the uneven deposition of Zn ions as well as the preferential deposition at the protruded surfaces. As a consequence, the BTO@Zn-symmetric cell showed cycling stability for more than 2000 h (1000 cycles) at 1 mA cm^{-2}, 1 mAh cm^{-2} (Figure 17c) and for 1500 cycles at 5 mA cm^{-2}, 2.5 mAh cm^{-2} (Figure 17d). After 300 cycles at 2 A g^{-1}, the BTO@Zn–MnO$_2$ full cell battery showed high rate capability and approximately 100% CE in a mild aqueous electrolyte.

Zhang et al. investigated CuO nanowires grown on a Cu mesh (CM@CuO) by anodic oxidation, followed by thermal annealing in air [126]. Because of the decreased energy barrier and increased number of active sites, Zn^{2+} was preferentially absorbed by CuO and dispersed evenly on the Zn surface, which was advantageous for Zn nucleation. Furthermore, as illustrated in Figure 18a, the Cu generated from CuO enhanced the electrical conductance, which in turn increased the electric field during the Zn nucleation, resulting in a more ordered Zn growth. Because of the increased zincophilicity and uniform Zn^{2+} distribution on the surface, small Zn nanosheets were consistently developed on the CuO nanowires, as shown in the SEM image (Figure 18b). Zn remained within the CuO nanowire structure even when the capacity was increased to 5 mAh cm^{-2} (Figure 18c). In contrast to the dendrite-free plating behavior of CM@CuO@Zn, the early Zn growth on the pristine CM was uneven, and micron-sized Zn dendrites (5 μm) were observed in the CM@Zn anode (Figure 18d,e). This work demonstrated that CM@CuO@Zn shows excellent electrochemical performance owing to its uniform electric field and 3D structure. The as-prepared Zn anode showed excellent cycling stability (340 h) and low voltage hysteresis (20 mV) in symmetric cells at 1 mA cm^{-2} and 1 mAh cm^{-2}. Meanwhile, a 3D hierarchical N-doped carbon cloth (NC) was produced using magnetron sputtering as a scaffold combining a 3D architecture and an interfacial control to efficiently promote the uniform nucleation of Zn metal and reduce the growth of Zn dendrites [127]. The dynamics of Zn^{2+} transfer and deposition could be accelerated by the desolvation process of Zn^{2+} with surface chemistry control. The 3D hierarchical conductive carbon scaffold uniformly distributed the electric fields and lowered the local current density. The N-containing functional groups in NC served as nucleation sites to distribute the Zn nuclei uniformly across the electrode surface. As Zn was constantly deposited, an electric charge was generated in the uneven nucleation sites. The carbon scaffold doped with N atoms encouraged the homogenous nucleation and growth of Zn, thus constraining the formation of Zn dendrites. The N atoms generated by the addition of pyrrole lowered the migration energy barrier and promoted the redistribution of Zn ions. Furthermore, the zincophilicity of the NC scaffold was enhanced by the strong electrostatic interaction between the negatively charged pyrrole N sites in the carbon lattice and the Zn atoms, resulting in the homogeneous nucleation and growth of Zn. The negative state of the pyrrole N site can be explained by the formation of σ-bonds between the doped N atoms and the adjacent carbons from which electrons moved toward pyrrole N through the inductive effect. As a result, the assembled half-cell showed a high zinc stripping/plating CE of 98.8%. The NC-Zn symmetrical cell showed stable operation at a high current density of 5 mA cm^{-2} with an overpotential as low as 11 mV after 210 cycles.

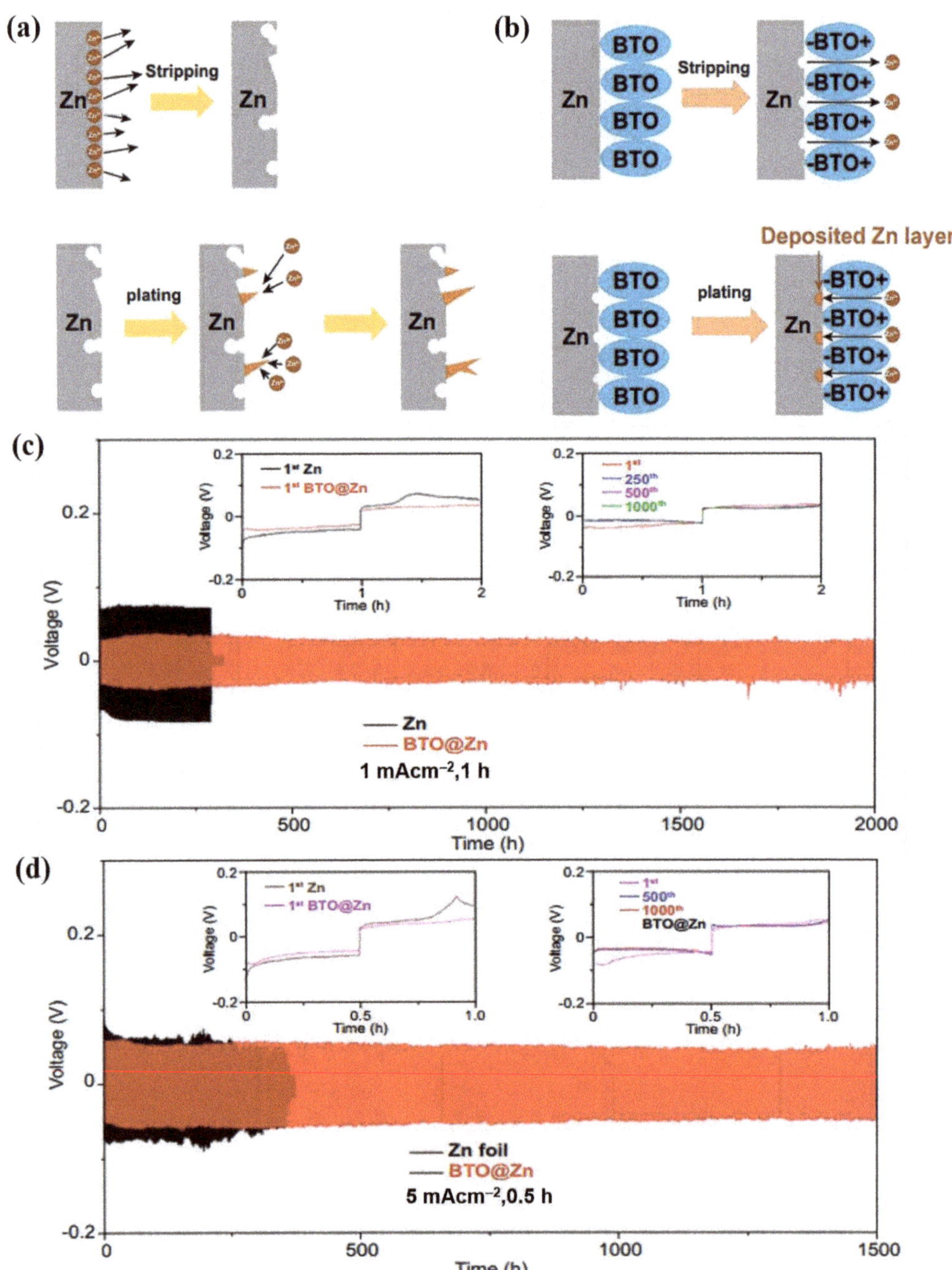

Figure 17. Schematic of Zn ion transport during Zn stripping/plating for (**a**) bare Zn, and (**b**) BTO@ Zn foil. Cyclic performances of the symmetric cells with Zn and BTO@Zn at (**c**) 1 mA cm^{-2} (1 mAh cm^{-2}), and (**d**) 5 mA cm^{-2} (2.5 mAh cm^{-2}). Reprinted with permission from [125].

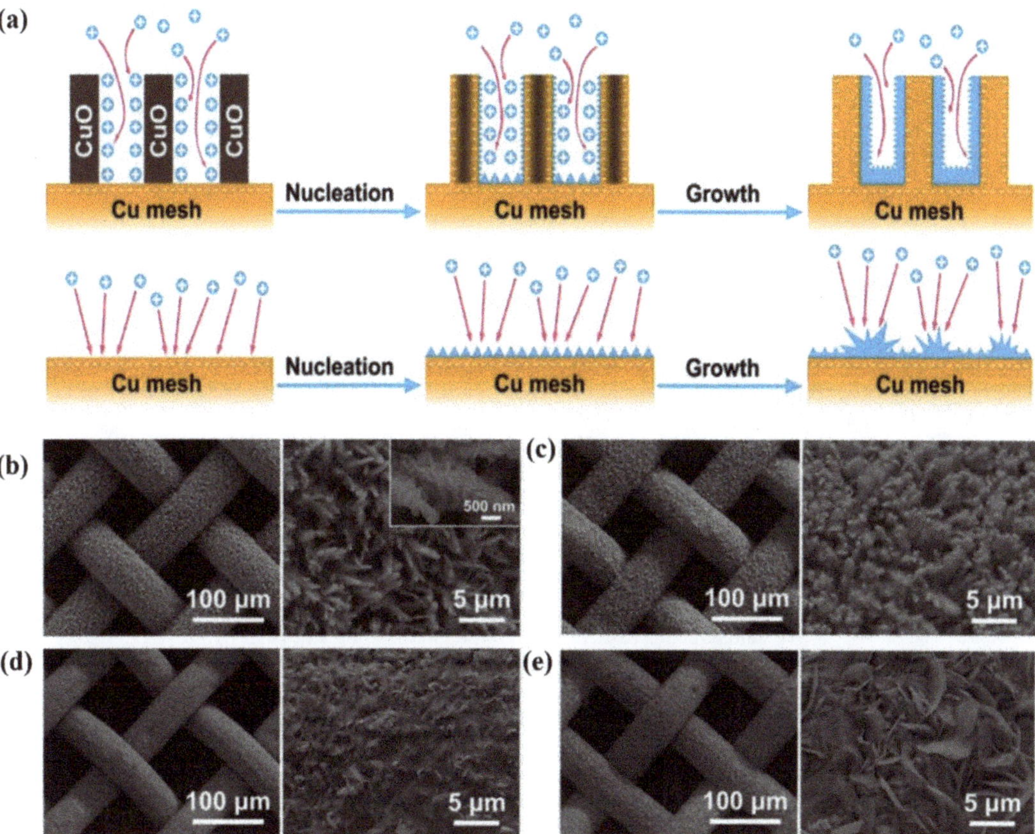

Figure 18. (**a**) Schematic illustration of Zn deposition on CM@CuO and CM. SEM images of CM@CuO@Zn with the capacities of (**b**) 1 and (**c**) 5 mAh cm^{-2}. SEM images of CM@Zn with the capacities of (**d**) 1 and (**e**) 5 mAh cm^{-2}. Reprinted with permission from [126]. Copyright 2020 Wiley-VCH GmbH.

The various modification principles used for modifying Zn anodes and their performances (symmetric and full cell) for AZIBs in mildly acidic electrolytes are summarized in Table 2.

According to the strategies discussed above for modifying the Zn metal surface, the drawbacks must be addressed concurrently to achieve a practical improvement of Zn metal anodes. As for the dendrites, they can be effectively suppressed by increasing the Zn nucleation sites, redistributing the ionic flux of Zn^{2+}, and controlling the electric field. Corrosion and HER can be resolved by reducing the chemical activity of H$_2$O and introducing the protective layer on the Zn metal anode. As can be seen from Table 2, these strategies improve the performance of the Zn anodes compared to the pristine state, but some of them are not still viable enough for practical applications in the perspectives of overall performance. For example, the use of carbon-based materials with a high specific surface area such as rGO can inhibit the dendrite formation, but can accelerate the corrosion, which may result in the degraded full-cell performance. Therefore, the effectiveness of modification methods should be strictly evaluated considering various performance metrics together to realize the high-performance Zn anode.

Table 2. Modification mechanisms and electrochemical performances of recently reported Zn metal anodes in mildly acidic electrolytes for AZIBs.

Anode	Mechanism	Corrosion Potential (V) (vs. Ag/AgCl)	Symmetrical Cell Performance:Lifespan (h), Capacity, and Current Density	Full-cell Performance:Capacity (mAh g^{-1})/Cycle/Capacity Retention (%)/Current Density (mA g^{-1})	Ref
TiO$_2$@Zn	Shielding Zn surface	−0.89	150 h 1 mA cm^{-2}, 1 mAhcm^{-2}	134 mAh g^{-1}/1000 cycles/85%/1 A g^{-1}	[116]
Al$_2$O$_3$@Zn	Shielding Zn surface	−0.88	500 h 1 mA cm^{-2}, 1 mAh cm^{-2}	158 mAh g^{-1}/1000 cycles/89%/1 A g^{-1}	[117]
Cu-Zn/Zn	Shielding Zn surface	−0.96	1500 h 1 mA cm^{-2}, 0.5 mAh cm^{-2}	-	[101]
rGO@Zn	Shielding Zn surface	-	300 h 1 mA cm^{-2}, 0.5 mAh cm^{-2}	61 mAh g^{-1}/5000 cycles/86%/1 A g^{-1}	[119]
NA-Zn	Controlling nucleation sites	-	2000 h 0.25 mA cm^{-2}, 0.05 mAh cm^{-2}	67 mAh g^{-1}/2000 cycles/-/0.5 A g^{-1}	[120]
ZrO$_2$@ Zn	Controlling nucleation sites	-	2100 h 5 mA cm^{-2}, 1 mAh cm^{-2}	52 mAh g^{-1}/3000 cycles/42%/1 A g^{-1}	[121]
CaCO3@Zn	Controlling nucleation sites	-	900 h 0.25 mA cm^{-2}, 0.05 mAh cm^{-2}	177 mAh g^{-1}/1000 cycles/86%/1 A g^{-1}	[92]
CNT@Zn	Controlling nucleation sites	-	200 h 2 mA cm^{-2}, 2 mAh cm^{-2}	167 mAh g^{-1}/1000 cycles/89%/-	[122]
PA@Zn	Redistributing Zn^{2+} ion flux	−0.96	8000 h 0.5 mA cm^{-2}, 0.25 mAh cm^{-2}	154 mAh g^{-1}/1000 cycles/88%/0.6 A g^{-1}	[94]
PVDF@Zn	Redistributing Zn^{2+} ion flux	-	2000 h 0.25 mA cm^{-2}, 0.05 mAh cm^{-2}	57 mAh g^{-1}/2000 cycles/-/1 A g^{-1}	[123]
MOF-PVDF@Zn	Redistributing Zn^{2+} ion flux	-	500 h 1 mA cm^{-2}, 0.5 mAh cm^{-2}	-	[124]
BTO@Zn	Creating uniform electric field	-	1500 h 5 mA cm^{-2}, 0.5 mAh cm^{-2}	74 mAh g^{-1}/300 cycles/67%/2 A g^{-1}	[125]
CM@CuO@Zn	Creating uniform electric field	-	350 h 1 mA cm^{-2}, 1 mAh cm^{-2}	-	[126]
NC@Zn	Creating uniform electric field	-	300 h 1 mA cm^{-2}, 0.5 mAh cm^{-2}	-	[127]

6. Conclusions and Outlook

In summary, the major challenges associated with the application of Zn metal anodes in mild aqueous electrolytes and the modification strategies employed to overcome these challenges are discussed to provide an insight into the development of high-performance AZIBs. AZIBs are regarded as promising alternatives to LIBs (especially for large-scale energy storage systems) owing to their safety, cost-effectiveness, environmental benignity, and high energy density. The main challenges associated with the use of Zn metal anodes in neutral or mildly acidic media are dendrite growth, corrosion, and H$_2$ evolution. The presence of uneven nucleation sites on Zn metal facilitates uncontrollable charge transfer, which accelerates the generation of an uneven electric field, causing the growth of dendrites. The corrosion of Zn metal anodes can be categorized as self-corrosion or corrosion by an electrochemical reaction. The latter refers to the irreversible Zn consumption due to the removal of Zn from the electrode as well as the side reactions between the electrode and electrolyte. H$_2$ evolution is a complicated process that is affected by the reduction potential, overpotential, surface area, electrolyte pH, and operating temperature. To address these issues, researchers have proposed various modification strategies, which can largely be categorized as (i) shielding the Zn metal from side reactions, (ii) regulating the Zn deposition behavior (controlling the nucleation sites and redistributing the Zn^{2+} ion flux), and (iii) establishing a uniform electric field.

Despite the considerable progress made in this field, there is still a long way to go to address the complex challenges discussed thus far. An efficient approach to produce a dendrite-free and highly reversible Zn anode is to strategically tailor the 3D structure of the electrode. The construction of 3D nanostructured Zn metal anodes has not been fully explored yet. Fabricating Zn anodes with hierarchical 3D architectures provides new opportunities for increasing the rate capability and service life of AZIBs because it allows a more uniform distribution of the electric current and provides a large number of reaction sites. The fabrication of 3D Zn architecture can be realized by the coating of Zn metal (e.g., electrodeposition) on a well-defined conductive 3D framework (e.g., metal form, porous carbon, carbon mesh, etc.) or by the selective etching of the sacrificial components in Zn composites. The caveat here is that while the increase in the surface area of the electrode can improve the energy density and suppress the dendrite development, it can accelerate the corrosion of Zn metal. This can be avoided by coating 3D Zn metal with corrosion inhibitors or by introducing organic/inorganic additives in the plating electrolyte. In addition, although the behavior of Zn electrodes in mild aqueous electrolytes differs from that in alkaline electrolytes, the knowledge of the operation of Zn batteries in alkaline aqueous media can provide time-saving guidance for the study of AZIBs. Furthermore, more progress can be made by investigating the previously reported strategies for modifying metal anodes (e.g., in Li/Na metal batteries).

In addition to the Zn metal anode, the other components (cathode and electrolyte) of the cell should also be taken into consideration to improve the performance of the anode. The type of electrolyte, electrolyte concentration, and additives are important factors that strongly influence the electrochemical reactions on the Zn metal anode. In particular, in mild aqueous electrolytes, $ZnSO_4$ and $Zn(CF_3SO_3)_2$ are mainly used within a certain concentration range. In most cases, $ZnSO_4$ is preferable because of its relatively low price as compared to that of $Zn(CF_3SO_3)_2$. However, $ZnSO_4$ suffers from the formation of hydroxide sulfate by-products ($Zn_4(OH)_6SO_4 \cdot nH_2O$). Exploring other Zn salts for application under mild aqueous conditions and some appropriate $ZnSO_4$-based additives is another potential approach. Mn-based materials are the most widely used cathode materials. As revealed by many studies on AZIBs cathodes, Mn-based oxides can be operated at high working voltages, but it is still difficult to fully understand their reaction mechanism because of the diverse phase transitions occurring during the reactions. Apart from Mn-based cathode materials, V-based materials and Prussian blue analogs can be potential candidates for application in conjunction with Zn metal anodes for AZIBs. However, in this case, the limitations of these materials (V-based oxides: low operation voltage, Prussian blue analogs: low specific capacity) should be considered.

Finally, most studies on Zn metal anodes have mainly focused on coin cells or pouch cells. To meet the market demands (especially in large-scale energy storage), these studies need to be expanded to larger cells (e.g., cylindrical and prismatic cells). In practical applications, unknown challenges that have not been encountered in coin cells can be encountered. The know-how of various cell configurations can narrow the gap between the lab-scale results and the commercialization of AZIBs. Finally, the development of flexible and wearable AZIBs can be an intriguing research topic because of their intrinsic safety, low cost, and facile fabrication (without the need for a glovebox). To realize this, not only should a sufficiently thin Zn metal electrode (beyond the commercial Zn foil) be developed, but appropriate components (cathode, solid electrolyte, and packing material) compatible with Zn anodes should also be carefully designed.

Author Contributions: Conceptualization, J.H., V.P.H.H., L.T.H.; validation, L.T.H.; investigation, V.P.H.H.; writing—original draft preparation, V.P.H.H. and L.T.H.; writing—review and editing, J.H.; supervision, J.H.; funding acquisition, J.H. All authors have read and agreed to the published version of the manuscript.

Funding: This work was supported by the Gachon University research fund of 2021 (GCU-202103450001) and the Korea Institute of Energy Technology Evaluation and Planning (KETEP) and the Ministry of Trade, Industry & Energy (MOTIE) of the Korea (No. 20194030202290).

Institutional Review Board Statement: Not applicable.

Informed Consent Statement: Not applicable.

Data Availability Statement: Not applicable.

Conflicts of Interest: The authors declare no competing interests.

References

1. Jiao, Y.; Kang, L.; Berry-Gair, J.; McColl, K.; Li, J.; Dong, H.; Jiang, H.; Wang, R.; Corà, F.; Brett, D.J.L.; et al. Enabling stable MnO_2 matrix for aqueous zinc-ion battery cathodes. *J. Mater. Chem. A* **2020**, *8*, 22075–22082. [CrossRef]
2. Pan, H.; Shao, Y.; Yan, P.; Cheng, Y.; Han, K.S.; Nie, Z.; Wang, C.; Yang, J.; Li, X.; Bhattacharya, P.; et al. Reversible aqueous zinc/manganese oxide energy storage from conversion reactions. *Nat. Energy* **2016**, *1*, 16039. [CrossRef]

3. Yang, Z.; Zhang, J.; Kintner-Meyer, M.C.W.; Lu, X.; Choi, D.; Lemmon, J.P.; Liu, J. Electrochemical Energy Storage for Green Grid. *Chem. Rev.* **2011**, *111*, 3577–3613. [CrossRef] [PubMed]

4. Dong, H.; Li, J.; Zhao, S.; Zhao, F.; Xiong, S.; Brett, D.J.L.; He, G.; Parkin, I.P. An anti-aging polymer electrolyte for flexible rechargeable zinc-ion batteries. *J. Mater. Chem. A* **2020**, *8*, 22637–22644. [CrossRef]

5. Dong, H.; Li, J.; Zhao, S.; Jiao, Y.; Chen, J.; Tan, Y.; Brett, D.J.L.; He, G.; Parkin, I.P. Investigation of a Biomass Hydrogel Electrolyte Naturally Stabilizing Cathodes for Zinc-Ion Batteries. *ACS Appl. Mater. Interfaces* **2021**, *13*, 745–754. [CrossRef] [PubMed]

6. Qiu, C.; Zhu, X.; Xue, L.; Ni, M.; Zhao, Y.; Liu, B.; Xia, H. The function of Mn^{2+} additive in aqueous electrolyte for Zn/δ-MnO_2 battery. *Electrochim. Acta* **2020**, *351*, 136445. [CrossRef]

7. Tarascon, J.M.; Armand, M. Issues and challenges facing rechargeable lithium batteries. *Nature* **2001**, *414*, 359–367. [CrossRef]

8. Fuller, T.F. Batteries: Bigger and better. *Nat. Energy* **2016**, *1*, 16003. [CrossRef]

9. Nitta, N.; Wu, F.; Lee, J.T.; Yushin, G. Li-ion battery materials: Present and future. *Mater. Today* **2015**, *18*, 252–264. [CrossRef]

10. Abdin, Z.; Khalilpour, K.R. Chapter 4—Single and Polystorage Technologies for Renewable-Based Hybrid Energy Systems. In *Polygeneration with Polystorage for Chemical and Energy Hubs*; Khalilpour, K.R., Ed.; Elsevier Inc.: Amsterdam, The Netherlands, 2019; pp. 77–131. [CrossRef]

11. Zhang, Z.J.; Ramadass, P.; Fang, W. 18-Safety of Lithium-Ion Batteries. In *Lithium-Ion Batteries*; Pistoia, G., Ed.; Elsevier B.V.: Amsterdam, The Netherlands, 2014; pp. 409–435. [CrossRef]

12. Yoshino, A. 1-Development of the Lithium-Ion Battery and Recent Technological Trends. In *Lithium-Ion Batteries*; Pistoia, G., Ed.; Elsevier: Amsterdam, The Netherlands, 2014; pp. 1–20. [CrossRef]

13. Mun, Y.S.; Pham, T.N.; Hoang Bui, V.K.; Tanaji, S.T.; Lee, H.U.; Lee, G.-W.; Choi, J.S.; Kim, I.T.; Lee, Y.-C. Tin oxide evolution by heat-treatment with tin-aminoclay (SnAC) under argon condition for lithium-ion battery (LIB) anode applications. *J. Power Sources* **2019**, *437*, 226946. [CrossRef]

14. Nguyen, T.L.; Park, D.; Kim, I.T. $Fe_xSn_yO_z$ Composites as Anode Materials for Lithium-Ion Storage. *J. Nanosci. Nanotechnol.* **2019**, *19*, 6636–6640. [CrossRef]

15. Nguyen, T.L.; Kim, J.H.; Kim, I.T. Electrochemical Performance of $Sn/SnO/Ni_3Sn$ Composite Anodes for Lithium-Ion Batteries. *J. Nanosci. Nanotechnol.* **2019**, *19*, 1001–1005. [CrossRef]

16. Pham, T.N.; Tanaji, S.T.; Choi, J.-S.; Lee, H.U.; Kim, I.T.; Lee, Y.-C. Preparation of Sn-aminoclay (SnAC)-templated Fe_3O_4 nanoparticles as an anode material for lithium-ion batteries. *RSC Adv.* **2019**, *9*, 10536–10545. [CrossRef]

17. Nguyen, T.P.; Kim, I.T. Ag Nanoparticle-Decorated MoS_2 Nanosheets for Enhancing Electrochemical Performance in Lithium Storage. *Nanomaterials* **2021**, *11*, 626. [CrossRef] [PubMed]

18. Nguyen, T.P.; Kim, I.T. Self-Assembled Few-Layered MoS_2 on SnO_2 Anode for Enhancing Lithium-Ion Storage. *Nanomaterials* **2020**, *10*, 2558. [CrossRef]

19. Preman, A.N.; Lee, H.; Yoo, J.; Kim, I.T.; Saito, T.; Ahn, S.-k. Progress of 3D network binders in silicon anodes for lithium ion batteries. *J. Mater. Chem. A* **2020**, *8*, 25548–25570. [CrossRef]

20. Nguyen, T.P.; Kim, I.T. W2C/WS2 Alloy Nanoflowers as Anode Materials for Lithium-Ion Storage. *Nanomaterials* **2020**, *10*, 1336. [CrossRef]

21. Kim, W.S.; Vo, T.N.; Kim, I.T. GeTe-TiC-C Composite Anodes for Li-Ion Storage. *Materials* **2020**, *13*, 4222. [CrossRef]

22. Vo, T.N.; Kim, D.S.; Mun, Y.S.; Lee, H.J.; Ahn, S.-k.; Kim, I.T. Fast charging sodium-ion batteries based on Te-P-C composites and insights to low-frequency limits of four common equivalent impedance circuits. *Chem. Eng. J.* **2020**, *398*, 125703. [CrossRef]

23. Larcher, D.; Tarascon, J.M. Towards greener and more sustainable batteries for electrical energy storage. *Nat. Chem.* **2015**, *7*, 19–29. [CrossRef] [PubMed]

24. Wang, S.; Yang, Y.; Dong, Y.; Zhang, Z.; Tang, Z. Recent progress in Ti-based nanocomposite anodes for lithium ion batteries. *J. Adv. Ceram.* **2019**, *8*, 1–18. [CrossRef]

25. Li, M.; Lu, J.; Ji, X.; Li, Y.; Shao, Y.; Chen, Z.; Zhong, C.; Amine, K. Design strategies for nonaqueous multivalent-ion and monovalent-ion battery anodes. *Nat. Rev. Mater.* **2020**, *5*, 276–294. [CrossRef]

26. Li, X.; Chen, G.; Le, Z.; Li, X.; Nie, P.; Liu, X.; Xu, P.; Wu, H.B.; Liu, Z.; Lu, Y. Well-dispersed phosphorus nanocrystals within carbon via high-energy mechanical milling for high performance lithium storage. *Nano Energy* **2019**, *59*, 464–471. [CrossRef]

27. Chao, D.; Ouyang, B.; Liang, P.; Huong, T.T.T.; Jia, G.; Huang, H.; Xia, X.; Rawat, R.S.; Fan, H.J. C-Plasma of Hierarchical Graphene Survives SnS Bundles for Ultrastable and High Volumetric Na-Ion Storage. *Adv. Mater.* **2018**, *30*, e1804833. [CrossRef] [PubMed]

28. Zhang, W.; Liu, Y.; Guo, Z. Approaching high-performance potassium-ion batteries via advanced design strategies and engineering. *Sci. Adv.* **2019**, *5*, eaav7412. [CrossRef] [PubMed]

29. Abraham, K.M. How Comparable Are Sodium-Ion Batteries to Lithium-Ion Counterparts? *ACS Energy Lett.* **2020**, *5*, 3544–3547. [CrossRef]

30. Karuppasamy, K.; Jothi, V.R.; Nichelson, A.; Vikraman, D.; Tanveer, W.H.; Kim, H.-S.; Yi, S.-C. Chapter 14—Nanostructured transition metal sulfide/selenide anodes for high-performance sodium-ion batteries. In *Nanostructured, Functional, and Flexible Materials for Energy Conversion and Storage Systems*; Pandikumar, A., Rameshkumar, P., Eds.; Elsevier Inc.: Amsterdam, The Netherlands, 2020; pp. 437–464. [CrossRef]

31. Jana, A.; Paul, R.; Roy, A.K. Chapter 2—Architectural design and promises of carbon materials for energy conversion and storage: In laboratory and industry. In *Carbon Based Nanomaterials for Advanced Thermal and Electrochemical Energy Storage and Conversion*; Paul, R., Etacheri, V., Wang, Y., Lin, C.-T., Eds.; Elsevier Inc.: Amsterdam, The Netherlands, 2019; pp. 25–61. [CrossRef]

32. Anoopkumar, V.; John, B.; Mercy, T.D. Potassium-Ion Batteries: Key to Future Large-Scale Energy Storage? *ACS Appl. Energy Mater.* **2020**, *3*, 9478–9492. [CrossRef]
33. Min, X.; Xiao, J.; Fang, M.; Wang, W.; Zhao, Y.; Liu, Y.; Abdelkader, A.M.; Xi, K.; Kumar, R.V.; Huang, Z. Potassium-ion batteries: Outlook on present and future technologies. *Energy Environ. Sci.* **2021**, *14*, 2186–2243. [CrossRef]
34. Liu, S.; Kang, L.; Hu, J.; Jung, E.; Zhang, J.; Jun, S.C.; Yamauchi, Y. Unlocking the Potential of Oxygen-Deficient Copper-Doped Co3O4 Nanocrystals Confined in Carbon as an Advanced Electrode for Flexible Solid-State Supercapacitors. *ACS Energy Lett.* **2021**, *6*, 3011–3019. [CrossRef]
35. Liu, S.; Kang, L.; Kim, J.M.; Chun, Y.T.; Zhang, J.; Jun, S.C. Recent Advances in Vanadium-Based Aqueous Rechargeable Zinc-Ion Batteries. *Adv. Energy Mater.* **2020**, *10*, 2000477. [CrossRef]
36. Kang, L.; Zhang, M.; Zhang, J.; Liu, S.; Zhang, N.; Yao, W.; Ye, Y.; Luo, C.; Gong, Z.; Wang, C.; et al. Dual-defect surface engineering of bimetallic sulfide nanotubes towards flexible asymmetric solid-state supercapacitors. *J. Mater. Chem. A* **2020**, *8*, 24053–24064. [CrossRef]
37. Liu, S.; Kang, L.; Jun, S.C. Challenges and Strategies toward Cathode Materials for Rechargeable Potassium-Ion Batteries. *Adv. Mater.* **2021**, 2004689. [CrossRef] [PubMed]
38. Wang, J.; Yamada, Y.; Sodeyama, K.; Watanabe, E.; Takada, K.; Tateyama, Y.; Yamada, A. Fire-extinguishing organic electrolytes for safe batteries. *Nat. Energy* **2018**, *3*, 22–29. [CrossRef]
39. Chen, L.; Cao, L.; Ji, X.; Hou, S.; Li, Q.; Chen, J.; Yang, C.; Eidson, N.; Wang, C. Enabling safe aqueous lithium ion open batteries by suppressing oxygen reduction reaction. *Nat. Commun.* **2020**, *11*, 2638. [CrossRef] [PubMed]
40. Yan, C.; Lv, C.; Wang, L.; Cui, W.; Zhang, L.; Dinh, K.N.; Tan, H.; Wu, C.; Wu, T.; Ren, Y.; et al. Architecting a Stable High-Energy Aqueous Al-Ion Battery. *J. Am. Chem. Soc.* **2020**, *142*, 15295–15304. [CrossRef]
41. Kumar, S.; Verma, V.; Arora, H.; Manalastas, W.; Srinivasan, M. Rechargeable Al-Metal Aqueous Battery Using NaMnHCF as a Cathode: Investigating the Role of Coated-Al Anode Treatments for Superior Battery Cycling Performance. *ACS Appl. Energy Mater.* **2020**, *3*, 8627–8635. [CrossRef]
42. Tian, H.; Li, Z.; Feng, G.; Yang, Z.; Fox, D.; Wang, M.; Zhou, H.; Zhai, L.; Kushima, A.; Du, Y.; et al. Stable, high-performance, dendrite-free, seawater-based aqueous batteries. *Nat. Commun.* **2021**, *12*, 237. [CrossRef]
43. Kim, H.; Hong, J.; Park, K.-Y.; Kim, H.; Kim, S.-W.; Kang, K. Aqueous Rechargeable Li and Na Ion Batteries. *Chem. Rev.* **2014**, *114*, 11788–11827. [CrossRef]
44. Chao, D.; Zhou, W.; Xie, F.; Ye, C.; Li, H.; Jaroniec, M.; Qiao, S.-Z. Roadmap for advanced aqueous batteries: From design of materials to applications. *Sci. Adv.* **2020**, *6*, eaba4098. [CrossRef]
45. Shin, J.; Lee, J.; Park, Y.; Choi, J.W. Aqueous zinc ion batteries: Focus on zinc metal anodes. *Chem. Sci.* **2020**, *11*, 2028–2044. [CrossRef]
46. Konarov, A.; Voronina, N.; Jo, J.H.; Bakenov, Z.; Sun, Y.-K.; Myung, S.-T. Present and Future Perspective on Electrode Materials for Rechargeable Zinc-Ion Batteries. *ACS Energy Lett.* **2018**, *3*, 2620–2640. [CrossRef]
47. Song, M.; Tan, H.; Chao, D.; Fan, H.J. Recent Advances in Zn-Ion Batteries. *Adv. Funct. Mater.* **2018**, *28*, 1802564. [CrossRef]
48. Lu, Y.; Zhu, T.; van den Bergh, W.; Stefik, M.; Huang, K. A High Performing Zn-Ion Battery Cathode Enabled by In Situ Transformation of V(2) O(5) Atomic Layers. *Angew. Chem. Int. Ed.* **2020**, *59*, 17004–17011. [CrossRef] [PubMed]
49. Zhang, X.; Wang, L.; Fu, H. Recent advances in rechargeable Zn-based batteries. *J. Power Sources* **2021**, *493*, 229677. [CrossRef]
50. Li, C.; Xie, X.; Liang, S.; Zhou, J. Issues and Future Perspective on Zinc Metal Anode for Rechargeable Aqueous Zinc-ion Batteries. *Energy Environ. Mater.* **2020**, *3*, 146–159. [CrossRef]
51. Jia, H.; Wang, Z.; Tawiah, B.; Wang, Y.; Chan, C.-Y.; Fei, B.; Pan, F. Recent advances in zinc anodes for high-performance aqueous Zn-ion batteries. *Nano Energy* **2020**, *70*, 104523. [CrossRef]
52. Han, C.; Li, W.; Liu, H.K.; Dou, S.; Wang, J. Principals and strategies for constructing a highly reversible zinc metal anode in aqueous batteries. *Nano Energy* **2020**, *74*, 104880. [CrossRef]
53. Xie, C.; Li, Y.; Wang, Q.; Sun, D.; Tang, Y.; Wang, H. Issues and solutions toward zinc anode in aqueous zinc-ion batteries: A mini review. *Carbon Energy* **2020**, *2*, 540–560. [CrossRef]
54. Guo, L.; Guo, H.; Huang, H.; Tao, S.; Cheng, Y. Inhibition of Zinc Dendrites in Zinc-Based Flow Batteries. *Front. Chem.* **2020**, *8*, 557. [CrossRef]
55. Du, W.; Ang, E.H.; Yang, Y.; Zhang, Y.; Ye, M.; Li, C.C. Challenges in the material and structural design of zinc anode towards high-performance aqueous zinc-ion batteries. *Energy Environ. Sci.* **2020**, *13*, 3330–3360. [CrossRef]
56. Li, Q.; Zhao, Y.; Mo, F.; Wang, D.; Yang, Q.; Huang, Z.; Liang, G.; Chen, A.; Zhi, C. Dendrites issues and advances in Zn anode for aqueous rechargeable Zn-based batteries. *EcoMat* **2020**, *2*, e12035. [CrossRef]
57. Hao, J.; Li, X.; Zeng, X.; Li, D.; Mao, J.; Guo, Z. Deeply understanding the Zn anode behaviour and corresponding improvement strategies in different aqueous Zn-based batteries. *Energy Environ. Sci.* **2020**, *13*, 3917–3949. [CrossRef]
58. Yi, Z.; Chen, G.; Hou, F.; Wang, L.; Liang, J. Zinc-Ion Batteries: Strategies for the Stabilization of Zn Metal Anodes for Zn-Ion Batteries (Adv. Energy Mater. 1/2021). *Adv. Energy Mater.* **2021**, *11*, 2170001. [CrossRef]
59. Zhao, C.; Wang, X.; Shao, C.; Li, G.; Wang, J.; Liu, D.; Dong, X. The strategies of boosting the performance of highly reversible zinc anodes in zinc-ion batteries: Recent progress and future perspectives. *Sustain. Energy Fuels* **2021**, *5*, 332–350. [CrossRef]
60. Wang, J.; Yang, Y.; Zhang, Y.; Li, Y.; Sun, R.; Wang, Z.; Wang, H. Strategies towards the challenges of zinc metal anode in rechargeable aqueous zinc ion batteries. *Energy Storage Mater.* **2021**, *35*, 19–46. [CrossRef]

61. Zheng, J.; Archer, L.J.S.A. Controlling electrochemical growth of metallic zinc electrodes: Toward affordable rechargeable energy storage systems. *Sci. Adv.* **2021**, *7*, eabe0219. [CrossRef] [PubMed]
62. Hu, L.; Xiao, P.; Xue, L.; Li, H.; Zhai, T. The rising zinc anodes for high-energy aqueous batteries. *EnergyChem* **2021**, *3*, 100052. [CrossRef]
63. Zhao, Z.; Fan, X.; Ding, J.; Hu, W.; Zhong, C.; Lu, J. Challenges in Zinc Electrodes for Alkaline Zinc–Air Batteries: Obstacles to Commercialization. *ACS Energy Lett.* **2019**, *4*, 2259–2270. [CrossRef]
64. Wu, T.H.; Zhang, Y.; Althouse, Z.D.; Liu, N. Nanoscale design of zinc anodes for high-energy aqueous rechargeable batteries. *Mater. Today Nano* **2019**, *6*, 100032. [CrossRef]
65. Tan, P.; Chen, B.; Xu, H.; Cai, W.; He, W.; Zhang, H.; Liu, M.; Shao, Z.; Ni, M. Integration of Zn–Ag and Zn–Air Batteries: A Hybrid Battery with the Advantages of Both. *ACS Appl. Mater. Interfaces* **2018**, *10*, 36873–36881. [CrossRef]
66. Li, P.-C.; Hu, C.-C.; You, T.-H.; Chen, P.-Y. Development and characterization of bi-functional air electrodes for rechargeable zinc-air batteries: Effects of carbons. *Carbon* **2017**, *111*, 813–821. [CrossRef]
67. Zhou, W.; Zhu, D.; He, J.; Li, J.; Chen, H.; Chen, Y.; Chao, D. A scalable top-down strategy toward practical metrics of Ni–Zn aqueous batteries with total energy densities of 165 W h kg^{-1} and 506 W h L^{-1}. *Energy Environ. Sci.* **2020**, *13*, 4157–4167. [CrossRef]
68. Sun, W.; Wang, F.; Hou, S.; Yang, C.; Fan, X.; Ma, Z.; Gao, T.; Han, F.; Hu, R.; Zhu, M.; et al. Zn/MnO$_2$ Battery Chemistry With H$^+$ and Zn^{2+} Coinsertion. *J. Am. Chem. Soc.* **2017**, *139*, 9775–9778. [CrossRef] [PubMed]
69. Zhang, N.; Dong, Y.; Jia, M.; Bian, X.; Wang, Y.; Qiu, M.; Xu, J.; Liu, Y.; Jiao, L.; Cheng, F. Rechargeable Aqueous Zn–V$_2$O$_5$ Battery with High Energy Density and Long Cycle Life. *ACS Energy Lett.* **2018**, *3*, 1366–1372. [CrossRef]
70. Xiaowei, C.; Yanliang, L.; Yan, Y. Electrolyte dictated materials design for beyond lithium ion batteries. In *Energy Harvesting and Storage: Materials, Devices, and Applications VIII*; International Society for Optics and Photonics: Bellingham, WA, USA, 2018; Volume 10663.
71. Liang, Y.; Jing, Y.; Gheytani, S.; Lee, K.-Y.; Liu, P.; Facchetti, A.; Yao, Y. Universal quinone electrodes for long cycle life aqueous rechargeable batteries. *Nat. Mater.* **2017**, *16*, 841–848. [CrossRef] [PubMed]
72. Bischoff, C.F.; Fitz, O.S.; Burns, J.; Bauer, M.; Gentischer, H.; Birke, K.P.; Henning, H.-M.; Biro, D. Revealing the Local pH Value Changes of Acidic Aqueous Zinc Ion Batteries with a Manganese Dioxide Electrode during Cycling. *J. Electrochem. Soc.* **2020**, *167*, 020545. [CrossRef]
73. Higashi, S.; Lee, S.W.; Lee, J.S.; Takechi, K.; Cui, Y. Avoiding short circuits from zinc metal dendrites in anode by backside-plating configuration. *Nat. Commun.* **2016**, *7*, 11801. [CrossRef]
74. Blanc, L.E.; Kundu, D.; Nazar, L.F. Scientific Challenges for the Implementation of Zn-Ion Batteries. *Joule* **2020**, *4*, 771–799. [CrossRef]
75. Zeng, X.; Hao, J.; Wang, Z.; Mao, J.; Guo, Z. Recent progress and perspectives on aqueous Zn-based rechargeable batteries with mild aqueous electrolytes. *Energy Storage Mater.* **2019**, *20*, 410–437. [CrossRef]
76. Wippermann, K.; Schultze, J.W.; Kessel, R.; Penninger, J. The inhibition of zinc corrosion by bisaminotriazole and other triazole derivatives. *Corros. Sci.* **1991**, *32*, 205–230. [CrossRef]
77. Hwang, B.; Oh, E.-S.; Kim, K. Observation of electrochemical reactions at Zn electrodes in Zn-air secondary batteries. *Electrochim. Acta* **2016**, *216*, 484–489. [CrossRef]
78. Toussaint, G.; Stevens, P.; Akrour, L.; Rouget, R.; Fourgeot, F. Development of a Rechargeable Zinc-Air Battery. *ECS Trans.* **2019**, *28*, 25–34. [CrossRef]
79. Zhang, W.; Zhai, X.; Zhang, Y.; Wei, H.; Ma, J.; Wang, J.; Liang, L.; Liu, Y.; Wang, G.; Ren, F.; et al. Application of Manganese-Based Materials in Aqueous Rechargeable Zinc-Ion Batteries. *Front. Energy Res.* **2020**, *8*, 195. [CrossRef]
80. Zhao, Y.; Zhu, Y.; Zhang, X. Challenges and perspectives for manganese-based oxides for advanced aqueous zinc-ion batteries. *InfoMat* **2020**, *2*, 237–260. [CrossRef]
81. Yang, H.; Chang, Z.; Qiao, Y.; Deng, H.; Mu, X.; He, P.; Zhou, H. Constructing a Super-Saturated Electrolyte Front Surface for Stable Rechargeable Aqueous Zinc Batteries. *Angew. Chem. Int. Ed.* **2020**, *59*, 9377–9381. [CrossRef]
82. Yuksel, R.; Buyukcakir, O.; Seong, W.K.; Ruoff, R.S. Metal-Organic Framework Integrated Anodes for Aqueous Zinc-Ion Batteries. *Adv. Energy Mater.* **2020**, *10*, 1904215. [CrossRef]
83. Hou, Z.; Gao, Y.; Tan, H.; Zhang, B. Realizing high-power and high-capacity zinc/sodium metal anodes through interfacial chemistry regulation. *Nat. Commun.* **2021**, *12*, 3083. [CrossRef]
84. Yu, Y.; Xie, J.; Zhang, H.; Qin, R.; Liu, X.; Lu, X. High-Voltage Rechargeable Aqueous Zinc-Based Batteries: Latest Progress and Future Perspectives. *Small Sci.* **2021**, *1*, 2000066. [CrossRef]
85. Li, G.; Liu, Z.; Huang, Q.; Gao, Y.; Regula, M.; Wang, D.; Chen, L.-Q.; Wang, D. Stable metal battery anodes enabled by polyethylenimine sponge hosts by way of electrokinetic effects. *Nat. Energy* **2018**, *3*, 1076–1083. [CrossRef]
86. Wang, Z.; Huang, J.; Guo, Z.; Dong, X.; Liu, Y.; Wang, Y.; Xia, Y. A Metal-Organic Framework Host for Highly Reversible Dendrite-free Zinc Metal Anodes. *Joule* **2019**, *3*, 1289–1300. [CrossRef]
87. Pei, A.; Zheng, G.; Shi, F.; Li, Y.; Cui, Y. Nanoscale Nucleation and Growth of Electrodeposited Lithium Metal. *Nano Lett.* **2017**, *17*, 1132–1139. [CrossRef] [PubMed]

88. Zhang, Q.; Luan, J.; Tang, Y.; Ji, X.; Wang, S.; Wang, H. A facile annealing strategy for achieving in situ controllable Cu_2O nanoparticle decorated copper foil as a current collector for stable lithium metal anodes. *J. Mater. Chem. A* **2018**, *6*, 18444–18448. [CrossRef]

89. Sagane, F.; Ikeda, K.-i.; Okita, K.; Sano, H.; Sakaebe, H.; Iriyama, Y. Effects of current densities on the lithium plating morphology at a lithium phosphorus oxynitride glass electrolyte/copper thin film interface. *J. Power Sources* **2013**, *233*, 34–42. [CrossRef]

90. Yang, Q.; Liang, G.; Guo, Y.; Liu, Z.; Yan, B.; Wang, D.; Huang, Z.; Li, X.; Fan, J.; Zhi, C. Do Zinc Dendrites Exist in Neutral Zinc Batteries: A Developed Electrohealing Strategy to In Situ Rescue In-Service Batteries. *Adv. Mater.* **2019**, *31*, 1903778. [CrossRef] [PubMed]

91. Zhang, Q.; Luan, J.; Tang, Y.; Ji, X.; Wang, H. Interfacial Design of Dendrite-Free Zinc Anodes for Aqueous Zinc-Ion Batteries. *Angew. Chem. Int. Ed.* **2020**, *59*, 13180–13191. [CrossRef]

92. Kang, L.; Cui, M.; Jiang, F.; Gao, Y.; Luo, H.; Liu, J.; Liang, W.; Zhi, C. Nanoporous $CaCO_3$ Coatings Enabled Uniform Zn Stripping/Plating for Long-Life Zinc Rechargeable Aqueous Batteries. *Adv. Energy Mater.* **2018**, *8*, 1801090. [CrossRef]

93. Huang, S.; Zhu, J.; Tian, J.; Niu, Z. Recent Progress in the Electrolytes of Aqueous Zinc-Ion Batteries. *Chem. A Eur. J.* **2019**, *25*, 14480–14494. [CrossRef] [PubMed]

94. Zhao, Z.; Zhao, J.; Hu, Z.; Li, J.; Li, J.; Zhang, Y.; Wang, C.; Cui, G. Long-life and deeply rechargeable aqueous Zn anodes enabled by a multifunctional brightener-inspired interphase. *Energy Environ. Sci.* **2019**, *12*, 1938–1949. [CrossRef]

95. Kim, D.; Lee, C.; Jeong, S. A concentrated electrolyte for zinc hexacyanoferrate electrodes in aqueous rechargeable zinc-ion batteries. *IOP Conf. Ser. Mater. Sci. Eng.* **2018**, *284*, 012001. [CrossRef]

96. Kasiri, G.; Trócoli, R.; Bani Hashemi, A.; La Mantia, F. An electrochemical investigation of the aging of copper hexacyanoferrate during the operation in zinc-ion batteries. *Electrochim. Acta* **2016**, *222*, 74–83. [CrossRef]

97. Zeng, Y.; Zhang, X.; Meng, Y.; Yu, M.; Yi, J.; Wu, Y.; Lu, X.; Tong, Y. Achieving Ultrahigh Energy Density and Long Durability in a Flexible Rechargeable Quasi-Solid-State $Zn–MnO_2$ Battery. *Adv. Mater.* **2017**, *29*, 1700274. [CrossRef]

98. Lee, B.; Seo, H.R.; Lee, H.R.; Yoon, C.S.; Kim, J.H.; Chung, K.Y.; Cho, B.W.; Oh, S.H. Critical Role of pH Evolution of Electrolyte in the Reaction Mechanism for Rechargeable Zinc Batteries. *ChemSusChem* **2016**, *9*, 2948–2956. [CrossRef]

99. Jiang, B.; Xu, C.; Wu, C.; Dong, L.; Li, J.; Kang, F. Manganese Sesquioxide as Cathode Material for Multivalent Zinc Ion Battery with High Capacity and Long Cycle Life. *Electrochim. Acta* **2017**, *229*, 422–428. [CrossRef]

100. Zhao, S.; Han, B.; Zhang, D.; Huang, Q.; Xiao, L.; Chen, L.; Ivey, D.G.; Deng, Y.; Wei, W. Unravelling the reaction chemistry and degradation mechanism in aqueous Zn/MnO_2 rechargeable batteries. *J. Mater. Chem. A* **2018**, *6*, 5733–5739. [CrossRef]

101. Cai, Z.; Ou, Y.; Wang, J.; Xiao, R.; Fu, L.; Yuan, Z.; Zhan, R.; Sun, Y. Chemically resistant Cu–Zn/Zn composite anode for long cycling aqueous batteries. *Energy Storage Mater.* **2020**, *27*, 205–211. [CrossRef]

102. Chao, D.; Zhou, W.; Ye, C.; Zhang, Q.; Chen, Y.; Gu, L.; Davey, K.; Qiao, S.-Z. An Electrolytic $Zn–MnO_2$ Battery for High-Voltage and Scalable Energy Storage. *Angew. Chem. Int. Ed.* **2019**, *58*, 7823–7828. [CrossRef] [PubMed]

103. Zhang, T.; Tang, Y.; Guo, S.; Cao, X.; Pan, A.; Fang, G.; Zhou, J.; Liang, S. Fundamentals and perspectives in developing zinc-ion battery electrolytes: A comprehensive review. *Energy Environ. Sci.* **2020**, *13*, 4625–4665. [CrossRef]

104. Lu, J.; Xiong, T.; Zhou, W.; Yang, L.; Tang, Z.; Chen, S. Metal Nickel Foam as an Efficient and Stable Electrode for Hydrogen Evolution Reaction in Acidic Electrolyte under Reasonable Overpotentials. *ACS Appl. Mater. Interfaces* **2016**, *8*, 5065–5069. [CrossRef] [PubMed]

105. Barton, G.W.; Scott, A.C. Industrial applications of a mathematical model for the zinc electrowinning process. *J. Appl. Electrochem.* **1994**, *24*, 377–383. [CrossRef]

106. Selvakumaran, D.; Pan, A.; Liang, S.; Cao, G. A review on recent developments and challenges of cathode materials for rechargeable aqueous Zn-ion batteries. *J. Mater. Chem. A* **2019**, *7*, 18209–18236. [CrossRef]

107. Ibrahim, M.A.M. Improving the throwing power of acidic zinc sulfate electroplating baths. *J. Chem. Technol. Biotechnol.* **2000**, *75*, 745–755. [CrossRef]

108. Kühne, H.M.; Schefold, J. Tafel Plots from Illuminated Photoelectrodes: A New Insight Into Charge Transfer Mechanism. *J. Electrochem. Soc.* **1990**, *137*, 568–575. [CrossRef]

109. Beverskog, B.; Puigdomenech, I. Revised pourbaix diagrams for zinc at 25–300 °C. *Corros. Sci.* **1997**, *39*, 107–114. [CrossRef]

110. Dong, H.; Lei, T.; He, Y.; Xu, N.; Huang, B.; Liu, C.T. Electrochemical performance of porous Ni_3Al electrodes for hydrogen evolution reaction. *Int. J. Hydrog. Energy* **2011**, *36*, 12112–12120. [CrossRef]

111. Miles, M.H.; Kissel, G.; Lu, P.W.T.; Srinivasan, S. Effect of Temperature on Electrode Kinetic Parameters for Hydrogen and Oxygen Evolution Reactions on Nickel Electrodes in Alkaline Solutions. *J. Electrochem. Soc.* **1976**, *123*, 332–336. [CrossRef]

112. Zhang, Q.B.; Hua, Y.X.; Dong, T.G.; Zhou, D.G. Effects of temperature and current density on zinc electrodeposition from acidic sulfate electrolyte with [BMIM]HSO4 as additive. *J. Appl. Electrochem.* **2009**, *39*, 1207. [CrossRef]

113. Pan, D.; Ma, L.; Xie, Y.; Jen, T.C.; Yuan, C. On the physical and chemical details of alumina atomic layer deposition: A combined experimental and numerical approach. *J. Vac. Sci. Technol. A* **2015**, *33*, 021511. [CrossRef]

114. Johnson, R.W.; Hultqvist, A.; Bent, S.F. A brief review of atomic layer deposition: From fundamentals to applications. *Mater. Today* **2014**, *17*, 236–246. [CrossRef]

115. Kim, H.; Lee, H.-B.-R.; Maeng, W.J. Applications of atomic layer deposition to nanofabrication and emerging nanodevices. *Thin Solid Film.* **2009**, *517*, 2563–2580. [CrossRef]

116. Zhao, K.; Wang, C.; Yu, Y.; Yan, M.; Wei, Q.; He, P.; Dong, Y.; Zhang, Z.; Wang, X.; Mai, L. Ultrathin Surface Coating Enables Stabilized Zinc Metal Anode. *Adv. Mater. Interfaces* **2018**, *5*, 1800848. [CrossRef]
117. He, H.; Tong, H.; Song, X.; Song, X.; Liu, J. Highly stable Zn metal anodes enabled by atomic layer deposited Al_2O_3 coating for aqueous zinc-ion batteries. *J. Mater. Chem. A* **2020**, *8*, 7836–7846. [CrossRef]
118. Xie, X.; Liang, S.; Gao, J.; Guo, S.; Guo, J.; Wang, C.; Xu, G.; Wu, X.; Chen, G.; Zhou, J. Manipulating the ion-transfer kinetics and interface stability for high-performance zinc metal anodes. *Energy Environ. Sci.* **2020**, *13*, 503–510. [CrossRef]
119. Xia, A.; Pu, X.; Tao, Y.; Liu, H.; Wang, Y. Graphene oxide spontaneous reduction and self-assembly on the zinc metal surface enabling a dendrite-free anode for long-life zinc rechargeable aqueous batteries. *Appl. Surf. Sci.* **2019**, *481*, 852–859. [CrossRef]
120. Cui, M.; Xiao, Y.; Kang, L.; Du, W.; Gao, Y.; Sun, X.; Zhou, Y.; Li, X.; Li, H.; Jiang, F.; et al. Quasi-Isolated Au Particles as Heterogeneous Seeds To Guide Uniform Zn Deposition for Aqueous Zinc-Ion Batteries. *ACS Appl. Energy Mater.* **2019**, *2*, 6490–6496. [CrossRef]
121. Liang, P.; Yi, J.; Liu, X.; Wu, K.; Wang, Z.; Cui, J.; Liu, Y.; Wang, Y.; Xia, Y.; Zhang, J. Highly Reversible Zn Anode Enabled by Controllable Formation of Nucleation Sites for Zn-Based Batteries. *Adv. Funct. Mater.* **2020**, *30*, 1908528. [CrossRef]
122. Zeng, Y.; Zhang, X.; Qin, R.; Liu, X.; Fang, P.; Zheng, D.; Tong, Y.; Lu, X. Dendrite-Free Zinc Deposition Induced by Multifunctional CNT Frameworks for Stable Flexible Zn-Ion Batteries. *Adv. Mater.* **2019**, *31*, 1903675. [CrossRef]
123. Hieu, L.T.; So, S.; Kim, I.T.; Hur, J. Zn anode with flexible β-PVDF coating for aqueous Zn-ion batteries with long cycle life. *Chem. Eng. J.* **2021**, *411*, 128584. [CrossRef]
124. Liu, M.; Yang, L.; Liu, H.; Amine, A.; Zhao, Q.; Song, Y.; Yang, J.; Wang, K.; Pan, F. Artificial Solid-Electrolyte Interface Facilitating Dendrite-Free Zinc Metal Anodes via Nanowetting Effect. *ACS Appl. Mater. Interfaces* **2019**, *11*, 32046–32051. [CrossRef]
125. Wu, K.; Yi, J.; Liu, X.; Sun, Y.; Cui, J.; Xie, Y.; Liu, Y.; Xia, Y.; Zhang, J. Regulating Zn Deposition via an Artificial Solid–Electrolyte Interface with Aligned Dipoles for Long Life Zn Anode. *Nano-Micro Lett.* **2021**, *13*, 79. [CrossRef]
126. Zhang, Q.; Luan, J.; Huang, X.; Zhu, L.; Tang, Y.; Ji, X.; Wang, H. Simultaneously Regulating the Ion Distribution and Electric Field to Achieve Dendrite-Free Zn Anode. *Small* **2020**, *16*, 2000929. [CrossRef]
127. He, M.; Shu, C.; Zheng, R.; Xiang, W.; Hu, A.; Yan, Y.; Ran, Z.; Li, M.; Wen, X.; Zeng, T.; et al. Manipulating the ion-transference and deposition kinetics by regulating the surface chemistry of zinc metal anodes for rechargeable zinc-air batteries. *Green Energy Environ.* **2021**. [CrossRef]

Review

Review of ZnO Binary and Ternary Composite Anodes for Lithium-Ion Batteries

Vu Khac Hoang Bui [1], Tuyet Nhung Pham [2], Jaehyun Hur [3,*] and Young-Chul Lee [1,*]

1   Department of BioNano Technology, Gachon University, Seongnam-si 13120, Korea; hoangvu2101@gachon.ac.kr
2   Phenikaa University Nano Institute (PHENA), PHENIKAA University, Hanoi 12116, Vietnam; nhungpham240694@gmail.com
3   Department of Chemical and Biological Engineering, Gachon University, Seongnam-si 13120, Korea
*   Correspondence: jhhur@gachon.ac.kr (J.H.); dreamdbs@gachon.ac.kr (Y.-C.L.); Tel.: +82-31-750-5593 (J.H.); +82-31-750-8751 (Y.-C.L.)

Abstract: To enhance the performance of lithium-ion batteries, zinc oxide (ZnO) has generated interest as an anode candidate owing to its high theoretical capacity. However, because of its limitations such as its slow chemical reaction kinetics, intense capacity fading on potential cycling, and low rate capability, composite anodes of ZnO and other materials are manufactured. In this study, we introduce binary and ternary composites of ZnO with other metal oxides (MOs) and carbon-based materials. Most ZnO-based composite anodes exhibit a higher specific capacity, rate performance, and cycling stability than a single ZnO anode. The synergistic effects between ZnO and the other MOs or carbon-based materials can explain the superior electrochemical characteristics of these ZnO-based composites. This review also discusses some of their current limitations.

Keywords: ZnO; composites; binary; ternary; LIBs; anode

Citation: Bui, V.K.H.; Pham, T.N.; Hur, J.; Lee, Y.-C. Review of ZnO Binary and Ternary Composite Anodes for Lithium-Ion Batteries. *Nanomaterials* **2021**, *11*, 2001. https://doi.org/10.3390/nano 11082001

Academic Editor: Christian M. Julien

Received: 28 June 2021
Accepted: 1 August 2021
Published: 4 August 2021

Publisher's Note: MDPI stays neutral with regard to jurisdictional claims in published maps and institutional affiliations.

1. Introduction

Rechargeable lithium-ion batteries (LIBs) are widely used owing to their high specific energy, high electrochemical performance, and extended lifetime [1]. They are extensively adopted to power various electronic appliances, such as laptops and mobile phones. However, graphite, which is used as an anode in LIBs, limits their practical applications because of its low theoretical capacity (372 mAh g^{-1}) [2]. Semiconductor metal oxides (MOs) have the potential to enhance the performance of LIBs because of their higher theoretical capacity and safety than the traditional materials, such as carbon materials [3]. In the past ten years, the literature and patents on MOs and their composites for LIB applications have drastically increased (Figure 1), which is a trend that is expected to continue. Among MOs, zinc oxide (ZnO) (978 mAh g^{-1}) has an excellent theoretical capacity (Table 1), which is only slightly lower than that of ferric oxide (Fe$_2$O$_3$). Compared to other MOs, ZnO not only has a higher theoretical capacity but also a lower cost, ease of synthesis, various synthesis methods, chemical stability, and different morphologies [4]. Therefore, in this study, ZnO-based composites were selected as the target.

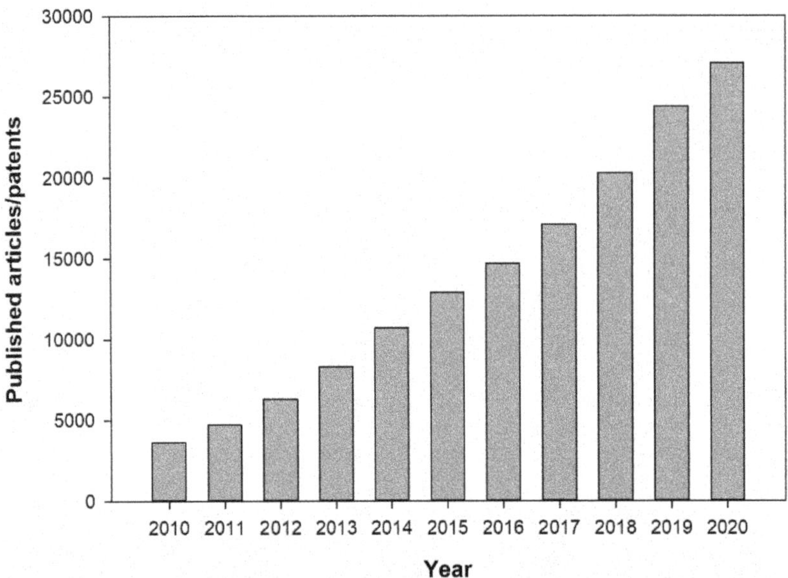

Figure 1. Published articles/patents related to MOs and their composites for LIB applications. Data collected from Google Scholar (https://scholar.google.com) (accessed on 15 June 2021) database with keyword: "Metal oxide lithium-ion batteries".

As mentioned above, ZnO is a promising anode material candidate for LIBs owing to its high theoretical capacity (978 mAh g^{-1}) [5]. ZnO is a low-band gap semiconductor (3.37 eV) with unique properties such as a high exciton binding energy (60 meV), photoelectric response, and electron mobility. The above features along with its good thermal and chemical stability render ZnO useful for various applications [6]. The general electrochemical mechanism for a ZnO anode in an LIB is as follows [7]:

$$ZnO + 2Li^+ + 2e^- \leftrightarrow Zn + Li_2O \tag{1}$$

$$Zn + Li^+ + e^- \leftrightarrow LiZn \tag{2}$$

Equation (1) is a conversion reaction, and (2) is an alloying–dealloying reaction. In (1), ZnO captures more lithium ions (Li$^+$) than traditional anodes, which proves beneficial during (2) [7]. Materials that undergo both conversion and alloying–dealloying reactions have higher capacities than those that involve only alloying reactions [8]. Unlike tin oxide (SnO$_2$), which undergoes irreversible conversion and reversible alloying–dealloying reactions [9], for ZnO, both reactions are reversible [7,10].

Table 1. Common MOs utilized as LIB anodes and their theoretical capacities.

Metal Oxide	Lithium Intercalation Method	Theoretical Capacity (mAh g^{-1})	References
Co$_3$O$_4$	Conversion	890	[9,11]
CoO	Conversion	716	[11]
CuO	Conversion	674	[11]
Fe$_2$O$_3$	Conversion	1006	[12]
NiO	Conversion	718	[9,13]
RuO$_2$	Conversion	806	[11,14]
SnO$_2$	Alloying	740	[15]
TiO$_2$	Intercalation	335	[9,16]
ZnCo$_2$O$_4$	Conversion	903	[9,17]
ZnFe$_2$O$_4$	Conversion	1000	[18,19]
ZnO	Alloying	978	[5,7,10]

However, ZnO has numerous limitations such as its slow chemical reaction kinetics, intense capacity fading on potential cycling, and low rate capability [20]. Moreover, it tends to aggregate and undergo a remarkable volume change (228%) during the charge/discharge cycles [7,21]. Although a thin layer formation occurs during the first cycle, it is extremely thin to be robust to the volume variation in ZnO. Thus, nanocracks can form inside ZnO and lead to continuous growth of the solid electrolyte interface (SEI) layer [22,23]. To realize LIB anodes with high reversible capacity, structural stability, and activity, materials with high conductivity are necessary [5].

Some strategies to increase the electrochemical performance of ZnO are to realize nanoscale-sized particles and to fabricate different nanoarchitectures, such as nanospheres, nanorods, nanotubes, and nanosheets [24–27]. Nanostructured MOs can decrease the Li^+ diffusion time and enhance the rate performance [9]. ZnO nanoarchitectures not only provide a barrier against the volume variation during the cycling process but also enhance the electrode/electrolyte contact area [28]. However, such different nanosized and morphological ZnO forms still have a low intrinsic electrical conductivity and exhibit a large volume change during the charge/discharge process [29]. Concurrently, composites of ZnO and other materials can be formed to increase its initial capacity and decrease the degradation during the charge/discharge process [7]. In addition, coating ZnO on materials such as carbon can alleviate the volume change problem [7,30]. Thus, ZnO-based nanomaterials offer a high electrochemical conductivity, a short transport path for Li^+, and an extended lifetime [7]. In this review, we focus on ZnO-based binary and ternary composites.

2. ZnO Binary Composites

The alloying reaction of MO anodes in LIBs can cause a large volume change and structural stress in the anodes, thus damaging them, such as through cracking and pulverization [31]. These problems not only lower the LIB performance but also threaten safety. One strategy to overcome these issues is to design composite anodes of carbon materials and MOs [20].

2.1. ZnO–MO Composites

In the binary MO composites, typically, Li^+ insertion/extraction occurs for both transition Mos during the charge/discharge process, which enhances the kinetics of the reduction–oxidation reactions and improves the electrical conductivity compared to the individual oxides; these, in turn, enhance the anodic performance in battery applications [32]. ZnO–SnO$_2$ composite anodes have been synthesized using various methods including hydrothermal methods, ball milling, layer-by-layer approaches, chemical vapor deposition, and physical vapor deposition [33–37]. Recently, Zhao et al. (2019) prepared two types of ZnO–SnO$_2$ composites by atomic layer deposition (ALD): (1) intermixed films in which Zn, Sn, and O were atomically mixed in a single amorphous layer, and (2) nanolaminated films with a well-defined interfacial formation between the ZnO and SnO$_2$ layers (Figure 2). The ALD processes were $100 \times (2\ ZnO + 3\ SnO_2)$ and $100 \times (20\ ZnO + 30\ SnO_2)$ for the intermixed (Figure 2a) and nanolaminated (Figure 2b) ZnO–SnO$_2$ composites, respectively. X-ray photoelectron spectroscopy (XPS) data confirmed the co-existence of ZnO and SnO$_2$ in the as-prepared composites. The Zn, Sn, and O atomic concentrations remained unchanged over the entire intermixed ZnO–SnO$_2$ thin film, which indicated uniform ZnO and SnO$_2$ mixing. In contrast, the Zn and Sn atomic concentrations alternated throughout the nanolaminated ZnO–SnO$_2$ film having a ZnO surface layer. Under a sputter depth of 1600 s of etching, Zn rapidly faded and was substituted by Sn, indicating an interface between the layers. From the cyclic voltammetry measurement, the electrochemical reactions for intermixed and nanomixed ZnO–SnO$_2$ and ZnO anodes can be described in the following equations:

$$\text{Conversion: } SnO_2 + 4Li^+ + 4e^- \rightarrow Sn + 2Li_2O$$

$$\text{Alloying-dealloying: } Sn + xLi^+ + xe^- \leftrightarrow Li_xSn\ (0 \leq x \leq 4.4)$$

$$\text{Conversion: } ZnO + 2Li^+ + 2e^- \rightarrow Zn + Li_2O$$

$$\text{Alloying-dealloying: } Zn + yLi^+ + ye^- \leftrightarrow Li_yZn \ (0 \leq y \leq 1)$$

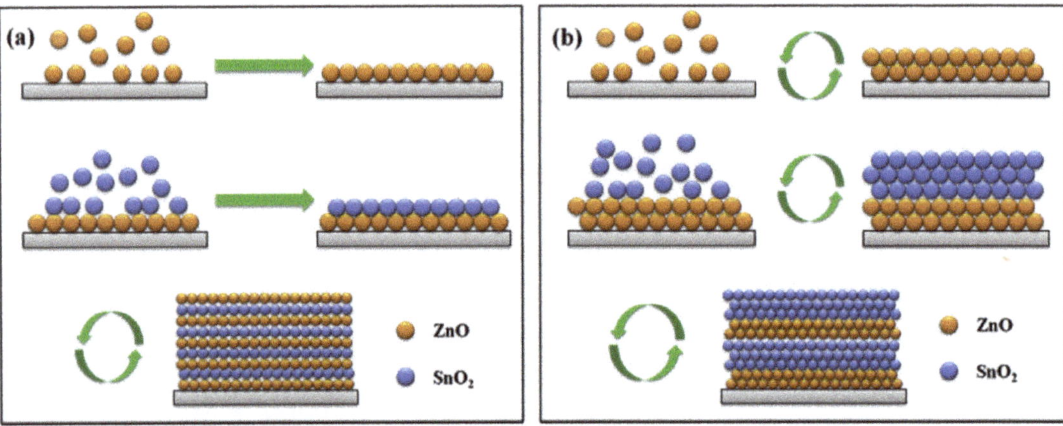

Figure 2. Synthesis of intermixed (**a**) and nanolaminated (**b**) ZnO–SnO$_2$ composite films by ALD. Reprinted with permission from [6]. Copyright 2019, Elsevier.

Although the CV curves of the intermixed and nanolaminated ZnO–SnO$_2$ composites were similar, in the cathodic scan, the latter showed a lower reversibility than the former. Compare to single ZnO and SnO$_2$ anodes, both intermixed and nanolaminated ZnO–SnO$_2$ composites have a higher capacity and initial Coulombic efficiency (CE). The first discharge capacity and Coulombic efficiency (CE) of the intermixed ZnO–SnO$_2$ were 2667 mAh g^{-1} and 80.2%, respectively, and for the nanolaminated ZnO–SnO$_2$, these were 2471 mAh g^{-1} and 71.4%, respectively. The higher CE indicated that intermixed ZnO–SnO$_2$ is more reversible. Both the intermixed and nanolaminated ZnO–SnO$_2$ composites presented a better cycling stability than pure SnO$_2$ and ZnO. For the nanolaminated composite, a capacity decrease occurred only after the 35th cycle. In comparison, the capacity of the intermixed composite was initially stable until the 10th cycle, subsequently increased to 1752 mAh g^{-1}, and remained almost constant up to the 50th cycle. After the 50th cycle, particle expansion occurred in the intermixed ZnO–SnO$_2$ electrode without disruption of their interconnections, resulting in a surface with fewer crevasses and more cycling stability than the nanolaminated composite surface. At the atomic scale, when mixing ZnO and SnO$_2$, ZnO reduction suppresses the alloying of Sn, and the produced reduced Zn^0 also assists SnO$_2$, which improves the morphological stability of the anode during the cycling process and increases its cyclability. For the nanolaminated composite, the interface formation between ZnO and SnO$_2$ leads to isolation intercalation and alloying reactions with the injected lithium, resulting in a severe volume change, specifically in the latter reaction. Moreover, the conductivity of the intermixed composite was improved. In view of these results, Zhao et al. (2019) explored the atomic ratio effect on the electrochemical properties of intermixed composites. For intermixed ZnO−2SnO$_2$ composites containing ZnO and SnO$_2$ in a 1:2 atomic ratio, the discharge capacities in the first, second, and thirtieth cycles were 2637, 2230, and 1771 mAh g^{-1}, respectively, and the initial CE was 84.5%. The corresponding capacities for intermixed 2ZnO–SnO$_2$ composites with ZnO and SnO$_2$ in a 2:1 atomic ratio were 2495, 2018, and 1492 mAh g^{-1}, and the first cycle CE was 80.9%. Both the ZnO–2SnO$_2$ and 2ZnO–SnO$_2$ composites had high cyclabilities of 1955 and 1794 mAh g^{-1} at 0.5 A g^{-1} after 50 cycles, respectively. It was concluded that the change in the atomic ratio did not significantly affect the electrochemical activity of these intermixed composites. Although the capacity could be tuned by changing the atomic ratio, the rate and cycling performance were minor. Interestingly, annealing the $ZnO–SnO_2$ composite film in ambient helium at 1000 °C for 2 h led to the formation of a Zn_2SnO_4 film, making the surface of the annealed composite rough with some pinholes. The first and second discharge capacities of the Zn_2SnO_4 film were 2363 and 1915 mAh g^{-1}, its CE was 81% after the first cycle, and its capacity remained ~1515 mAh g^{-1} at 0.5 A g^{-1} after 50 cycles. As the current density was increased in the order 0.5, 0.8, 1, 2, and 5 A g^{-1}, the discharge capacity decreased in sequence as 1652, 1331, 1074, 818, and 514 mAh g^{-1}. When the current density was reduced to 0.5 A g^{-1} again, the discharge capacity increased to reach 1448 mAh g^{-1}, indicating the good rate capability of the Zn_2SnO_4 film. Thus, annealing did not have a significant influence on the electrochemical potential. Therefore, the most influencing factor of the electrochemical performance of $ZnO–SnO_2$ composites is the degree of mixing, and not the crystallinity degree or the exact composition [6].

In addition to ZnO and SnO_2, nickel oxide (NiO) is an interesting MO owing to its high theoretical specific capacity, low cost, eco-friendliness, and good electrochemical activities [38]. SnO_2 undergoes reversible alloying–dealloying reactions, while NiO undergoes reversible conversion reactions. Transition MOs with unique structures such as multi-layer or yolk–shell structures present enhanced electrochemical performance [39,40]. Nanoscale materials provide enhanced surface areas compared to their bulk forms and thus contribute additional active sites for Li$^+$ insertion/desertion, ensuring rapid transfer of charged particles [41,42]. Concurrently, a yolk–shell structure with a hollow feature provides a short pathway for Li$^+$ and electron mobility, enhancing the kinetics of the electrochemical reaction [32,39]. Additionally, the presence of abundant channels on the porous shell allows important electrode/electrolyte interactions. Moreover, the hollow structure and the gap between the yolk and the shell can significantly prevent volume change and damage of the yolk–shell structure during the discharge/charge process [43–45]. Li et al. (2018) produced yolk–shell ZnO/NiO microspheres with a shell of nanorods and a microsphere yolk by a controlled thermal treatment of a bimetallic organic framework in air. Scanning electron microscopy (SEM) and high-resolution transmission electron microscopy (TEM) showed that the obtained microspheres had a diameter of 2 μm. There was a gap (~200 nm) between the yolk and the shell structure, which provided an additional barrier against the changes in the volume and structure during the cycling process [45]. The co-existence of Zn^{2+} and Ni^{2+} was critical for obtaining the yolk–shell structure. Without the introduction of the Ni^{2+} source (Ni(NO$_3$)$_2$•6H$_2$O), only ultrathin products were produced after the solvothermal process, and ZnO nanoparticles (NPs) with diameters $\leq$ 100 nm were obtained after calcination. In contrast, without the injection of the Zn^{2+} source (Zn(NO$_3$)$_2$•6H$_2$O), only layer stacking was achieved, and ultrathin NiO nanosheets were formed after calcination. The ZnO/NiO microspheres (0.276 g cm^{-3}) had a higher packing density than commercial graphite (0.198 g cm^{-3}), and a high-packing density electrode material is expected to result in high-energy density LIBs [46]. Thus, a large surface area along with mesopores can provide an additional barrier to penetration and prevent structural variation during the charge/discharge process. XPS analysis showed that the presence of amorphous carbon can increase the ZnO/NiO conductivity as well as offering a barrier to prevent structural change during the charge/discharge cycles [47]. The CV curves and X-ray powder diffraction (XRD) measurement indicated the reversible formation of ZnO and NiO after charging:

$$\text{Conversion: } ZnO + 2Li^+ + 2e^- \leftrightarrow Zn + Li_2O$$

$$\text{Alloying-dealloying: } Zn + xLi^+ + xe^- \leftrightarrow Li_xZn$$

$$\text{Conversion: } NiO + 2Li^+ \leftrightarrow Ni + Li_2O$$

Due to the presence of NiO, the capacity of the composites is much higher than single ZnO and NiO. The first discharge/charge capacity of the ZnO/NiO microspheres was 1221.7/769.2 mAh g^{-1}, and their initial CE was 62.9% at 0.1 A g^{-1}. However, unlike the mentioned $ZnO–SnO_2$ composite, the presence of NiO did not improve the initial CE.

Therefore, different performances of the obtained composites can be found depending on the chosen MO partner. The electrolyte decomposition, SEI formation, occurrence of unclear irreversible reactions, and Li^+ confinement in the electrode explain the low CE. The authors suggested that electrolyte optimization can enhance the initial CE of the MOs [48,49]. After 200 cycles, the specific capacity of the microspheres remained approximately 1008.6 mAh g^{-1}. The ZnO/NiO microspheres demonstrated long-term cycling stability (592.4 mAh g^{-1} at 0.5 A g^{-1} after 1000 cycles) and a high rate capability (437.1 mAh g^{-1} at 2 A g^{-1}). The high electrochemical activity of the ZnO/NiO microspheres can be explained based on their morphology and the ZnO and NiO synergetic effect. Additionally, the specific capacity of the ZnO/NiO microspheres increased during the charge/discharge cycles. For MO-based electrodes, the increase in the specific capacity can be attributed to the following: (1) generation of a polymeric layer on electrolyte degradation, which is called "pseudo-capacitance-type behavior" [50,51], (2) high accessibility of the host materials for Li^+ insertion/desertion originating from the enhanced kinetics of Li diffusion by the activation process [52], (3) formation of defects in the charge/discharge process, which improves the reaction kinetics and contributes additional active sites for Li^+ insertion/desertion [53], and (4) storage of interfacial lithium [54]. Electrochemical impedance spectroscopy (EIS) measurements showed that the charge transfer resistance (R_{ct}) decreased on charge/discharge cycling, which can be ascribed to the following reasons: (1) improved charge particle mobilization owing to formation of the yolk–shell structure and amorphous-phase carbon composite [55,56], (2) increased electrolyte penetration in the electrode by the electrochemical reactions, which reduces the interfacial electrode/electrolyte impedance [38,51], and (3) enhanced Li^+ diffusion kinetics owing to the repeated lithiation/delithiation on cycling. At 0.5 A g^{-1} and the 1000th cycle, the ZnO/NiO microsphere electrode still delivered a reversible capacity of 592.4 mAh g^{-1}; however, some minor pulverization was observed. In addition, both capacitive and diffusion-controlled processes occurred in the ZnO/NiO electrode. The diffusion-controlled processes, i.e., intercalation, conversion, and alloying, resulted in the high capacity of the electrode, whereas the capacitive behavior ensured rapid charge movement [57,58]. The optimized temperature for the preparation of the ZnO/NiO microspheres was 450 °C (heating rate: 2 °C min^{-1}) for 20 min [59].

Recently, Tu et al. (2021) introduced carbon cloth with a WO_3/ZnO heterostructure film as an anode for LIBs. Tungsten trioxide (WO_3) is another candidate to combine with ZnO to form WO_3/ZnO composites. WO_3 has a high theoretical capacity (~700 mAh g^{-1}) [60,61] and similar limitations to other MOs such as quick capacity fading at a high current density, aggregation, and low conductivity. The carbon cloth (CC)-supported WO_3/ZnO was prepared via the repeated immersion of CC-supported WO_3 (previously prepared by hydrothermal and thermal treatment) into a ZnO QD solution, dried, and calcinated at 450 °C in air for 1 hour. Amorphous ZnO QDs (~2 nm) were found to be uniformly deposited on the surface of CC-supported WO_3. According to the CV curve and XRD, this WO_3/ZnO electrode undergoes both intercalation and conversion reactions during the discharge process. The obtained composite has a high discharge capacity (~1500 mAh g^{-1} at 0.28 C), rate performance (~500 mAh g^{-1} at 9.0 C), and cycling stability (1100 mAh g^{-1} at 1 C after 300 cycles). The high rate performance of the composites can be explained by the fast ion transfer due to the formation of the directional internal electric field at the interface of WO_3 and ZnO in the WO_3/ZnO heterostructures. In addition, the initial CE of the composites is 79.9%. However, the thick ZnO QD layer can lead to a decrease in the electrochemical performance due to the aggregation of ZnO into the gaps between WO_3. The R_{ct} of WO_3/ZnO is lower than that for WO_3, thus confirming that the combination of WO_3 and ZnO can increase the interfacial charge transfer kinetics. The increase in the reversible capacity of the WO_3/ZnO electrode can be explained by both diffusion-controlled and surface-controlled processes, with a greater contribution belonging to the latter. The surface-controlled process resulted in the rapid lithiation/delithiation process at high rates, high stability, and high reversibility [62].

In another study, Karunakaran et al. (2018) synthesized a ZnO/Cu_2MgO_3 hollow porous nanocage as an anode for LIBs using a one-step cost-effective ultrasonic spray pyrolysis method. The nanocage had a diameter of ~400–700 nm, specific surface area of 46.373 m^2g^{-1}, pore size of 2–6 nm, and pore volume of 0.103 $cm^3 g^{-1}$. The porous structured shell resulted in easier electrolyte diffusion and more rapid Li^+ transport and provided a buffer matrix to relieve the volume change during the cycling. The first discharge/charge capacity of the ZnO/Cu_2MgO_3 electrode at 0.3 A g^{-1} was 990.75 mAh g^{-1}/663.45 mAh g^{-1}, and the initial CE was 66.98%. The as-prepared composite anode had a high discharge capacity (528 and 441 mAh g^{-1} after 400 cycles at 0.3 and 0.5 A g^{-1}, respectively), cycle stability, and rate capability. The charge transfer resistance of an as-assembled cell with ZnO/Cu_2MgO_3 as the anode was ~170 Ω, which decreased to approximately 146 Ω after four cycles, indicating a fading process after the cycling [63]. Field emission SEM images of the ZnO/Cu_2MgO_3 surface after 400 cycles at 0.5 A g^{-1} showed its robust and stable microstructure [64].

Similar to the yolk−shell structure composites, the composites of micrometer/submicrometer dimensions are promising as anodes for LIBs. Hou et al. (2015) synthesized hierarchical mesoporous $ZnO/ZnFe_2O_4$ (ZZFO) sub-microcubes (SMCs) as LIB anodes. The electrochemical reaction of $ZnFe_3O_4$ can be described by the following equation [19]:

$$\text{Conversion: } ZnFe_2O_4 + 9Li^+ + 9e^- \leftrightarrow LiZn + 2Fe + 4Li_2O$$

ZZFO was prepared by post-calcination (500 °C for 2 h in air) of the Prussian blue analog (PBA) of $Zn_3[Fe(CN)_6]_2$. Prussian blue and PBA can be employed as templates for preparing porous transition MOs [65–69]. X-ray fluorescence spectrometry confirmed that the ZnO/ZZFO molar ratio in the composite was 2:1. A simple chemical etching of ZZFO using 6 M NaOH for 5 h formed $ZnFe_2O_4$ (ZFO) (Figure 3). NaOH was used in the etching process because of ZnO dissolution in high-pH aqueous solutions [70]. The obtained products were cube-shaped, with dimensions in the range of ~500–700 nm. The ZZFO SMCs were composed of 10–15 nm NPs and contained many pores of ~3–7 nm between neighbor nanocrystallites. The results indicated that the ZZFO SMCs had a better initial discharge and reversible capacity, cycling performance, and rate capability than the single ZFO SMCs. The first discharge capacities of ZZFO and ZFO were ~1892 mAh g^{-1} and 854 mAh g^{-1}, respectively, at 1 A g^{-1}, and their initial CEs were ~70% and ~72%, respectively. In addition, at 1 A g^{-1}, the discharge capacities of ZZFO and ZFO remained stable after 200 cycles. The good cycling properties of ZZFO and ZFO can be partially explained by the porous SMC structure, which provides additional free space against the volume change occurring during the repeated Li^+ insertion/extraction process. Although ZFO had a higher specific surface area (SSA) than ZZFO, it had a lower specific capacity than ZZFO. Thus, it could be concluded that the Li storage performance of the ZZFO SMCs is dependent on the structure and the components, instead of the SSA. The electrochemical performance of the ZZFO SMCs can be attributed to the synergistic effect between ZnO and ZFO as well as the good dispersion of these nanophases. The homogeneous dispersion of ZnO prevents the self-aggregation of ZFO during the charge/discharge cycles and enhances the cycling performance of ZZFO. In addition, the ZnO and ZFO composite relieves the volume expansion during the charge/discharge process [71].

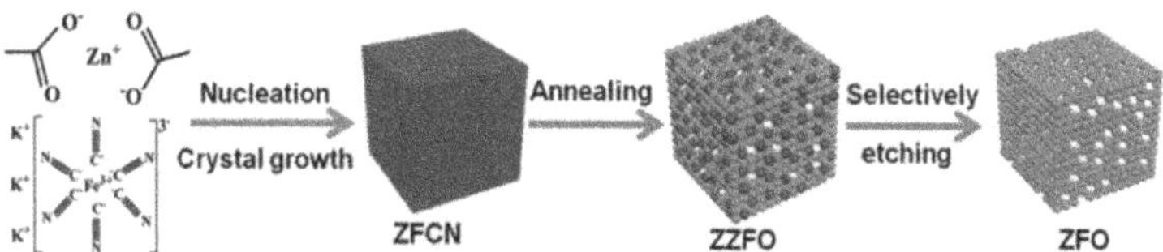

Figure 3. Scheme of ZZFO and ZFO synthesis process. Reprinted with permission from [71]. Copyright 2015, John Wiley and Sons.

2.2. ZnO–Carbon-Based Composites

Forming MO composites, such as ZnO, with porous nanostructured carbon materials can enhance the specific capacity, mechanical stability/flexibility, and electronic conductivity of the MOs [7]. The carbon layer/matrix enhances the performance of MO LIB anodes by two mechanisms: (1) by serving as a buffer matrix to relieve the volume expansion, preventing its pulverization, and increasing the electrical conductivity [72–75], and (2) by offering additional beneficial effects such as increased Li storage sites, high electrical conductivity, and improved electrode/electrolyte wettability, leading to an enhanced specific capacity, cycling stability, and rate performance [76]. However, in many composites, because the porous carbon (PC) material can account for >50% of the total electrode weight, it only slightly improves the capacity. Consequently, the total energy density is significantly reduced, and PC materials are not highly suitable for actual LIB applications [77].

2.2.1. ZnO-Coated Carbon-Based Material Composites

As mentioned above, porous nanostructured carbon-based materials can improve the specific capacity, cycling stability, and rate performance of MO anodes. They also offer a good conductive network and a buffer matrix for strain accommodation [5]. Hsieh et al. (2013) anchored ZnO nanocrystals (80–100 nm) with graphene nanosheets (GNs) using a method consisting of microwave heating, the modified Hummers method, and dispersion homogenization. A uniform dispersion of ZnO nanocrystals was found on both sides of the GNs and even in their interspacing layers. These ZnO nanocrystals not only acted as spacers to support the stereo GN framework but also as reduction–oxidation sites for improving the lithium storage capacity. In addition, the improvement in the rate capacity can be explained by the increase in the interlayer distance (i.e., d_{002}) of the GNs due to the insertion of the ZnO nanocrystals. A large d_{002} value yielded a high Li^+ diffusion rate in the 3D GN framework. This can be explained by the migration of Li^+ ions to bond to these sites as well as the active sites. When there were no ZnO spacers, the GNs tended to stack to form aggregates, decreased the accessible sites for the accommodation of Li^+, and reduced the ionic diffusion rate performance [78]. The GNs served as a buffer against the volume change related to the strain during cycling [79], thereby increasing the cycling performance. Thus, the Zn@GN anode presented an improved performance compared to ZnO, with a Li storage capacity of 850 mAh g^{-1} (at 0.1 C), an 82% CE (at the first cycle), a good rate capacity (capacity retention ~60% at 5 C), and increased cycling stability (capacity decay ~8% at the 50th cycle at 1 C). In comparison, bare ZnO had a storage capacity of 606 mAh g^{-1} (at 0.1 C), a 48.4% CE (at first cycle), and a relatively poorer cyclic performance [20].

Owing to its nanoporous structure, good conductive network, and strain accommodation, PC has been utilized as a dispersion medium for many MOs, such as ZnO, SnO_2, TiO_2, and Fe_3O_4 [5,80,81]. It protects MOs from aggregation and pulverization [5]. Shen et al. (2013) prepared ZnO/PC composites using the solvothermal method. The first discharge/charge capacity of ZnO (54 wt%)/PC was 2017.4/1062.9 mAh g^{-1} (CE: 52.68%), and it had a high reversible capacity of 653.7 mAh g^{-1} at 0.1 A g^{-1} after 100 cycles. The high lithium storage potential of this ZnO/PC composite can be attributed to its nanoporous structure and interconnected network. Lowering the ZnO loading reduced the discharge capacity of the ZnO/PC composite, whereas increasing the ZnO loading blocked the PC pores. In addition, at a high ZnO loading, the increase in the ZnO particle size increased the mechanical stress in the composite. The morphology of pure ZnO severely deteriorated after the lithiation/delithiation cycling, which led to a low reversible capacity. In contrast, the ZnO/PC composite maintained its original morphology without significant pulverization or cracks after the charge/discharge cycles. It can be concluded that the PC host provided space for the volume variation in the ZnO particles during the intercalation/deintercalation process, thereby preventing electrode degradation [5].

Carbon fibers (CFs) can also be used as supporting and conducting materials [82]. Han et al. (2019) prepared CFs@pore–ZnO as anodes for LIBs. The shell thickness of ZnO was approximately 100 nm. After thermal treatment, the ZnO particles became porous and their surface was rougher, which can be attributed to the thermal decomposition of the carbon material in the metal–organic framework structure and the emission of CO_2 and H_2O. The rough surface of the CFs@pore–ZnO composite increased the electrode/electrolyte contact area, whereas the induced porosity was beneficial for Li^+ mobility, electrolyte penetration, and addition of space against the volume expansion during the repeated Li^+ insertion/desertion. Accordingly, the first discharge/charge capacities of the CFs@pore–ZnO composite and pure CFs were 955/533 mAh g^{-1} and 230/221 mAh g^{-1} at 0.1 A g^{-1}, respectively. At the 300th cycle, the reversible capacity of the CFs@pore–ZnO composite was approximately 510 mAh g^{-1}, whereas the discharge capacity of the pure CFs gradually reduced to 149 mAh g^{-1} after only 150 cycles. The discharge capacity of the composite was even higher than the estimated theoretical capacity, which can be explained by the synergistic effect between ZnO and the CFs as well as the Li^+ storage in the voids between the ZnO polyhedra. The CFs@pore–ZnO composite exhibited good rates at 0.1, 0.2, 0.5, 1, and 2 A g^{-1}, with discharge capacities of 462, 378, 313, 270, and 240 mAh g^{-1}, respectively. In addition, when the current density was again decreased to 0.1 A g^{-1} after 60 cycles, the discharge capacity increased to 451 mAh g^{-1}, which gradually increased to 510 mAh g^{-1} after 100 cycles. After 50 cycles, the pure CFs were pulverized, which did not occur in the CFs@pore–ZnO composite [83].

Three-dimensional graphene aerogels (GAs) have different applications such as catalysis, gas chromatography, gas storage and separation, and sensing, owing to their mechanical stability and high mass and electron transfer rates [84]. Fan et al. (2016) prepared GAs with anchored sub-micrometer mulberry-like ZnO (ZnO@GAs) as anodes for LIBs using the solvothermal method. In the XRD patterns, the diffraction peak at approximately $2\theta = 25°$ disappeared, indicating the distribution of ZnO on both sides of the GA and prevention of stacking. The SEM images indicated that in the ZnO@GA composite, ZnO had a uniform diameter of ~500 nm and was composed of many small particles that aggregated into a mulberry-like morphology. Brunauer–Emmett–Teller (BET) analysis showed that the ZnO@GA composite had an SSA of 45.3 m^2 g^{-1} and a well-defined 10.7 nm mesopore. These nanoporous voids are beneficial for the rapid diffusion of the electrolyte to the active sites. XPS analysis suggested the presence of C–O–Zn linkages (285.6 eV in C 1s and 532.1 in O 1s). Their formation can be attributed to the substitution of hydrogen in the hydroxyl groups or a possible ring-opening reaction of the epoxy groups by Zn^{2+} in ZnO [85–87]. The first discharge/charge specific capacity of the ZnO@GA composite was 1001/713 mAh g^{-1} at 1.6 A g^{-1} (initial CE: 71.2%). The reversible capacities of the as-prepared composite were 365, 320, and 230 mAh g^{-1} at 1, 2, and 10 A g^{-1}, respectively. At 1.6 A g^{-1} and after 500 cycles, the reversible capacity remained at approximately 445 mAh g^{-1}, with a CE of approximately 100%. The morphology of the nanoclusters was well preserved, and the spherical structure was clearly retained. The robustness of the spherical structure can be explained by the generation of an SEI layer around the polymer/gel-like coating, which maintains the structural integrity and thus enhances the electrochemical performance. EIS measurements confirmed that the ZnO@GA electrode had smaller charge transfer and interface layer resistances than the ZnO electrode. The high electrochemical performance is contributed by the strong oxygen bridges (C–O–Zn) resulting from the interaction of graphene and ZnO, providing a good pathway for electron transfer during the cycling process. In addition, the hierarchical structure of the ZnO microballs prevents the stacking of the graphene layers, allowing the GAs to facilitate Li^+ transfer. The GA framework can also increase the electrical conductivity and relieve the electrode volume variation. The above synergistic effects improve the electrochemical performance [87].

The specific capacity of LIBs can be increased by the preparation of freestanding composites. In this type of composite, the use of binders and current collectors is rendered unnecessary [88]. It should be noted that a high mass loading is beneficial for a high energy density [77]. Nevertheless, ZnO-based composites frequently have low ZnO mass loadings (<1 mg cm^{-2}) [89]. Concurrently, a thick electrode with a high mass density typically leads to a poor electrochemical performance owing to the high charge transport resistance and the electrode removal from the current collector [90–92]. To overcome these limitations, inactive materials, such as conductive agents, binders, and metallic current collectors, are required [77]. Therefore, composites with a high mass loading and that avoid inactive additives while maintaining their high electrochemical properties should be developed. In addition, the 3D structure of composites enhances the effective electrode surface area and promotes the diffusion of lithium ions [88]. Using this approach, Zhao et al. (2018) synthesized a composite of a 3D interconnected carbon foam anchored with a two-dimensional (2D) ZnO nanomembrane (C/ZnO NM foam) as an anode for LIBs. The advantage of a 2D NM is that its deformation to a wrinkled structure allows remarkable strain accommodation during lithiation without damage, such as cracking [12,93,94]. Concurrently, a 3D interconnected structure offers sufficient voids for high mass loading, prevents peeling off from thick active materials, and eliminates the utilization of organic binders [95,96]. In the above study, the carbon foam had a pore diameter of 200 µm and a surface area of ~50 m^2 g^{-1}, and it was freestanding and could be folded or compressed. Owing to the large contact area, ZnO NMs could be anchored on the carbon foam surface by physisorption. The composite presented a high surface area originating from the high ZnO loading as well as the high porosity of the carbon foam. The as-prepared anode maintained 92% capacity at 2 A g^{-1} and 5 A g^{-1} after 700 and 500 cycles, respectively, and achieved an areal capacity of 4.3 mAh cm^{-2} at 80 mA g^{-1}, which is close to the acceptable capacity for practical applications (4 mAh cm^{-2}) [77]. The C/ZnO NM foam realized discharge capacities of 450, 375, 288, 175, and 80 mAh g^{-1} at 0.25, 0.5, 1, 2, and 4 A g^{-1}, respectively. When the current density was returned to 0.25 mA g^{-1}, a capacity of ~450 mAh g^{-1} was recovered. After the first and second cycles, the CE was maintained at approximately 99.5%. This can be attributed to the role of the carbon foam structure of providing sufficient voids for the ZnO NM to release the strain by deformation without cracking. The carbon foam also prevents the generation of an SEI layer during the following cycles. The interconnected framework and the open pores allow carbon to enhance the electron and ionic transport over the electrode, assisting lithiation/delithiation reversibility. After 100 cycles at 0.64 A g^{-1}, the starting structure of the carbon foam remained intact, and ZnO continued to be tightly anchored to the carbon foam framework. In contrast, the CE of the pure ZnO NMs was low because of the ZnO fracture and excessive formation of the SEI layer. To use C/ZnO NM foam anodes in full cells, cathode additives or extra cathodes are necessary to recover the capacity losses due to the SEI generation, which will sacrifice the energy density of the battery. In addition, specially designed electrolytes, such as gel-like electrolytes, and surface modification by thin oxides are effective approaches for improving the initial CE [77].

Li et al. (2017) formed another new freestanding composite, ZnO NM/expanded graphite (EG), using ALD. EG is a 3D material with an ultra-high SSA and large pores [97–99]. The large pores in EG provide sufficient voids for Li$^+$ storage as well as volume expansion [100–105]. EG also provides mechanical support and conductive channels for active materials such as ZnO, thereby avoiding the electrode pulverization originating from the volume change during the cycling process. In the above study, the thickness of the ZnO layer was controlled by adjusting the ALD cycles (from 100 to 800 cycles). ALD prepares thin films by a surface chemical reaction. To prevent the occurrence of gas-phase reactions, precursors with surface-saturating concentrations are alternatively pumped into the reaction chamber. ALD achieves a conformal growth of NPs with precise thickness control and good adhesion. ZnO NM/EG could be compressed into a flexible and self-standing film and be embedded into a battery, without the requirement of a conductive agent, binder, and current collector. The resulting anode had a capacity of 438 mAh g^{-1} at 0.2 A g^{-1} after 500 cycles. The combination of the high capacity of ZnO and the support provided by EG explained the performance of the anode [106].

2.2.2. Carbon-Based Material-Coated ZnO Composites

Coating carbon on ZnO leads to rapid electron mobility over its entire surface during cycling, thus enhancing the reversibility and kinetics of the Li^+ insertion/extraction [107,108]. Carbon coating also protects MOs from dissolving in the electrolyte, protects composite deformation, and maintains high conductivity [7,109]. In 2016, Quartarone et al. synthesized graphite-coated ZnO nanosheets as binder-free anodes for LIBs. The ZnO nanosheets were prepared by the hydrothermal process and coated with graphite by thermal evaporation. The graphite-coated ZnO nanosheet composite having a graphite thickness of 350 Å presented the first discharge/charge capacity of 1470/968 mAh g^{-1} (CE: 65.85%) and a specific capacity of 600 mAh g^{-1} after 100 cycles at 1 A g^{-1}. The specific capacity of the uncoated ZnO nanosheet was 400 mAh g^{-1} at 1 A g^{-1} after 100 cycles. The formation of micropores (pore diameter < 1 nm) within the composite and the enhancement of the exposed surface area were ascribed to the ZnO NPs' nucleation. The ZnO NPs were 15 nm in size, and the nanosheet width and length were approximately 0.8–1.5 µm. The small sizes of the ZnO NPs and their nanostructure not only increased the electrode/electrolyte contact area and the electrical contact but also achieved high strain accommodation. Concurrently, the graphite coating acted as a buffer against the change in the composite morphology during the charge/discharge process [110].

Gan et al. (2017) prepared N-doped carbon-coated ZnO nanorods (ZnO/NC-Z NRDs) using a solvent-free method as anodes for LIBs. N-doped carbon was synthesized by thermal treatment of zeolitic imidazolate framework-8 (ZIF-8), which was previously in situ grown on ZnO NRD surfaces (Figure 4). The study claimed that by using the in situ strategy, the adhesive force from the interaction between the carbon and ZnO NRDs was stronger than that with conventional ex situ methods, which resulted in a limited confinement effect in the latter [29]. Thus, compared to conventional ex situ methods, in situ methods realized a tighter coating of carbon on the ZnO NRD surfaces. TEM results showed that the carbon layer thickness was approximately 15–20 nm. Based on the XRD patterns, ZIF-8 was completely converted into amorphous carbon and ZnO after carbonization in Ar atmosphere. From the XPS analysis, the nitrogen concentration in the ZnO/NC-Z NRDs was approximately 2.3% and considered to enhance the composite electrical conductivity [29]. The carbon prepared from ZIF-8 exhibited specific properties such as a large surface area, tunable porosity, and structural stability and flexibility [29]. Owing to these characteristics, ZIF-8 can be employed in different energy storage systems, including supercapacitors, LIBs, and sodium-ion batteries [111–114]. Similar to the above-mentioned carbon-based materials, N-doped carbon increased the NRD conductivity and, consequently, enhanced the Li^+ diffusion rate. It also acted as a buffer layer, alleviating the volume variation during the cycling process. ZnO/NC-Z NRDs had a specific BET surface area of 135.9 m^2 g^{-1} (higher than ZnO NRDs, 47.9 m^2 g^{-1}) and a mesoporous structure (pore size of 3.85–5.53 nm). The presence of N-doped carbon also increased the ZnO/NC-Z NRD pore volume. The enhancements in the SSA and pore volume can be attributed to the decomposition of the ZIF-8 framework during the sintering process. The as-prepared composite had a capacity of 1439 mAh g^{-1} at 0.2 A g^{-1} and an initial CE of 76%; the latter was higher than that of the ZnO/C-P NRDs (71%) (sample prepared by ex situ method) and ZnO NRDs (67%). The higher CE of the ZnO/NC-Z NRD composite can be explained by the coated carbon layer preventing the detrimental reactions between the electrolyte and the ZnO NRDs. After 200 and 850 cycles, the capacity of the composite was reduced to 1011 mAh g^{-1} at 0.2 A g^{-1} and to 544 mAh g^{-1} at 1 A g^{-1} (capacity retention: 87.7%). The CE of the ZnO/NC-Z NRDs was found to be approximately 100% after the first cycle, indicating the ease of Li^+ insertion/extraction and efficient electron and ion transfer. Using the four-point probe method showed that the electrical conductivity of the ZnO/NC-Z NRDs was higher than that of the ZnO/C-P NRDs and ZnO NRDs. After 150 cycles, the NRD morphology of the ZnO/NC-Z NRDs was preserved without any significant structural damage. The charge transfer resistance ($R_{(sf+ct)}$) of the ZnO/NC-Z NRDs (61 Ω) was lower than that of the ZnO/C-P NRDs (102 Ω) and ZnO NRDs (175 Ω).

Moreover, the ZnO/NC-Z NRDs could be additionally used as precursors to synthesize N-doped carbon nanotubes (CNTs) (Figure 4), which had a capacity of 1001.1 mAh g^{-1} after 100 cycles at 0.2 A g^{-1} and a capacity retention of 99.1% [29].

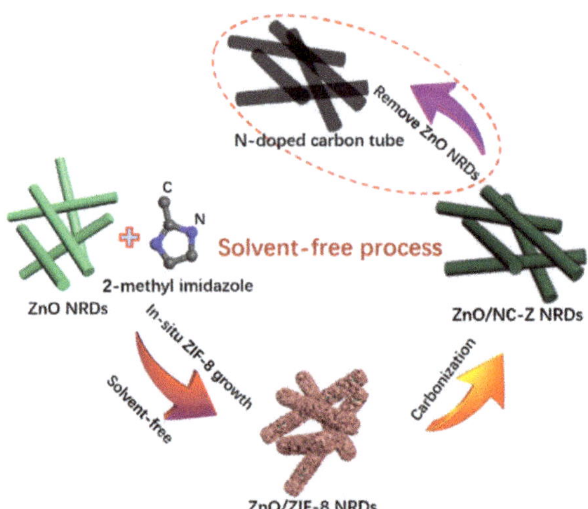

Figure 4. Scheme of synthesis of ZnO/NC-Z NRDs and N-doped carbon nanotubes. Reprinted with permission from [29]. Copyright 2017, Elsevier.

Recently, Thauer et al. (2021) introduced ZnO/C composites prepared via thermal treatment. As anodes for LIBs, ZnO/C composites calcinated at 700 °C showed the first discharge/charge capacities of 1061/671 mAh g^{-1} at 0.1 A g^{-1} (initial CE: 63.24%). The carbon content in the optimized composite was found to be around 5.7 wt.%. XRD measurement confirmed the two-step reaction mechanism, including the conversion and alloying processes (Equations (1) and (2)). After the 100th cycle at 0.1 A g^{-1}, the discharge capacity of the ZnO/C composite was reduced to 212 mAh g^{-1}. The irreversible conversion reaction can explain the rapid capacity. The presence of carbon is believed to improve the electronic conductivity and Li$^+$ diffusion [115]. Eisenmann et al. (2021) indicated that the carbon coating on Mn-doped ZnO could partially reduce manganese, reallocate the crystal structure, and increase the specific capacity of Mn-doped ZnO [116]. The particle size of carbon-coated Mn-doped ZnO and Mn-doped ZnO was similar (~ 20 nm), indicating that the carbon layer did not affect the size of particles. Through measurement with thermal gravimetric analyzers, the carbon content was ~20 wt.%. The lithium reaction mechanism was nearly the same between carbon-coated Mn-doped ZnO and Mn-doped ZnO. However, more reversible alloying–dealloying and much more reversible conversion were found in the case of carbon-coated Mn-doped ZnO. At 0.1 C, carbon-coated Mn-doped ZnO had higher specific capacities (~200 mAh g^{-1}). The authors found that a carbon-coated layer is most likely to improve the capacity retention and rate capability rather than result in a capacity improvement. When the current density was increased from 0.1 to 1 C, the specific capacity decreased by only 138 mAh g^{-1} from 740 to 602 mAh g^{-1}. Overall, the presence of a coated carbon layer can increase the electrochemical performance of Mn-doped ZnO via the decreased volume change upon cycling and the improvement in the reversible conversion reaction [116].

3. ZnO Ternary Composites

Besides binary composites, ternary composites have been investigated in different studies. Most of these ternary composites comprised ZnO, another MO, and PC-based materials. The synergistic effects between these materials can explain the improved electrochemical performance of the corresponding composites. For example, the PC materials serve as a buffer matrix to alleviate the volume expansion and capacity fading problems, whereas the ZnO and MO composite can protect the active materials from aggregation and enhance the electrode conductivity [117]. However, the electrochemical characteristics of ternary composites are less remarkable than those of the abovementioned binary composites.

Kose et al. (2016) prepared a freestanding ZnO/SnO_2/multi-walled CNT (ZnO/SnO_2/MWCNT) buckypaper composite using the sol–gel coating method. The MWCNT structure provided the mechanical stability offered by active materials. At the first cycle and 0.2 C, the specific capacities of ZnO/SnO_2/MWCNT, ZnO/MWCNT, and SnO_2/MWCNT were 1584, 1152, and 1491 mAh g^{-1}, respectively, which, after 100 cycles, were reduced to 487, 460, and 441 mAh g^{-1}, respectively. The MWCNTs act as a buffer matrix by interacting with the MO cluster and preventing the volume variation and capacity fading [118]. The higher discharge capacity of ZnO/SnO_2/MWCNT than that of the binary composites originates from the compositing of ZnO and SnO_2 as well as the high conductivity provided by the MWCNTs. In the binary composites, aggregation of the MOs is one of the significant problems [118,119]. In ZnO/SnO_2/MWCNT ternary composites, the uniform dispersion of the MOs as well as the diffusion barriers between ZnO and SnO_2 prevents the aggregation of Zn and Sn atoms [117]. Recently, Zhang et al. (2021) prepared SnO_2/ZnO@Polypyrrole (PPy) via an electrospinning technology. SnO_2 SnO_2/ZnO was prepared with polyvinylpyrrolidone (PVP) and then coated with PPy via the electrospinning process. PVP with a high molecular weight (Mw = 1,300,000 g mol^{-1}) resulted in a smaller diameter, dense structure, and higher electrochemical performance than PVP with a lower molecular weight (Mw = 58,000 g mol^{-1}). The obtained composites had an initial discharge/charge capacity of 1861.8/1138.1 mAh g^{-1} at 0.2 C (initial CE: 61.12%). After 100 cycles at 0.2 C, the discharge capacity of the SnO_2/ZnO@PPy composite was reduced to 626.1 mAh g^{-1}. Ppy can increase the capacity of composites by enhancing the conductivity of the composites and alleviating the change in the electrode during electrochemical cycling [120–123]. Similar to porous carbon materials, the porosity of PPy also improved the contact area with the electrolyte, accelerated the ion/electron diffusion rate, and increased the Li^+ reversibility during the cycling process. EIS measurement additionally indicated that SnO_2/ZnO@PPy has a higher charge transfer rate than SnO_2/ZnO due to the presence of PPy [124].

In addition to SnO_2, SnO, which is formed by the reduction of SnO_2, has a high theoretical capacity (approximately 880 mAh g^{-1}) [125]. Joshi et al. (2016) examined binder-free SnO_x–ZnO/carbon nanofiber (CNF) composites as LIB anodes. The optimal Sn/Zn ratio for the SnO_x–ZnO CNF composites was found to be 75:25 (wt.%). Increasing the ZnO content decreased the electrochemical performance of the composite owing to its low electrochemical activity [126]. At the optimal condition, the first discharge/charge capacity of the SnO_x–ZnO CNF composite was 1910/1400 mAh g^{-1} (CE: 73.3%) at 0.1 A g^{-1}. After 55 cycles, its reversible capacity was 963 mAh g^{-1} at 0.1 A g^{-1}. The CNFs enhanced the Li_2O decomposition and thus increased the reversible capacity. In addition, ZnO protected Sn from aggregation, which resulted in a cell with a high discharge capacity and increased stability. Amorphous SnO_x and ZnO were embedded in the CNFs, and the morphology of the resultant SnO_x–ZnO CNFs was uniform, smooth, long, and free of agglomerated particles. Moreover, there was no significant deterioration of the SnO_x–ZnO CNF composite morphology after 55 cycles [127].

Another ternary composite can be formed with ZnO and NiO. Similar to binary ZnO/MO and SnO_2/MO composites, the limitations of NiO/MO binary composites originate from their poor electronic conductivity and structural change during the repeated cycling, which result in a poor rate capability and rapid capacity fading. In 2018, Ma et al. prepared a NiO–ZnO/reduced graphene oxide (NiO–ZnO/RGO) composite by a process consisting of ultrasonic, freeze drying, and thermal treatments. The SEM results showed that the synthesized NiO–ZnO nanoflakes were uniformly distributed on the RGO sheet. The as-prepared electrodes had a first discharge capacity of 1393 mAh g^{-1} (CE: 66.3%) and high reversible capacities of 1017 mAh g^{-1} at 0.1 A g^{-1} after 200 cycles and 458 mAh g^{-1} at 0.5 A g^{-1} after 400 cycles. The reversible capacity of the ternary composite (1017 mAh g^{-1}) was higher than the theoretical capacity (833 mAh g^{-1}). After 15 cycles, the NiO–ZnO/RGO composite electrode had a higher CE (98%) than the NiO–ZnO hybrid anode. The reversible capacities of NiO and NiO–ZnO were reduced to 212 mAh g^{-1} at 180 cycles and 247 mAh g^{-1} at 150 cycles, respectively. The enhancement in the capacity of the NiO–ZnO/RGO composite was contributed by the formation of a reversible polymeric gel-like film with a high material viscosity provided by ZnO, which enhances the adhesion between the active material layer and the current collector [128]. NiO–ZnO/RGO had a smaller charge transfer resistance than the NiO–ZnO binary composite. Similar to MWCNTs and other carbon-based materials, the RGO prevents NiO–ZnO agglomeration and the volume variation during the charge/discharge cycles. NiO–ZnO nanoflakes were considered to provide abundant electrochemical reaction sites and decrease the Li$^+$ diffusion length, whereas the role of RGO was to enhance the Li$^+$ and electron transfer rates during the cycling process [129]. In the above study, the synergistic effect between NiO–ZnO and RGO was similar to that in the abovementioned binary composites. However, the combination effects of NiO and ZnO were not clearly explained.

Germanium oxide (GeO$_2$) is a promising anode material because of its high theoretical reversible capacity (1125 mAh g^{-1}), low operating voltage, and good thermal stability [130–132]. He et al. (2019) reported freestanding mesoporous foldable GeO$_x$/ZnO/C (FGCZ) composite nanofibers with uniform distributions of GeO$_x$ and ZnO. A solution of polyacrylonitrile (PAN), zinc acetate (Zn(Ac)$_2$), CNTs, and GeO$_2$ NPs was used to fabricate nanofibers by electrospinning. The precursor nanofibers were stabilized by 2 h of annealing in air at 250 °C and were subsequently carbonized by 6 h of annealing in air at 700 °C to yield the FGCZ nanofibers. The resulting nanofibers possessed uniform diameters of approximately 300 nm, longer than those of the GeO$_x$ sample. The above can be explained based on the presence of Zn(Ac)$_2$ causing plasticization of PAN via the formation of N–Zn coordinative bonds. XPS analysis confirmed the existence of Ge$_2$O$_3$, formed by the reduction of GeO$_2$ by carbon at 700 °C, as well as ZnO in the composite fibers. Raman spectra, via I$_D$/I$_G$, indicated FGCZ was more disordered than GC (the sample prepared in the absence of Zn(Ac)$_2$) owing to the presence of Zn(Ac)$_2$. FGCZ also had a larger surface area (532.56 m^{-2} g^{-1}) than GC (236.33 m^{-2} g^{-1}). Mesopores with widths of 4–7 nm were also found in FGCZ, whose highly porous structure enhanced the electron transmission, provided more Li storage sites, and increased the rate of Li ion transport, thus improving the electrochemical performance. In their study, besides serving as the ZnO precursor and promoting the formation of mesopores, the added Zn(Ac)$_2$ in the electrospun solution achieved the following: (i) enhancement in the mechanical properties and flexibility of the composites (Figure 5), and (ii) assisting in the dispersion of GeO$_2$ NPs. Thus, the as-prepared FGCZ composite presented good electrochemical characteristics with a first discharge/charge capacity of 1000/890 mAh g^{-1} at 0.2 A g^{-1} (CE: 66.9%). After the 200th cycle at 0.2 A g^{-1} and 500th cycle at 1 A g^{-1}, the FGCZ composite achieved discharge capacities of 617 and 464 mAh g^{-1}, respectively. Owing to the presence of amorphous active materials, their uniform dispersion, and the good conductivity of CNTs, the FGCZ composite showed rapid Li ion diffusion and therefore exhibited higher reversible capacities than the GC sample at the same high current density. When the FGCZ composite was assembled into full cells using a commercial flexible LiCoO$_2$/CNT as the cathode, it displayed a discharge capacity of 417 mAh g^{-1} at 0.1 A g^{-1}. A ten-fold bent full battery had a discharge capacity of 391 mAh g^{-1}, which confirmed the good mechanical stability of the FGCZ composite. He et al. (2019) found that their full battery with FGCZ powered a light-emitting diode under different bending conditions. Specifically, under different bending angles, there was no significant difference in the EIS plots of the FGCZ composite [133].

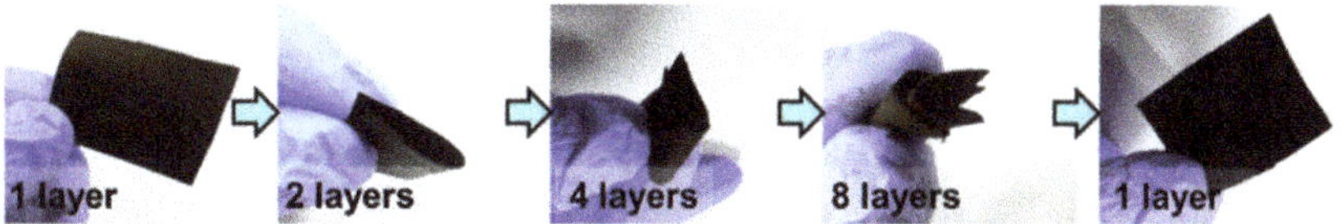

Figure 5. Folding process of FGCZ composite. Reprinted with permission from [133]. Copyright 2019, Elsevier.

Co_3O_4 is also considered as an anode candidate for LIBs owing to its high theoretical capacity [134]. However, it is limited by its high cost and toxicity [28]. To reduce its cost and improve its eco-friendliness, Co can be replaced by different metals, such as Fe, Mn, Ni, and Zn, to form a ternary structure [134]. Among the various combinations, $ZnCo_2O_4$ is a promising candidate because of its high specific capacity (975 mAh g^{-1}), inexpensive cost, low operating voltage, and environmental friendliness [135–138]. Ge et al. (2015) synthesized a $ZnO/ZnCo_2O_4/C$ PC/shell composite comprising $ZnCo_2O_4$ as the shell, ZnO as the core, and a homogeneously carbon-coated $ZnCo_2O_4$ shell surface (Figure 6). The $ZnO/ZnCo_2O_4/C$ composite had a diameter of 800 nm and a mesoporous structure with an SSA of 27.9 m^2 g^{-1}, an average pore size of 14 nm, and a pore volume of 0.146 cm^3 g^{-1}. The initial discharge/charge capacity of the $ZnO/ZnCo_2O_4/C$ composite was 1278.8/974 mAh g^{-1} at 0.5 A g^{-1}, and the initial CE was 76.2%. The as-prepared ternary composite maintained a reversible capacity of 669 mAh g^{-1} after 250 cycles at 0.5 A g^{-1} and 715 mAh g^{-1} after 50 cycles at a high rate of 1.6 A g^{-1}. In contrast, at 0.5 A g^{-1}, the reversible capacity of the $ZnO/ZnCo_2O_4$ composite rapidly reduced to 524.4 mAh g^{-1} only after 110 cycles. The structure of the $ZnO/ZnCo_2O_4/C$ composite was maintained well after 250 cycles, without any collapse and shedding. The EIS of $ZnO/ZnCo_2O_4/C$ had a smaller diameter than that of $ZnO/ZnCo_2O_4$, indicating that the former had a smaller charge transfer resistance and more rapid reaction throughout the charging process owing to the existence of the carbon layer. Overall, hierarchically porous core/shell structures provide abundant active sites, improve the electrode/electrolyte contact area, and offer abundant channels for the penetration of the electrolyte. Moreover, they relieve the structure pulverization caused by the Li^+ insertion/desertion. Concurrently, the carbon layer effectively enhances the composite conductivity, therefore improving the electron transfer rate, efficiently protects $ZnCo_2O_4$ from agglomeration and pulverization, and partially alleviates the strain resulting from the volume variation during the cycling process [28].

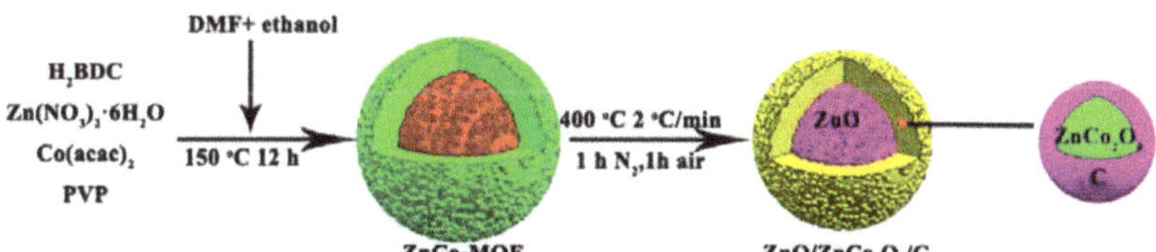

Figure 6. Scheme of preparation of $ZnO/ZnCo_2O_4/C$ composite. Reprinted with permission from [28]. Copyright 2015, American Chemical Society.

Similar to $ZnCo_2O_4$, Ma et al. (2017) introduced $ZnO/ZnFe_2O_4$/N-doped C-micro-polyhedra (ZZFO-C) as anodes for LIBs. The composites were prepared by thermal treatment of ZIF–ZnFe (molar ratio of 3:1) for 2 h at 500 °C (Figure 7). XRD results confirmed the co-existence of $ZnFe_2O_4$ and ZnO. XPS analysis identified various nitrogen-doped carbon species and oxygen functional groups. The ZZFO-C composites had an average size of ~420 nm, slightly smaller than that of the ZIF–ZnFe precursor, owing to the partial framework decomposition and contraction throughout the calcination. The ZZFO-C composites possessed rougher surfaces than their precursors and comprised clustered MOs/carbon NPs (size of ~20 nm) and many small holes. The formation of the hollow structure is a consequence of the decomposition of the inner ZIF–ZnFe to form gaseous products, such as CO_2, H_2O, and NO_2 [66,75,139–141]. The $ZnO/ZnFe_2O_4$ NPs and the carbon layers overlapped with each other in the composite. Ma et al. (2017) found that when the annealing temperature was increased to 650 °C, the carbon in the composite was removed (noted as ZZFO) (Figure 7). The surface area and pore size of ZZFO-C were 84.3 m^2 g^{-1} and 13 nm, respectively, and for ZZFO, they were 31.2 m^2 g^{-1} and 39 nm, respectively. The mesoporous structures of ZZFO-C and ZZFO were beneficial for the Li^+ transport and could alleviate the volume change during insertion/desertion cycles. In terms of the electrochemical performance, ZZFO-C had a first discharge capacity of 1751 mAh g^{-1} (CE: 67.4%). The higher CE of ZZFO-C than that of ZZFO (60.8%) can be explained by the presence of the N-doped carbon matrix, which improves the reversibility of the electrode. After 100 cycles at 0.2 A g^{-1} and 2.0 A g^{-1}, the obtained ZZFO-C composite presented reversible capacities of 1000 mAh g^{-1} and 620 mAh g^{-1}, respectively. It also displayed a good rate capability with specific capacities of 1075, 1052, 1024, 928, 842, and 787 mAh g^{-1} at 0.05, 0.1, 0.2, 0.5, 1, and 2 A g^{-1}, respectively. After returning to 0.1 A g^{-1} at the 75th cycle, the ZZFO-C composite recovered its initial specific capacity (1190 mAh g^{-1}), which continuously increased to reach 1328 mAh g^{-1} at the 90th cycle. Compared to ZZFO-C, the ZZFO sample showed poorer electrochemical activity. The good electrochemical performance of the ZZFO-C composite can be attributed to the synergistic effect between the N-doped carbon matrix and the two active components as well as its distinct hierarchical hollow structure [142].

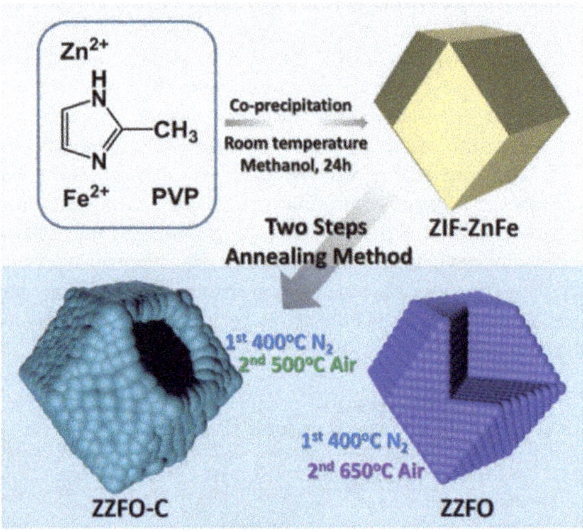

Figure 7. Scheme of synthesis process of ZZFO-C and ZZFO. Reprinted with permission from [142]. Copyright 2017, Elsevier.

Zhang et al. (2016) prepared a freestanding 3D ZnO/graphene/CNT ternary composite as an anode for LIBs by the sol–gel technique followed by vacuum-assisted filtration. Based on thermogravimetric analysis, the ZnO NPs (average size of 8 nm) accounted for approximately 84 wt% and were uniformly distributed on the composite. The ZnO/graphene/CNT composite had a first discharge capacity of 1503 mAh g^{-1} at 0.1 A g^{-1} (CE: 60%) and high cyclability and rate capability, with a reversible discharge capacity of 620 mAh g^{-1} after 100 cycles at 0.1 A g^{-1}. The CE was approximately 100% from the seventh cycle. Similar to the previously mentioned ZnO NPs and PC binary composites, the large surface area and high conductive network of the graphene/CNT structure maintain good electronic contact between the particles and suppress the aggregation and volume variation in the ZnO NPs during the cycling process. In addition, the

CNTs not only adopt the role of a graphene modifier, change the surface characteristics, and prevent the agglomeration of active materials but also serve as a link between the graphene layers [88]. These effects contribute to the high first discharge capacity and good cycle/rate performance of the ternary composite [88].

4. Conclusions and Perspectives

ZnO is a promising anode candidate for LIBs owing to its high theoretical capacity (978 mAh g^{-1}) [5]. However, because of its limitations, such as its slow chemical reaction kinetics, rapid capacity fading, and poor rate capability [20], composites of ZnO must be formed with other materials. The highlighted studies on ZnO-based binary and ternary composites with different synthesis methods are summarized in Table 2, Table 3. Most ZnO-based composites had higher first discharge capacities than only ZnO. Moreover, most of the composites discussed in this review showed a good cycling stability and rate performance. The large electrode/electrolyte contact area, abundant charge storage reaction sites, short Li$^+$ diffusion path, improved conductivity, stability structure, and potential to relieve the volume expansion could explain the good electrical performance of these anode composites. To synthesize high-performance anodes, different factors such as size, morphology, crystallinity, phase composition, and porosity should be considered [9]. Some limitations of ZnO-based composites have also been found and should be overcome rapidly. The synergistic effects between the components of the composites as well as the correlation between the composite structure and the electrical performance should be investigated in more detail. In general, the initial CEs of ZnO-based composites were low and should be increased, for which electrolyte optimization, surface modification, and coating of ZnO are possible solutions [9,48,49]. In the case of ZnO–MO binary composites, the choice of MO partner to prepare ZnO–MOs may increase the initial CE. Both conversion and alloying MOs can increase the capacity of composites. However, between reversible conversion and alloying–dealloying reactions, the reversible alloying–dealloying reaction may have more benefits in the increase in the initial CE. For example, the presence of alloying materials (SnO$_2$) in the composite can increase the initial CE, while the presence of conversion materials (NiO) has less impact on the improvement in the initial CE [6,59]. Moreover, the relationships between active materials, binders, and electrolyte additives should be further investigated in order to improve the SEI layer and thus result in a better CE and cycle performance [9]. Strategies to develop ZnO-based composites that have high energy densities should be identified. For example, in ZnO–carbon-based composites, the carbon component contributes slightly to the capacity but accounts for more than 50% of the total electrode weight. Therefore, the total energy density of these composites is significantly lowered, which reduces the potential of PC materials in real LIB applications. In addition, there are considerable efforts to synthesize ZnO ternary composites with different structures. However, their electrochemical performance is not significantly higher when compared to similar binary composite anodes for LIBs. For example, the ZnO/SnO$_2$/MWCNT and NiO–ZnO/RGO ternary composites mentioned in this paper did not have a better initial discharge capacity and cycling performance when compared to ZnO–SnO$_2$ and ZnO–NiO. In addition, most of the studies mentioned in this review focused on developing composites with a high specific capacity instead of producing composites with a high packing density and a high energy density. Thus, we recommend aiming at generating novel composites or optimizing ZnO-based composites that not only have a high specific capacity, cycling stability, and rate performance but also a high initial CE and energy density. In this review, some studies such as Li et al. (ZnO–NiO microspheres, 2018) and Zhao et al. (C/ZnO NMs, 2018) focused on improvements not only in the electrochemical performance of their studied composites but also in their energy density [59,77]. Moreover, a more synthetic process than the current ones should be developed to reduce the production cost to meet industrial requirements.

Table 2. Summary of highlighted ZnO binary composite studies.

Anode	Synthesis Method	Electrochemical Performance				Reference
		Initial Discharge Capacity (mAh g^{-1})	Initial CE (%)	Cycling Performance (mAh g^{-1}/Cycles)	Current Density	
Intermixed ZnO–SnO$_2$	ALD	2667	80.2	1752/50	0.5 A g^{-1}	[6]
Nanolaminated ZnO–SnO$_2$	ALD	2471	71.4	~1250/50	0.5 A g^{-1}	[6]
ZnO–NiO microspheres	Controlled calcination treatment	1221.7	62.9	1008.6/200	0.1 A g^{-1}	[59]
WO$_3$/ZnO film	Hydrothermal and thermal treatment	~1500	79.9	~1100/300 at 1 C	0.28 C	[62]
ZnO/Cu$_2$MgO$_3$	One-step cost-effective ultrasonic spray pyrolysis	990.75	66.98	528/400	0.3 A g^{-1}	[64]
ZnO/ZnFe$_2$O$_4$ SMCs	Thermal treatment	1892	~70	837/200 cycles	1 A g^{-1}	[71]
ZnO@GNs	High-performance homogenizing	850	82	Capacity decay of ~8% after 50 cycles at 1 C	0.1 C	[20]
ZnO/PC	Solvothermal	2017.4	52.68	653.7/100	0.1 A g^{-1}	[5]
CF@pore-ZnO	Thermal treatment	955	55.81	510/300	0.1 A g^{-1}	[83]
C/ZnO NMs	Pyrolysis and immersion coating	~1100	59	260/700 at 2 A g^{-1}	0.08 A g^{-1}	[77]
ZnO/EG	ALD	~1000	~60	438/500	0.2 A g^{-1}	[106]
ZnO@GAs	Solvothermal	1001	71.2	445/500	1.6 A g^{-1}	[87]
Graphene-coated ZnO nanosheet	Hydrothermal and thermal evaporation process	1470	65.8	600/100	1 A g^{-1}	[110]
ZnO/NC-Z NRDs	Thermal treatment	1439	76	1011/200	0.2 A g^{-1}	[29]
ZnO/C	Thermal treatment	1061	63.24	212/100	0.1 A g^{-1}	[115]
Carbon-coated Mn-doped ZnO	Thermal treatment	~1200	~61.6%	~740/80	0.1 C	[116]

Table 3. Summary of highlighted ZnO ternary composite studies.

Anode	Synthesis Method	Electrochemical Performance				Reference
		Initial Discharge Capacity (mAh g^{-1})	Initial CE (%)	Cycling Performance (mAh g^{-1}/Cycles)	Current Density	
ZnO/SnO$_2$/MWCNT	Sol–gel coating	1584	~57	487/100	0.2 C	[117]
SnO$_2$/ZnO@PPy	Electrospinning and thermal treatment	1861	61.12	626.1/100	0.2 C	[124]
SnO$_x$–ZnO/CNF fiber	Electrospinning and thermal treatment	1910	73.3	963/55	0.1 A g^{-1}	[127]
NiO–ZnO/RGO	Ultrasonic, freeze drying, and thermal treatment	1393	66.3	1017/200	0.1 A g^{-1}	[129]
GeO$_x$/ZnO/C	Electrospinning and thermal treatment	1000	66.9	617/200	0.2 A g^{-1}	[133]
ZnO/ZnCo$_2$O$_4$/C	Thermal treatment	1278	76.2	669/250	0.5 A g^{-1}	[28]
3D ZnO/graphene/CNTs	Sol–gel technique following by vacuum-assisted filtration	1503	60	620/100 cycles	0.1 A g^{-1}	[88]

Author Contributions: V.K.H.B.: conceptualization, methodology, investigation, resources, visualization, writing—original draft, writing—review and editing; T.N.P.: resources, writing—review and editing; J.H.: conceptualization, resources, writing—review and editing, supervision, funding acquisition; Y.-C.L.: conceptualization, resources, writing—review and editing, supervision, funding acquisition. All authors have read and agreed to the published version of the manuscript.

Funding: This research was supported by the Basic Science Research Capacity Enhancement Project through a Korea Basic Science Institute (National Research Facilities and Equipment Center) grant funded by the Ministry of Education (2019R1A6C1010016). This research was also supported by the Korea Institute of Energy Technology Evaluation and Planning (KETEP) and the Ministry of Trade, Industry and Energy (MOTIE) of the Republic of Korea (No. 20194030202290).

Institutional Review Board Statement: Not applicable.

Informed Consent Statement: Not applicable.

Data Availability Statement: Statistics about published articles/patents related to MOs and their composites for LIB applications are archived from Google Scholar (https://scholar.google.com) (accessed on 15 June 2021) database with keyword: "Metal oxide lithium-ion batteries".

Conflicts of Interest: The authors declare no conflict of interest.

References

1. Kang, S.; Li, Y.; Wu, M.; Cai, M.; Shen, P.K. Synthesis of hierarchically flower-like FeWO$_4$ as high performance anode materials for Li-ion batteries by a simple hydrothermal process. *Int. J. Hydrog. Energy* **2014**, *39*, 16081–16087. [CrossRef]
2. Nguyen, Q.H.; Phung, V.D.; Kidanu, W.G.; Ahn, Y.N.; Nguyen, T.L.; Kim, I.T. Carbon-free Cu/Sb$_x$O$_y$/Sb nanocomposites with yolk-shell and hollow structures as high-performance anodes for lithium-ion storage. *J. Alloys Compd.* **2021**, *878*, 160447. [CrossRef]
3. Zhu, J.; Zhang, G.; Gu, S.; Lu, B. SnO$_2$ nanorods on ZnO nanofibers: A new class of hierarchical nanostructures enabled by electrospinning as anode material for high-performance lithium-ion batteries. *Electrochim. Acta* **2014**, *150*, 308–313. [CrossRef]
4. Li, H.; Wei, Y.; Zhang, Y.; Yin, F.; Zhang, C.; Wang, G.; Bakenov, Z. Synthesis and electrochemical investigation of highly dispersed ZnO nanoparticles as anode material for lithium-ion batteries. *Ionics* **2016**, *22*, 1387–1393. [CrossRef]

5. Shen, X.; Mu, D.; Chen, S.; Wu, B.; Wu, F. Enhanced electrochemical performance of ZnO-loaded/porous carbon composite as anode materials for lithium ion batteries. *ACS Appl. Mater. Interfaces* **2013**, *5*, 3118–3125. [CrossRef]
6. Zhao, B.; Mattelaer, F.; Kint, J.; Werbrouck, A.; Henderick, L.; Minjauw, M.; Dendooven, J.; Detavernier, C. Atomic layer deposition of ZnO–SnO$_2$ composite thin film: The influence of structure, composition and crystallinity on lithium-ion battery performance. *Electrochim. Acta* **2019**, *320*, 134604. [CrossRef]
7. Zhang, J.; Gu, P.; Xu, J.; Xue, H.; Pang, H. High performance of electrochemical lithium storage batteries: ZnO-based nanomaterials for lithium-ion and lithium-sulfur batteries. *Nanoscale* **2016**, *8*, 18578–18595. [CrossRef] [PubMed]
8. Aravindan, V.; Jinesh, K.B.; Prabhakar, R.R.; Kale, V.S.; Madhavi, S. Atomic layer deposited (ALD) SnO$_2$ anodes with exceptional cycleability for Li-ion batteries. *Nano Energy* **2013**, *2*, 720–725. [CrossRef]
9. Chen, Y.; Chen, X.; Zhang, Y. A Comprehensive Review on Metal-Oxide Nanocomposites for High-Performance Lithium-Ion Battery Anodes. *Energy Fuels* **2021**, *35*, 6420–6442. [CrossRef]
10. Yuan, G.; Wang, G.; Wang, H.; Bai, J. Synthesis and electrochemical investigation of radial ZnO microparticles as anode materials for lithium-ion batteries. *Ionics* **2015**, *21*, 365–371. [CrossRef]
11. Fang, S.; Bresser, D.; Passerini, S. Transition Metal Oxide Anodes for Electrochemical Energy Storage in Lithium- and Sodium-Ion Batteries. *Adv. Energy Mater.* **2020**, *10*, 1902485. [CrossRef]
12. Liu, X.; Si, W.; Zhang, J.; Sun, X.; Deng, J.; Baunack, S.; Oswald, S.; Liu, L.; Yan, C.; Schmidt, O.G. Free-standing Fe$_2$O$_3$ nanomembranes enabling ultra-long cycling life and high rate capability for Li-ion batteries. *Sci. Rep.* **2014**, *4*, 1–8. [CrossRef] [PubMed]
13. Song, Y.; Hwang, J.; Lee, S.; Thirumalraj, B.; Kim, J.H.; Jenei, P.; Gubicza, J.; Choe, H. Synthesis of a High-Capacity NiO/Ni Foam Anode for Advanced Lithium-Ion Batteries. *Adv. Energy Mater.* **2020**, *22*, 2000351. [CrossRef]
14. Hassan, A.S.; Moyer, K.; Ramachandran, B.R.; Wick, C.D. Comparison of Storage Mechanisms in RuO$_2$, SnO$_2$, and SnS$_2$ for Lithium-Ion Battery Anode Materials. *J. Phys. Chem. C* **2016**, *120*, 2036–2046. [CrossRef]
15. Nguyen, T.P.; Kim, I.T. Self-assembled few-layered MoS$_2$ on SnO$_2$ anode for enhancing lithium-ion storage. *Nanomaterials* **2020**, *10*, 1–13. [CrossRef] [PubMed]
16. He, H.; Gan, Q.; Wang, H.; Xu, G.L.; Zhang, X.; Huang, D.; Fu, F.; Tang, Y.; Amine, K.; Shao, M. Structure-dependent performance of TiO$_2$/C as anode material for Na-ion batteries. *Nano Energy* **2018**, *44*, 217–227. [CrossRef]
17. Ai, C.; Yin, M.; Wang, C.; Sun, J. Synthesis and characterization of spinel type ZnCo$_2$O$_4$ as a novel anode material for lithium ion batteries. *J. Mater. Sci.* **2004**, *39*, 1077–1079. [CrossRef]
18. Guo, H.; Zhang, Y.; Marschilok, A.C.; Takeuchi, K.J.; Takeuchi, E.S.; Liu, P. A first principles study of spinel ZnFe$_2$O$_4$ for electrode materials in lithium-ion batteries. *Phys. Chem. Chem. Phys.* **2017**, *19*, 26322–26329. [CrossRef]
19. Yao, W.; Xu, Z.; Xu, X.; Xie, Y.; Qiu, W.; Xu, J.; Zhang, D. Two-dimensional holey ZnFe$_2$O$_4$ nanosheet/reduced graphene oxide hybrids by self-link of nanoparticles for high-rate lithium storage. *Electrochim. Acta* **2018**, *292*, 390–398. [CrossRef]
20. Hsieh, C.T.; Lin, C.Y.; Chen, Y.F.; Lin, J.S. Synthesis of ZnO@Graphene composites as anode materials for lithium ion batteries. *Electrochim. Acta* **2013**, *111*, 359–365. [CrossRef]
21. Lu, S.; Wang, H.; Zhou, J.; Wu, X.; Qin, W. Atomic layer deposition of ZnO on carbon black as nanostructured anode materials for high-performance lithium-ion batteries. *Nanoscale* **2017**, *9*, 1184–1192. [CrossRef] [PubMed]
22. Liu, N.; Lu, Z.; Zhao, J.; McDowell, M.T.; Lee, H.W.; Zhao, W.; Cui, Y. A pomegranate-inspired nanoscale design for large-volume-change lithium battery anodes. *Nat. Nanotechnol.* **2014**, *9*, 187–192. [CrossRef]
23. Gao, H.; Zhou, W.; Jang, J.H.; Goodenough, J.B. Cross-Linked Chitosan as a Polymer Network Binder for an Antimony Anode in Sodium-Ion Batteries. *Adv. Energy Mater.* **2016**, *6*, 1502130. [CrossRef]
24. Li, F.; Yang, L.; Xu, G.; Xiaoqiang, H.; Yang, X.; Wei, X.; Ren, Z.; Shen, G.; Han, G. Hydrothermal self-assembly of hierarchical flower-like ZnO nanospheres with nanosheets and their application in Li-ion batteries. *J. Alloys Compd.* **2013**, *577*, 663–668. [CrossRef]
25. Park, K.T.; Xia, F.; Kim, S.W.; Kim, S.B.; Song, T.; Paik, U.; Park, W.I. Facile synthesis of ultrathin ZnO nanotubes with well-organized hexagonal nanowalls and sealed layouts: Applications for lithium ion battery anodes. *J. Phys. Chem. C* **2013**, *117*, 1037–1043. [CrossRef]
26. Huang, X.H.; Guo, R.Q.; Wu, J.B.; Zhang, P. Mesoporous ZnO nanosheets for lithium ion batteries. *Mater. Lett.* **2014**, *122*, 82–85. [CrossRef]
27. Zhang, G.; Hou, S.; Zhang, H.; Zeng, W.; Yan, F.; Li, C.C.; Duan, H. High-performance and ultra-stable lithium-ion batteries based on MOF-derived ZnO@ZnO quantum dots/C core-shell nanorod arrays on a carbon cloth anode. *Adv. Mater.* **2015**, *27*, 2400–2405. [CrossRef]
28. Ge, X.; Li, Z.; Wang, C.; Yin, L. Metal-organic frameworks derived porous core/shell structured ZnO/ZnCo$_2$O$_4$/C hybrids as anodes for high-performance lithium-ion battery. *ACS Appl. Mater. Interfaces* **2015**, *7*, 26633–26642. [CrossRef]
29. Gan, Q.; Zhao, K.; Liu, S.; He, Z. Solvent-free synthesis of N-doped carbon coated ZnO nanorods composite anode via a ZnO support-induced ZIF-8 in-situ growth strategy. *Electrochim. Acta* **2017**, *250*, 292–301. [CrossRef]
30. Bresser, D.; Mueller, F.; Fiedler, M.; Krueger, S.; Kloepsch, R.; Baither, D.; Winter, M.; Paillard, E.; Passerini, S. Transition-metal-doped zinc oxide nanoparticles as a new lithium-ion anode material. *Chem. Mater.* **2013**, *25*, 4977–4985. [CrossRef]
31. Wang, X.; Zhou, X.; Yao, K.; Zhang, J.; Liu, Z. A SnO$_2$/graphene composite as a high stability electrode for lithium ion batteries. *Carbon* **2011**, *49*, 133–139. [CrossRef]

32. Guo, W.; Sun, W.; Wang, Y. Multilayer CuO@NiO Hollow Spheres: Microwave-Assisted Metal-Organic-Framework Derivation and Highly Reversible Structure-Matched Stepwise Lithium Storage. *ACS Nano* **2015**, *9*, 11462–11471. [CrossRef] [PubMed]
33. Belliard, F.; Irvine, J.T.S. Electrochemical performance of ball-milled ZnO-SnO$_2$ systems as anodes in lithium-ion battery. *J. Power Sources* **2001**, *97–98*, 219–222. [CrossRef]
34. Franken, R.H.; Van Der Werf, C.H.M.; Löffler, J.; Rath, J.K.; Schropp, R.E.I. Beneficial effects of sputtered ZnO:Al protection layer on SnO$_2$:F for high-deposition rate hot-wire CVD p-i-n solar cells. *Thin Solid Films* **2006**, *501*, 47–50. [CrossRef]
35. Wang, J.; Du, N.; Zhang, H.; Yu, J.; Yang, D. Layer-by-layer assembly synthesis of ZnO/SnO$_2$ composite nanowire arrays as high-performance anode for lithium-ion batteries. *Mater. Res. Bull.* **2011**, *46*, 2378–2384. [CrossRef]
36. Ahmad, M.; Yingying, S.; Sun, H.; Shen, W.; Zhu, J. SnO$_2$/ZnO composite structure for the lithium-ion battery electrode. *J. Solid State Chem.* **2012**, *196*, 326–331. [CrossRef]
37. Huang, Y.; Liu, X.; Lu, L.; Fang, J.; Ni, H.; Ji, Z. Preparation and characterization of ZnO/SnO$_2$ composite thin films as high-capacity anode for lithium-ion batteries. *Appl. Phys. A* **2015**, *120*, 519–524. [CrossRef]
38. Xie, Q.; Ma, Y.; Zeng, D.; Wang, L.; Yue, G.; Peng, D.L. Facile fabrication of various zinc-nickel citrate microspheres and their transformation to ZnO-NiO hybrid microspheres with excellent lithium storage properties. *Sci. Rep.* **2014**, *5*, 1–9. [CrossRef]
39. Pan, Q.; Zheng, F.; Ou, X.; Yang, C.; Xiong, X.; Liu, M. MoS$_2$ encapsulated SnO$_2$-SnS/C nanosheets as a high performance anode material for lithium ion batteries. *Chem. Eng. J.* **2017**, *316*, 393–400. [CrossRef]
40. Zhang, H.; Huang, X.; Noonan, O.; Zhou, L.; Yu, C. Tailored Yolk–Shell Sn@C Nanoboxes for High-Performance Lithium Storage. *Adv. Funct. Mater.* **2017**, *27*, 1606023. [CrossRef]
41. Wu, Z.S.; Ren, W.; Wen, L.; Gao, L.; Zhao, J.; Chen, Z.; Zhou, G.; Li, F.; Cheng, H.M. Graphene anchored with Co$_3$O$_4$ nanoparticles as anode of lithium ion batteries with enhanced reversible capacity and cyclic performance. *ACS Nano* **2010**, *4*, 3187–3194. [CrossRef]
42. Hu, Y.; Yan, C.; Chen, D.; Lv, C.; Jiao, Y.; Chen, G. One-dimensional Co$_3$O$_4$ nanonet with enhanced rate performance for lithium ion batteries: Carbonyl-B-cyclodextrin inducing and kinetic analysis. *Chem. Eng. J.* **2017**, *321*, 31–39. [CrossRef]
43. Yang, J.; Ouyang, Y.; Zhang, H.; Xu, H.; Zhang, Y.; Wang, Y. Novel Fe$_2$P/graphitized carbon yolk/shell octahedra for high-efficiency hydrogen production and lithium storage. *J. Mater. Chem. A* **2016**, *4*, 9923–9930. [CrossRef]
44. Yang, J.; Wu, Q.; Yang, X.; He, S.; Khan, J.; Meng, Y.; Zhu, X.; Tong, S.; Wu, M. Chestnut-Like TiO$_2$@α-Fe$_2$O$_3$ Core-Shell Nanostructures with Abundant Interfaces for Efficient and Ultralong Life Lithium-Ion Storage. *ACS Appl. Mater. Interfaces* **2017**, *9*, 354–361. [CrossRef]
45. Li, J.; Yan, D.; Lu, T.; Yao, Y.; Pan, L. An advanced CoSe embedded within porous carbon polyhedra hybrid for high performance lithium-ion and sodium-ion batteries. *Chem. Eng. J.* **2017**, *325*, 14–24. [CrossRef]
46. Lin, D.; Lu, Z.; Hsu, P.C.; Lee, H.R.; Liu, N.; Zhao, J.; Wang, H.; Liu, C.; Cui, Y. A high tap density secondary silicon particle anode fabricated by scalable mechanical pressing for lithium-ion batteries. *Energy Environ. Sci.* **2015**, *8*, 2371–2376. [CrossRef]
47. Wu, C.; Maier, J.; Yu, Y. Sn-Based Nanoparticles Encapsulated in a Porous 3D Graphene Network: Advanced Anodes for High-Rate and Long Life Li-Ion Batteries. *Adv. Funct. Mater.* **2015**, *25*, 3488–3496. [CrossRef]
48. Seng, K.H.; Li, L.; Chen, D.P.; Chen, Z.X.; Wang, X.L.; Liu, H.K.; Guo, Z.P. The effects of FEC (fluoroethylene carbonate) electrolyte additive on the lithium storage properties of NiO (nickel oxide) nanocuboids. *Energy* **2013**, *58*, 707–713. [CrossRef]
49. Wu, L.; Buchholz, D.; Bresser, D.; Gomes Chagas, L.; Passerini, S. Anatase TiO$_2$ nanoparticles for high power sodium-ion anodes. *J. Power Sources* **2014**, *251*, 379–385. [CrossRef]
50. Zou, F.; Chen, Y.M.; Liu, K.; Yu, Z.; Liang, W.; Bhaway, S.M.; Gao, M.; Zhu, Y. Metal organic frameworks derived hierarchical hollow NiO/Ni/graphene composites for lithium and sodium storage. *ACS Nano* **2016**, *10*, 377–386. [CrossRef]
51. Hu, H.; Zhang, J.; Guan, B.; Lou, X.W.D. Unusual Formation of CoSe@carbon Nanoboxes, which have an Inhomogeneous Shell, for Efficient Lithium Storage. *Angew. Chem. Int. Ed.* **2016**, *55*, 9514–9518. [CrossRef]
52. Sun, Y.; Hu, X.; Luo, W.; Huang, Y. Self-assembled hierarchical MoO$_2$/graphene nanoarchitectures and their application as a high-performance anode material for lithium-ion batteries. *ACS Nano* **2011**, *5*, 7100–7107. [CrossRef] [PubMed]
53. Guo, J.; Liu, Q.; Wang, C.; Zachariah, M.R. Interdispersed amorphous MnO$_x$-carbon nanocomposites with superior electrochemical performance as lithium-storage material. *Adv. Funct. Mater.* **2012**, *22*, 803–811. [CrossRef]
54. Jamnik, J.; Maier, J. Nanocrystallinity effects in lithium battery materials: Aspects of nano-ionics. Part IV. *Phys. Chem. Chem. Phys.* **2003**, *5*, 5215–5220. [CrossRef]
55. Xie, Q.; Li, F.; Guo, H.; Wang, L.; Chen, Y.; Yue, G.; Peng, D.L. Template-free synthesis of amorphous double-shelled zinc-cobalt citrate hollow microspheres and their transformation to crystalline ZnCo$_2$O$_4$ microspheres. *ACS Appl. Mater. Interfaces* **2013**, *5*, 5508–5517. [CrossRef]
56. Xie, Q.; Zeng, D.; Ma, Y.; Lin, L.; Wang, L.; Peng, D.L. Synthesis of ZnO-ZnCo$_2$O$_4$ hybrid hollow microspheres with excellent lithium storage properties. *Electrochim. Acta* **2015**, *169*, 283–290. [CrossRef]
57. Chao, D.; Zhu, C.; Yang, P.; Xia, X.; Liu, J.; Wang, J.; Fan, X.; Savilov, S.V.; Lin, J.; Fan, H.J.; et al. Array of nanosheets render ultrafast and high-capacity Na-ion storage by tunable pseudocapacitance. *Nat. Commun.* **2016**, *7*, 1–8. [CrossRef] [PubMed]
58. Chen, Z.; Wu, R.; Wang, H.; Jiang, Y.; Jin, L.; Guo, Y.; Song, Y.; Fang, F.; Sun, D. Construction of hybrid hollow architectures by in-situ rooting ultrafine ZnS nanorods within porous carbon polyhedra for enhanced lithium storage properties. *Chem. Eng. J.* **2017**, *326*, 680–690. [CrossRef]

59. Li, J.; Yan, D.; Hou, S.; Lu, T.; Yao, Y.; Chua, D.H.C.; Pan, L. Metal-organic frameworks derived yolk-shell ZnO/NiO microspheres as high-performance anode materials for lithium-ion batteries. *Chem. Eng. J.* **2018**, *335*, 579–589. [CrossRef]
60. Wang, C.; Zhao, Y.; Zhou, L.; Liu, Y.; Zhang, W.; Zhao, Z.; Hozzein, W.N.; Alharbi, H.M.S.; Li, W.; Zhao, D. Mesoporous carbon matrix confinement synthesis of ultrasmall WO_3 nanocrystals for lithium ion batteries. *J. Mater. Chem. A* **2018**, *6*, 21550–21557. [CrossRef]
61. Wu, X.; Yao, S. Flexible electrode materials based on WO_3 nanotube bundles for high performance energy storage devices. *Nano Energy* **2017**, *42*, 143–150. [CrossRef]
62. Tu, C.; Zhang, Z.; Shao, A.; Qi, X.; Zhu, C.; Li, C.; Yang, Z. Constructing a directional ion acceleration layer at WO_3/ZnO heterointerface to enhance Li-ion transfer and storage. *Compos. Part B Eng.* **2021**, *205*, 108511. [CrossRef]
63. Su, D.; Kim, H.S.; Kim, W.S.; Wang, G. Synthesis of tuneable porous hematites (α-Fe_2O_3) for gas sensing and lithium storage in lithium ion batteries. *Microporous Mesoporous Mater.* **2012**, *149*, 36–45. [CrossRef]
64. Karunakaran, G.; Kundu, M.; Kumari, S.; Kolesnikov, E.; Gorshenkov, M.V.; Maduraiveeran, G.; Sasidharan, M.; Kuznetsov, D. ZnO/Cu_2MgO_3 hollow porous nanocage: A new class of hybrid anode material for advanced lithium-ion batteries. *J. Alloys Compd.* **2018**, *763*, 94–101. [CrossRef]
65. Li, Z.; Li, B.; Yin, L.; Qi, Y. Prussion blue-supported annealing chemical reaction route synthesized double-shelled Fe_2O_3/Co_3O_4 hollow microcubes as anode materials for Lithium-Ion battery. *ACS Appl. Mater. Interfaces* **2014**, *6*, 8098–8107. [CrossRef]
66. Zhang, L.; Wu, H.B.; Madhavi, S.; Hng, H.H.; Lou, X.W. Formation of Fe_2O_3 microboxes with hierarchical shell structures from metal-organic frameworks and their lithium storage properties. *J. Am. Chem. Soc.* **2012**, *134*, 17388–17391. [CrossRef]
67. Yan, N.; Hu, L.; Li, Y.; Wang, Y.; Zhong, H.; Hu, X.; Kong, X.; Chen, Q. Co_3O_4 nanocages for high-performance anode material in lithium-ion batteries. *J. Phys. Chem. C* **2012**, *116*, 7227–7235. [CrossRef]
68. Nie, P.; Shen, L.; Luo, H.; Ding, B.; Xu, G.; Wang, J.; Zhang, X. Prussian blue analogues: A new class of anode materials for lithium ion batteries. *J. Mater. Chem. A* **2014**, *2*, 5852–5857. [CrossRef]
69. Huang, G.; Zhang, F.; Zhang, L.; Du, X.; Wang, J.; Wang, L. Hierarchical $NiFe_2O_4$/Fe_2O_3 nanotubes derived from metal organic frameworks for superior lithium ion battery anodes. *J. Mater. Chem. A* **2014**, *2*, 8048–8053. [CrossRef]
70. Zou, L.; Li, F.; Xiang, X.; Evans, D.G.; Duan, X. Self-generated template pathway to high-surface-area zinc aluminate spinel with mesopore network from a single-source inorganic precursor. *Chem. Mater.* **2006**, *18*, 5852–5859. [CrossRef]
71. Hou, L.; Lian, L.; Zhang, L.; Pang, G.; Yuan, C.; Zhang, X. Self-sacrifice template fabrication of hierarchical mesoporous bi-component-active ZnO/$ZnFe_2O_4$ sub-microcubes as superior anode towards high-performance lithium-ion battery. *Adv. Funct. Mater.* **2015**, *25*, 238–246. [CrossRef]
72. Yang, S.J.; Nam, S.; Kim, T.; Im, J.H.; Jung, H.; Kang, J.H.; Wi, S.; Park, B.; Park, C.R. Preparation and exceptional lithium anodic performance of porous carbon-coated ZnO quantum dots derived from a metal-organic framework. *J. Am. Chem. Soc.* **2013**, *135*, 7394–7397. [CrossRef]
73. Zou, F.; Hu, X.; Li, Z.; Qie, L.; Hu, C.; Zeng, R.; Jiang, Y.; Huang, Y. MOF-derived porous ZnO/$ZnFe_2O_4$/C octahedra with hollow interiors for high-rate lithium-ion batteries. *Adv. Mater.* **2014**, *26*, 6622–6628. [CrossRef] [PubMed]
74. Bresser, D.; Paillard, E.; Kloepsch, R.; Krueger, S.; Fiedler, M.; Schmitz, R.; Baither, D.; Winter, M.; Passerini, S. Carbon coated $ZnFe_2O_4$ nanoparticles for advanced lithium-ion anodes. *Adv. Energy Mater.* **2013**, *3*, 513–523. [CrossRef]
75. Zheng, F.; He, M.; Yang, Y.; Chen, Q. Nano electrochemical reactors of Fe_2O_3 nanoparticles embedded in shells of nitrogen-doped hollow carbon spheres as high-performance anodes for lithium-ion batteries. *Nanoscale* **2015**, *7*, 3410–3417. [CrossRef]
76. Wu, Z.S.; Ren, W.; Xu, L.; Li, F.; Cheng, H.M. Doped graphene sheets as anode materials with superhigh rate and large capacity for lithium ion batteries. *ACS Nano* **2011**, *5*, 5463–5471. [CrossRef] [PubMed]
77. Zhao, Y.; Huang, G.; Li, Y.; Edy, R.; Gao, P.; Tang, H.; Bao, Z.; Mei, Y. Three-dimensional carbon/ZnO nanomembrane foam as an anode for lithium-ion battery with long-life and high areal capacity. *J. Mater. Chem. A* **2018**, *6*, 7227–7235. [CrossRef]
78. Hsieh, C.T.; Lin, J.Y.; Mo, C.Y. Improved storage capacity and rate capability of Fe_3O_4-graphene anodes for lithium-ion batteries. *Electrochim. Acta* **2011**, *58*, 119–124. [CrossRef]
79. Zhang, M.; Lei, D.; Yin, X.; Chen, L.; Li, Q.; Wang, Y.; Wang, T. Magnetite/graphene composites: Microwave irradiation synthesis and enhanced cycling and rate performances for lithium ion batteries. *J. Mater. Chem.* **2010**, *20*, 5538–5543. [CrossRef]
80. Yu, Z.; Zhu, S.; Li, Y.; Liu, Q.; Feng, C.; Zhang, D. Synthesis of SnO_2 nanoparticles inside mesoporous carbon via a sonochemical method for highly reversible lithium batteries. *Mater. Lett.* **2011**, *65*, 3072–3075. [CrossRef]
81. Liu, J.; Zhou, Y.; Liu, F.; Liu, C.; Wang, J.; Pan, Y.; Xue, D. One-pot synthesis of mesoporous interconnected carbon-encapsulated Fe_3O_4 nanospheres as superior anodes for Li-ion batteries. *RSC Adv.* **2012**, *2*, 2262–2265. [CrossRef]
82. Liu, B.; Wang, X.; Liu, B.; Wang, Q.; Tan, D.; Song, W.; Hou, X.; Chen, D.; Shen, G. Advanced rechargeable lithium-ion batteries based on bendable $ZnCo_2O_4$-urchins-on-carbon-fibers electrodes. *Nano Res.* **2013**, *6*, 525–534. [CrossRef]
83. Han, Q.; Li, X.; Wang, F.; Han, Z.; Geng, D.; Zhang, W.; Li, Y.; Deng, Y.; Zhang, J.; Niu, S.; et al. Carbon fiber@ pore-ZnO composite as anode materials for structural lithium-ion batteries. *J. Electroanal. Chem.* **2019**, *833*, 39–46. [CrossRef]
84. Nardecchia, S.; Carriazo, D.; Ferrer, M.L.; Gutiérrez, M.C.; Monte, F.D. Three dimensional macroporous architectures and aerogels built of carbon nanotubes and/or graphene: Synthesis and applications. *Chem. Soc. Rev.* **2013**, *42*, 794–830. [CrossRef]
85. Son, D.I.; Kwon, B.W.; Park, D.H.; Seo, W.S.; Yi, Y.; Angadi, B.; Lee, C.L.; Choi, W.K. Emissive ZnO-graphene quantum dots for white-light-emitting diodes. *Nat. Nanotechnol.* **2012**, *7*, 465–471. [CrossRef]

86. Dou, Y.; Xu, J.; Ruan, B.; Liu, Q.; Pan, Y.; Sun, Z.; Dou, S.X. Atomic Layer-by-Layer Co_3O_4/Graphene Composite for High Performance Lithium-Ion Batteries. *Adv. Energy Mater.* **2016**, *6*, 1501835. [CrossRef]
87. Fan, L.; Zhang, Y.; Zhang, Q.; Wu, X.; Cheng, J.; Zhang, N.; Feng, Y.; Sun, K. Graphene Aerogels with Anchored Sub-Micrometer Mulberry-Like ZnO Particles for High-Rate and Long-Cycle Anode Materials in Lithium Ion Batteries. *Small* **2016**, *12*, 5208–5216. [CrossRef] [PubMed]
88. Zhang, Y.; Wei, Y.; Li, H.; Zhao, Y.; Yin, F.; Wang, X. Simple fabrication of free-standing ZnO/graphene/carbon nanotube composite anode for lithium-ion batteries. *Mater. Lett.* **2016**, *184*, 235–238. [CrossRef]
89. Sun, H.; Mei, L.; Liang, J.; Zhao, Z.; Lee, C.; Fei, H.; Ding, M.; Lau, J.; Li, M.; Wang, C.; et al. Three-dimensional holey-graphene/niobia composite architectures for ultrahigh-rate energy storage. *Science* **2017**, *356*, 599–604. [CrossRef] [PubMed]
90. Lv, D.; Zheng, J.; Li, Q.; Xie, X.; Ferrara, S.; Nie, Z.; Mehdi, L.B.; Browning, N.D.; Zhang, J.G.; Graff, G.L.; et al. High Energy Density Lithium-Sulfur Batteries: Challenges of Thick Sulfur Cathodes. *Adv. Energy Mater.* **2015**, *5*, 1402290. [CrossRef]
91. Gallagher, K.G.; Trask, S.E.; Bauer, C.; Woehrle, T.; Lux, S.F.; Tschech, M.; Lamp, P.; Polzin, B.J.; Ha, S.; Long, B.; et al. Optimizing Areal Capacities through Understanding the Limitations of Lithium-Ion Electrodes. *J. Electrochem. Soc.* **2016**, *163*, A138–A149. [CrossRef]
92. Peng, H.J.; Huang, J.Q.; Cheng, X.B.; Zhang, Q. Review on High-Loading and High-Energy Lithium–Sulfur Batteries. *Adv. Energy Mater.* **2017**, *7*, 1700260. [CrossRef]
93. Liu, J.; Liu, X.W. Two-dimensional nanoarchitectures for lithium storage. *Adv. Mater.* **2012**, *24*, 4097–4111. [CrossRef] [PubMed]
94. Si, W.; Mönch, I.; Yan, C.; Deng, J.; Li, S.; Lin, G.; Han, L.; Mei, Y.; Schmidt, O.G. A single rolled-up Si tube battery for the study of electrochemical kinetics, electrical conductivity, and structural integrity. *Adv. Mater.* **2014**, *26*, 7973–7978. [CrossRef]
95. Zhou, G.; Li, L.; Ma, C.; Wang, S.; Shi, Y.; Koratkar, N.; Ren, W.; Li, F.; Cheng, H.M. A graphene foam electrode with high sulfur loading for flexible and high energy Li-S batteries. *Nano Energy* **2015**, *11*, 356–365. [CrossRef]
96. Fang, R.; Zhao, S.; Hou, P.; Cheng, M.; Wang, S.; Cheng, H.M.; Liu, C.; Li, F. 3D Interconnected Electrode Materials with Ultrahigh Areal Sulfur Loading for Li-S Batteries. *Adv. Mater.* **2016**, *28*, 3374–3382. [CrossRef] [PubMed]
97. Afanasov, I.M.; Lebedev, O.I.; Kolozhvary, B.A.; Smirnov, A.V.; van Tendeloo, G. Nickel/Carbon composite materials based on expanded graphite. *New Carbon Mater.* **2011**, *26*, 335–340. [CrossRef]
98. Wang, L.; Zhu, Y.; Guo, C.; Zhu, X.; Liang, J.; Qian, Y. Ferric chloride-graphite intercalation compounds as anode materials for Li-ion batteries. *ChemSusChem* **2014**, *7*, 87–91. [CrossRef] [PubMed]
99. Ma, C.; Ma, C.; Wang, J.; Wang, H.; Shi, J.; Song, Y.; Guo, Q.; Liu, L. Exfoliated graphite as a flexible and conductive support for Si-based Li-ion battery anodes. *Carbon* **2014**, *72*, 38–46. [CrossRef]
100. Jiang, B.; Tian, C.; Zhou, W.; Wang, J.; Xie, Y.; Pan, Q.; Ren, Z.; Dong, Y.; Fu, D.; Han, J.; et al. In situ growth of TiO_2 in interlayers of expanded graphite for the fabrication of TiO_2-graphene with enhanced photocatalytic activity. *Chem. Eur. J.* **2011**, *17*, 8379–8387. [CrossRef]
101. Zhang, W.; Wan, W.; Zhou, H.; Chen, J.; Wang, X.; Zhang, X. In-situ synthesis of magnetite/expanded graphite composite material as high rate negative electrode for rechargeable lithium batteries. *J. Power Sources* **2013**, *223*, 119–124. [CrossRef]
102. Naderi, H.R.; Mortaheb, H.R.; Zolfaghari, A. Supercapacitive properties of nanostructured MnO_2/exfoliated graphite synthesized by ultrasonic vibration. *J. Electroanal. Chem.* **2014**, *719*, 98–105. [CrossRef]
103. Huang, Y.G.; Lin, X.L.; Zhang, X.H.; Pan, Q.C.; Yan, Z.X.; Wang, H.Q.; Chen, J.J.; Li, Q.Y. Fe_3C@carbon nanocapsules/expanded graphite as anode materials for lithium ion batteries. *Electrochim. Acta* **2015**, *178*, 468–475. [CrossRef]
104. Huang, Y.; Lin, X.; Pan, Q.; Li, Q.; Zhang, X.; Yan, Z.; Wu, X.; He, Z.; Wang, H. Al@C/Expanded Graphite Composite as Anode Material for Lithium Ion Batteries. *Electrochim. Acta* **2016**, *193*, 253–260. [CrossRef]
105. Hu, S.; Song, Y.; Yuan, S.; Liu, H.; Xu, Q.; Wang, Y.; Wang, C.X.; Xia, Y.Y. A hierarchical structure of carbon-coated Li_3VO_4 nanoparticles embedded in expanded graphite for high performance lithium ion battery. *J. Power Sources* **2016**, *303*, 333–339. [CrossRef]
106. Li, Y.; Zhao, Y.; Huang, G.; Xu, B.; Wang, B.; Pan, R.; Men, C.; Mei, Y. ZnO Nanomembrane/Expanded Graphite Composite Synthesized by Atomic Layer Deposition as Binder-Free Anode for Lithium Ion Batteries. *ACS Appl. Mater. Interfaces* **2017**, *9*, 38522–38529. [CrossRef] [PubMed]
107. Wu, S.; Wang, W.; Li, M.; Cao, L.; Lyu, F.; Yang, M.; Wang, Z.; Shi, Y.; Nan, B.; Yu, S.; et al. Highly durable organic electrode for sodium-ion batteries via a stabilized α-C radical intermediate. *Nat. Commun.* **2016**, *7*, 1–11. [CrossRef]
108. Wu, S.; Zhu, Y.; Huo, Y.; Luo, Y.; Zhang, L.; Wan, Y.; Nan, B.; Cao, L.; Wang, Z.; Li, M.; et al. Bimetallic organic frameworks derived CuNi/carbon nanocomposites as efficient electrocatalysts for oxygen reduction reaction. *Sci. China Mater.* **2017**, *60*, 654–663. [CrossRef]
109. Duan, Z.Q.; Liu, Y.T.; Xie, X.M.; Ye, X.Y.; Zhu, X.D. H-BN Nanosheets as 2D Substrates to Load 0D Fe_3O_4 Nanoparticles: A Hybrid Anode Material for Lithium-Ion Batteries. *Chem. Asian J.* **2016**, *11*, 828–833. [CrossRef]
110. Quartarone, E.; Dall'asta, V.; Resmini, A.; Tealdi, C.; Tredici, I.G.; Tamburini, U.A.; Mustarelli, P. Graphite-coated ZnO nanosheets as high-capacity, highly stable, and binder-free anodes for lithium-ion batteries. *J. Power Sources* **2016**, *320*, 314–321. [CrossRef]
111. Li, Z.; Yin, L. Sandwich-like reduced graphene oxide wrapped MOF-derived $ZnCo_2O_4$-ZnO-C on nickel foam as anodes for high performance lithium ion batteries. *J. Mater. Chem. A* **2015**, *3*, 21569–21577. [CrossRef]
112. Gan, Q.; Liu, S.; Zhao, K.; Wu, Y.; He, Z.; Zhou, Z. Graphene supported nitrogen-doped porous carbon nanosheets derived from zeolitic imidazolate framework for high performance supercapacitors. *RSC Adv.* **2016**, *6*, 78947–78953. [CrossRef]

113. Li, C.; Hu, Q.; Li, Y.; Zhou, H.; Lv, Z.; Yang, X.; Liu, L.; Guo, H. Hierarchical hollow Fe_2O_3 @MIL-101(Fe)/C derived from metal-organic frameworks for superior sodium storage. *Sci. Rep.* **2016**, *6*, 1–8. [CrossRef] [PubMed]

114. Yu, L.; Liu, J.; Xu, X.; Zhang, L.; Hu, R.; Liu, J.; Yang, L.; Zhu, M. Metal-organic framework-derived NiSb alloy embedded in carbon hollow spheres as superior lithium-ion battery anodes. *ACS Appl. Mater. Interfaces* **2017**, *9*, 2516–2525. [CrossRef] [PubMed]

115. Thauer, E.; Zakharova, G.S.; Andreikov, E.I.; Adam, V.; Wegener, S.A.; Nölke, J.H.; Singer, L.; Ottmann, A.; Asyuda, A.; Zharnikov, M.; et al. Novel synthesis and electrochemical investigations of ZnO/C composites for lithium-ion batteries. *J. Mater. Sci.* **2021**, *56*, 13227–13242. [CrossRef]

116. Eisenmann, T.; Birrozzi, A.; Mullaliu, A.; Giuli, G.; Trapananti, A.; Passerini, S.; Bresser, D. Effect of Applying a Carbon Coating on the Crystal Structure and De-/Lithiation Mechanism of Mn-Doped ZnO Lithium-Ion Anodes. *J. Electrochem. Soc.* **2021**, *168*, 030503. [CrossRef]

117. Köse, H.; Dombaycıoğlu, Ş.; Aydın, A.O.; Akbulut, H. Production and characterization of free-standing ZnO/SnO_2/MWCNT ternary nanocomposite Li-ion battery anode. *Int. J. Hydrog. Energy* **2016**, *41*, 9924–9932. [CrossRef]

118. Zhang, J.; Zhu, Y.; Cao, C.; Butt, F.K. Microwave-assisted and large-scale synthesis of SnO_2/carbon-nanotube hybrids with high lithium storage capacity. *RSC Adv.* **2015**, *5*, 58568–58573. [CrossRef]

119. Guler, M.O.; Cetinkaya, T.; Tocoglu, U.; Akbulut, H. Electrochemical performance of MWCNT reinforced ZnO anodes for Li-ion batteries. *Microelectron. Eng.* **2014**, *118*, 54–60. [CrossRef]

120. Kuwabata, S.; Masui, S.; Yoneyama, H. Charge–discharge properties of composites of $LiMn_2O_4$ and polypyrrole as positive electrode materials for 4 V class of rechargeable Li batteries. *Electrochim. Acta* **1999**, *44*, 4593–4600. [CrossRef]

121. Tang, W.; Liu, L.; Zhu, Y.; Sun, H.; Wu, Y.; Zhu, K. An aqueous rechargeable lithium battery of excellent rate capability based on a nanocomposite of MoO_3 coated with PPy and $LiMn_2O_4$. *Energy Environ. Sci.* **2012**, *5*, 6909–6913. [CrossRef]

122. Zhao, J.; Zhang, S.; Liu, W.; Du, Z.; Fang, H. Fe_3O_4/PPy composite nanospheres as anode for lithium-ion batteries with superior cycling performance. *Electrochim. Acta* **2014**, *121*, 428–433. [CrossRef]

123. Zhong, X.B.; Wang, H.Y.; Yang, Z.Z.; Jin, B.; Jiang, Q.C. Facile synthesis of mesoporous $ZnCo_2O_4$ coated with polypyrrole as an anode material for lithium-ion batteries. *J. Power Sources* **2015**, *296*, 298–304. [CrossRef]

124. Zhang, J.; Li, L.; Chen, J.; He, N.; Yu, K.; Liang, C. Controllable SnO_2/ZnO@PPy hollow nanotubes prepared by electrospinning technology used as anode for lithium ion battery. *J. Phys. Chem. Solids* **2021**, *150*, 109861. [CrossRef]

125. Köse, H.; Aydin, A.O.; Akbulut, H. Free-standing SnO_2/MWCNT nanocomposite anodes produced by different rate spin coatings for Li-ion batteries. *Int. J. Hydrog. Energy* **2014**, *39*, 21435–21446. [CrossRef]

126. Zhao, Y.; Li, X.; Dong, L.; Yan, B.; Shan, H.; Li, D.; Sun, X. Electrospun SnO_2-ZnO nanofibers with improved electrochemical performance as anode materials for lithium-ion batteries. *Int. J. Hydrog. Energy* **2015**, *40*, 14338–14344. [CrossRef]

127. Joshi, B.N.; An, S.; Jo, H.S.; Song, K.Y.; Park, H.G.; Hwang, S.; Al-Deyab, S.S.; Yoon, W.Y.; Yoon, S.S. Flexible, Freestanding, and Binder-free SnO_x-ZnO/Carbon Nanofiber Composites for Lithium Ion Battery Anodes. *ACS Appl. Mater. Interfaces* **2016**, *8*, 9446–9453. [CrossRef]

128. Qiao, L.; Wang, X.; Sun, X.; Li, X.; Zheng, Y.; He, D. Single electrospun porous NiO-ZnO hybrid nanofibers as anode materials for advanced lithium-ion batteries. *Nanoscale* **2013**, *5*, 3037–3042. [CrossRef]

129. Ma, L.; Pei, X.Y.; Mo, D.C.; Lyu, S.S.; Fu, Y.X. Fabrication of NiO-ZnO/RGO composite as an anode material for lithium-ion batteries. *Ceram. Int.* **2018**, *44*, 22664–22670. [CrossRef]

130. Wang, X.L.; Han, W.Q.; Chen, H.; Bai, J.; Tyson, T.A.; Yu, X.Q.; Wang, X.J.; Yang, X.Q. Amorphous hierarchical porous GeO_x as high-capacity anodes for Li ion batteries with very long cycling life. *J. Am. Chem. Soc.* **2011**, *133*, 20692–20695. [CrossRef]

131. Jin, S.; Li, N.; Cui, H.; Wang, C. Growth of the vertically aligned graphene@ amorphous GeO_x sandwich nanoflakes and excellent Li storage properties. *Nano Energy* **2013**, *2*, 1128–1136. [CrossRef]

132. Medvedev, A.G.; Mikhaylov, A.A.; Grishanov, D.A.; Yu, D.Y.W.; Gun, J.; Sladkevich, S.; Lev, O.; Prikhodchenko, P.V. GeO_2 Thin Film Deposition on Graphene Oxide by the Hydrogen Peroxide Route: Evaluation for Lithium-Ion Battery Anode. *ACS Appl. Mater. Interfaces* **2017**, *9*, 9152–9160. [CrossRef]

133. He, X.; Hu, Y.; Chen, R.; Shen, Z.; Wu, K.; Cheng, Z.; Pan, P. Foldable uniform GeO_x/ZnO/C composite nanofibers as a high-capacity anode material for flexible lithium ion batteries. *Chem. Eng. J.* **2019**, *360*, 1020–1029. [CrossRef]

134. Zhang, Y.Z.; Wang, Y.; Xie, Y.L.; Cheng, T.; Lai, W.Y.; Pang, H.; Huang, W. Porous hollow Co_3O_4 with rhombic dodecahedral structures for high-performance supercapacitors. *Nanoscale* **2014**, *6*, 14354–14359. [CrossRef] [PubMed]

135. Liu, B.; Zhang, J.; Wang, X.; Chen, G.; Chen, D.; Zhou, C.; Shen, G. Hierarchical three-dimensional $ZnCo_2O_4$ nanowire arrays/carbon cloth anodes for a novel class of high-performance flexible lithium-ion batteries. *Nano Lett.* **2012**, *12*, 3005–3011. [CrossRef] [PubMed]

136. Du, N.; Xu, Y.; Zhang, H.; Yu, J.; Zhai, C.; Yang, D. Porous $ZnCo_2O_4$ nanowires synthesis via sacrificial templates: High-performance anode materials of li-ion batteries. *Inorg. Chem.* **2011**, *50*, 3320–3324. [CrossRef]

137. Luo, W.; Hu, X.; Sun, Y.; Huang, Y. Electrospun porous $ZnCo_2O_4$ nanotubes as a high-performance anode material for lithium-ion batteries. *J. Mater. Chem.* **2012**, *22*, 8916–8921. [CrossRef]

138. Giri, A.K.; Pal, P.; Ananthakumar, R.; Jayachandran, M.; Mahanty, S.; Panda, A.B. 3D hierarchically assembled porous wrinkled-paper-like structure of $ZnCo_2O_4$ and Co-ZnO@C as anode materials for lithium-ion batteries. *Cryst. Growth Des.* **2014**, *14*, 3352–3359. [CrossRef]

139. Zhong, J.; Cao, C.; Liu, Y.; Li, Y.; Khan, W.S. Hollow core-shell η-Fe$_2$O$_3$ microspheres with excellent lithium-storage and gas-sensing properties. *Chem. Commun.* **2010**, *46*, 3869–3871. [CrossRef]
140. Zhou, L.; Xu, H.; Zhang, H.; Yang, J.; Hartono, S.B.; Qian, K.; Zou, J.; Yu, C. Cheap and scalable synthesis of α-Fe$_2$O$_3$ multi-shelled hollow spheres as high-performance anode materials for lithium ion batteries. *Chem. Commun.* **2013**, *49*, 8695–8697. [CrossRef] [PubMed]
141. Lin, H.B.; Rong, H.B.; Huang, W.Z.; Liao, Y.H.; Xing, L.D.; Xu, M.Q.; Li, X.P.; Li, W.S. Triple-shelled Mn$_2$O$_3$ hollow nanocubes: Force-induced synthesis and excellent performance as the anode in lithium-ion batteries. *J. Mater. Chem. A* **2014**, *2*, 14189–14194. [CrossRef]
142. Ma, Y.; Ma, Y.; Geiger, D.; Kaiser, U.; Zhang, H.; Kim, G.T.; Diemant, T.; Behm, R.J.; Varzi, A.; Passerini, S. ZnO/ZnFe$_2$O$_4$/N-doped C micro-polyhedrons with hierarchical hollow structure as high-performance anodes for lithium-ion batteries. *Nano Energy* **2017**, *42*, 341–352. [CrossRef]

 *nanomaterials*

Article

Microwave-Assisted Rapid Synthesis of $NH_4V_4O_{10}$ Layered Oxide: A High Energy Cathode for Aqueous Rechargeable Zinc Ion Batteries

Seokhun Kim, Vaiyapuri Soundharrajan , Sungjin Kim, Balaji Sambandam, Vinod Mathew, Jang-Yeon Hwang and Jaekook Kim *

Department of Materials Science and Engineering, Chonnam National University, Gwangju 61186, Korea; sin-tw@nate.com (S.K.); soundharajan.007@gmail.com (V.S.); babichunje1@hanmail.net (S.K.); matbalaji@gmail.com (B.S.); bethelvinod@gmail.com (V.M.); hjy@jnu.ac.kr (J.-Y.H.)
* Correspondence: jaekook@chonnam.ac.kr

Citation: Kim, S.; Soundharrajan, V.; Kim, S.; Sambandam, B.; Mathew, V.; Hwang, J.-Y.; Kim, J. Microwave-Assisted Rapid Synthesis of $NH_4V_4O_{10}$ Layered Oxide: A High Energy Cathode for Aqueous Rechargeable Zinc Ion Batteries. *Nanomaterials* **2021**, *11*, 1905. https://doi.org/ 10.3390/nano11081905

Academic Editor: Christophe Detavernier

Received: 25 June 2021
Accepted: 21 July 2021
Published: 24 July 2021

Publisher's Note: MDPI stays neutral with regard to jurisdictional claims in published maps and institutional affiliations.

Abstract: Aqueous rechargeable zinc ion batteries (ARZIBs) have gained wide interest in recent years as prospective high power and high energy devices to meet the ever-rising commercial needs for large-scale eco-friendly energy storage applications. The advancement in the development of electrodes, especially cathodes for ARZIB, is faced with hurdles related to the shortage of host materials that support divalent zinc storage. Even the existing materials, mostly based on transition metal compounds, have limitations of poor electrochemical stability, low specific capacity, and hence apparently low specific energies. Herein, $NH_4V_4O_{10}$ (NHVO), a layered oxide electrode material with a uniquely mixed morphology of plate and belt-like particles is synthesized by a microwave method utilizing a short reaction time (~0.5 h) for use as a high energy cathode for ARZIB applications. The remarkable electrochemical reversibility of Zn^{2+}/H^+ intercalation in this layered electrode contributes to impressive specific capacity (417 mAh g^{-1} at 0.25 A g^{-1}) and high rate performance (170 mAh g^{-1} at 6.4 A g^{-1}) with almost 100% Coulombic efficiencies. Further, a very high specific energy of 306 Wh Kg^{-1} at a specific power of 72 W Kg^{-1} was achieved by the ARZIB using the present NHVO cathode. The present study thus facilitates the opportunity for developing high energy ARZIB electrodes even under short reaction time to explore potential materials for safe and sustainable green energy storage devices.

Keywords: aqueous batteries; zinc-ion batteries; electrochemistry; electrode materials; ammonium vanadate

1. Introduction

The substantial features of being low-cost and highly safe have projected aqueous rechargeable zinc ion batteries (ARZIBs) for potential implementation in large-scale energy storage devices [1–3]. In specific, the abundance of natural zinc combined with the high capacity (~820 mAh g^{-1}) and low negative voltage (-0.76 V vs. standard hydrogen electrode (SHE)) of the zinc anode can facilitate the plausible use of ARZIBs as an efficient rechargeable device for smart grid applications [4–6]. Continual research efforts are witnessed, in this regard, to present ARZIBs as one of the remedies to the expected severe energy crisis [7–9]. In particular, such efforts have resulted in many good discoveries of suitable ARZIB cathodes, mainly based on manganese and vanadium-based materials, with structural and diffusion-related hurdles, however, researchers have identified ways for rectifying those snags. For instance, manganese-based cathodes experience poor cycling capability due to inherent Jahn-Teller distortion-induced manganese dissolution into the electrolyte during repeated cycling [10–12]. Tactically, this issue is addressed by ensuring the presence of enough manganese (II) ions in the electrolyte via the inclusion of $MnSO_4$ additive in the $ZnSO_4$ electrolyte, which results in reaching the reaction equilibrium faster and hence aids in the exceptional cyclability of the cathode [13–16]. Although, vanadium-based electrode materials have dominated the research activities in the lithium-ion battery application field for more than three decades, their use in ARZIBs was initiated only after 2016 [17–21]. In particular, layered vanadium oxide cathodes have gained wide attention for ARZIB applications due to their high capacity related to the wide operational potential range offered by the varied vanadium oxidation states, very stable cyclability, and extremely high rate performance arising from their structural durability [22–25]. For example, in 2016 Kundu et al. established a high-capacity bilayered Zn-stabilized V_2O_5 cathode for ARZIBs that exhibits a stable Zn electrochemical reversibility delivering 220 mAh g^{-1} specific capacity at 15 C with 80% retention over 1000 cycles [25]. Following this, several layered vanadates with their advantageous open framework structures and the multiple vanadium oxidation states leading to enhanced high capacity and supreme rate performances for prospective ARZIB applications were identified and reported [26–30]. Ammonium vanadium bronze, $NH_4V_4O_{10}$ (NHVO) with a monoclinic structure consisting of three V^{5+} and one V^{4+} states (variable oxidation states) has been reported to demonstrate reversible electrochemical Zn-intercalation under prolonged cycling due to its structural stability arising from the strong interaction between polyanionic ($V_3O_8^-$) and cationic layers by forming H-bonding network of N-H $\cdots$ O [17,21]. In addition, the existence of NH_4^+-ions as pillars between the vanadium interstitial layers offering increased interlayer distance results in enhanced ion diffusion rate and enriched electronic conductivity which will subdue the volume expansion due to divalent Zn^{2+} intercalation/de-intercalation [31]. For example, the larger interplanar spacing (~9.8 Å) and higher diffusion coefficient (~10^{-9}–10^{-10} cm^2 S^{-1}) of NHVO compared to that for $NH_4V_3O_8$ facilitates high zinc storage properties [32]. Recent first-principle calculations predicted that the lowest migration energy barrier (~0.63 eV) for Zn^{2+} diffusion and hence the feasible intercalation path is along the [010] direction in the interlayer region of the layered NHVO structure [33]. A majority of the reported NHVO cathode has been mostly synthesized by the hydrothermal reaction [33–35]. Herein, we present an $NH_4V_4O_{10}$ with a unique morphology consisting of plate-type particles and prepared by a microwave reaction for use as a cathode in an ARZIB. In combination with metallic zinc as the negative electrode, the prepared cathode delivers a very high specific capacity of ~417 mAh g^{-1} at 0.25 A g^{-1} and remarkably good cycling performance (more than 75% capacity retention after 1500 consecutive cycles at 2.5 A g^{-1} rate) in the presence of 3M $Zn(CF_3SO_3)_2$ electrolyte solution. Further, the electrochemical analyses aid in suggesting that the reaction mechanism is based on the reversible electrochemical intercalation of divalent Zn^{2+} and H^+ ions into the layered NHVO cathode. The results from the present study can expand the understanding of vanadate cathodes for the real-time development of various ARZIB cathodes with electrochemical behaviors analogous for safe and green large-scale energy storage devices.

2. Experimental Procedures

NHVO sample was prepared by using the microwave reaction method. In brief, appropriate amounts of ammonium vanadate (NH_4VO_3) and oxalic acid ($H_2C_2O_4 \cdot 2H_2O$) were initially dissolved in deionized water under stirring. Then, the resulting mixture was filled to 70% volume capacity of a 20 mL quartz tube and sealed. The mixture-contained vial was placed in a microwave reactor (Biotage Initiator Third generation, Biotage, Uppsala, Sweden) and heated to 200 °C for 0.5 h and allowed to cool naturally. The precipitate-containing solution was then washed with distilled water and ethanol several times before vacuum drying in an oven at 70 °C for 8 h.

2.1. Structural and Morphological Characterization

A Shimadzu X-ray diffractometer (Shimadzu, Kyoto, Japan) was used to record the powder X-ray powder diffraction (PXRD, Cu Kα radiation, with λ = 1.5406 Å) data for the prepared sample. Field-emission scanning electron microscopy (FE-SEM, Hitachi S-4700, Hitachi, Kerfeld, Germany) was used to study the surface morphology of the sample synthesized by the microwave reaction. Also, field-emission transmission electron microscopy (FE-TEM) analyses were performed using a Tecnai F20 instrument (Philips, Alemlo, The Netherlands) located at the Korea Basic Science Institue (KBSI), Chonnam National University) operating at 200 kV. The particle size distribution was calculated using dynamic light scattering method (DLS, ELS-8000, Otsuka electronics, Osaka, Japan). Prior to analysis, the powder sample was dispersed for 30 min in ethanol to obtain a homogenous solution. The surface area and pore size distribution of the samples were calculated using Brunauer–Emmett–Teller model (BET, BELSORP-mini II, MicrotracMRB, Osaka, Japan).

2.2. Electrochemical Characterization

A homogenous slurry was prepared by mixing the active material (80%), Super P carbon (10%), and polyacrylic acid (10%) binder in N-methyl-2-pyrrolidone. The doctor blade technique was followed to achieve a uniform coating of the formed slurry on the stainless steel foil current collector followed by vacuum drying at 80 °C. The coating-contained foil was then pressed between hot (~120 °C) stainless steel twin rollers and cut into circular discs to form the cathode. The active material loading was determined to be in the 1.2–1.6 mg range. A glass-fiber separator soaked in 3 M $Zn(CF_3SO_3)_2$ electrolyte was sandwiched between the prepared cathode and the zinc metal foil anode in a 2032-type coin cell under open-air conditions and aged for 24 h before electrochemical characterization. The electrochemical discharge/charge experiments were performed using a model 2004H battery testing system (BTS, Nagano Keiki Co., Ltd., Tokyo, Japan) at different current densities in the potential range 1.2–0.2 V vs. Zn^{2+}/Zn. Cyclic voltammetry (CV) scans were performed using a potentiostat workstation (PGSTAT302N, Autolab, Metrohm AG, Herisau Switzerland).

3. Results and Discussion

The NHVO sample powder was prepared by a facile microwave reaction that lasted for a short period of time (~0.5 h). The use of microwave irradiation in material synthesis is highly efficient in terms of maintaining uniform heating for the entire solution, ensuring reproducibility of materials synthesized due to better control of process parameters and facilitates less energy consumption because the irradiation directly heats the precursor solution and evades the heating of the apparatus [36–38]. Microwave irradiation is faster than conventional reactions as high reaction temperatures can be reached within a short time period. The rapid heating can help in accelerating the rate of reaction between the precursors in the solution. Moreover, side products are less formed during microwave irradiation. The structural evolution of the prepared powder was analyzed using PXRD and the results are given in Figure 1a. All the diffraction planes of (001), (200), (002), (110), (111), (003), (−112), (−311), (−312), (−205), (020), (−514) and (603) are well-indexed to the standard monoclinic NH_4VO_{10} with layered structure (JCPDS NO. 31-0075). As anticipated, the (001) line is the highly intense characteristic plane thereby implying that the growth of the particle most likely occurs in the [001] direction. FE-SEM was used to characterize the particle-size and morphology of the prepared sample and the results indicate that a mixed morphology of plate- and belt-shaped particles of a few hundred nanometers and micrometer dimensions, respectively, are distinguishable, as shown in Figure 1b. The high magnification image in Figure 1c shows a few plate-shaped particles conjoined to form a secondary flower-like structure. Further, the thickness of the plate-shaped particles is roughly estimated to be in the range of a few tens of nanometers. This implies that the unique morphology obtained by microwave reaction is within a short reaction time. To obtain further structural information, the TEM studies of the prepared sample were performed and the results (Figure 2a) confirmed the presence of a uniquely mixed morphology of plate-type and belt-type particles, thereby confirming the FE-SEM results. The high magnification TEM image reveals a portion of a single belt-shaped particle with ambiguous lattice fringes (Figure 2b). It is worth noting here that the crystallinity of the prepared NHVO sample appears to be sensitive to the prolonged exposure of the TEM instrument as the lattice fringes in the portion under the focus of the beam signal gradually began to undergo disintegration or slight amorphization thereby making it complicated to observe clear lattice fringes. However, almost clear lattice fringes are evolved in the high magnified TEM image shown in Figure 2b. The observed lattice fringe width of 0.96 nm corresponds to the characteristic (001) mother plane of the NHVO, as observed from the XRD pattern in Figure 1a. This clearly confirms that the particle growth occurs along the crystallographic [001] direction. In addition, the corresponding selected area electron diffraction (SAED) pattern recorded for the prepared NHVO revealed bright spots corresponding to almost single crystalline characteristics (Figure 2c). Furthermore, the matching of the bright spot from the SAED pattern (Figure 2c) corresponding to the (-205) plane in the XRD pattern (shown in Figure 1a) is also depicted thereby confirming the phase purity of the NHVO prepared by a microwave reaction. Moreover, elemental mapping images clearly reveal the uniform distribution of the corresponding elements throughout the area of study (Figure 2d–h).

The DLS method was used to determine that the particle-size distribution for the NHVO sample. The result shown in Figure 3a confirm that the average particle size was less than 600 nm and corroborated well with the electron microscopy results. BET N_2 adsorption-desorption analysis to determine the surface area confirmed that the layered NHVO sample demonstrates a type IV isotherm with H4 hysteresis (Figure 3b). Further, the surface area and total pore volume was determined to be 13.34 m^2 g^{-1} and 0.13 cm^3 g^{-1}, respectively. The existence of H4 hysteresis indicates that the NHVO sample consists of a microporous network with slit pore geometry. The steep increase of relative partial pressure (P/P_0) ~ 0.99 typically corresponds to a uniform particle size distribution in the NHVO sample. The Barrett–Joyner–Halenda (BJH) pore size distribution plot in Figure 3c is clearly indicative of the pore size distribution to be in the micrometre range. These results thus confirm that the determined surface area and porous characteristics can ultimately influence the electrochemical property of the prepared NHVO sample.

Electrochemical Zn-intercalation/de-intercalation behavior into the layered NHVO with a unique morphology was initially examined by cyclic voltammetry at 0.1 mV s^{-1} scan rate in the voltage range of 0.2–1.2 V vs. Zn^{2+}/Zn in $Zn(CF_3SO_3)_2$ electrolyte at ambient conditions and the results are given in Figure 4a. The well-observed cathodic shoulder at 0.91 V and peaks located at 0.6 V and 0.4 V corresponds to the multi-step reduction of vanadium to lower oxidation states ($V^{5+} \leftrightarrow V^{3+}$) due to the intercalation of Zn^{2+} ions into the layered NHVO host [22,39]. In the reverse anodic scan, 0.52, 0.71 and 1.01 V correspond to the continuous Zn^{2+} deintercalation across the layered structure of NHVO. This reveals a multi-step reaction mechanism usually known for vanadium-based cathodes. Interestingly, the cathodic peak around 0.8 V for the cycles can be associated with H^+/H_2O insertion, usually observed for vanadium-based cathodes demonstrating both Zn^{2+} and H^+ insertion [40,41]. From the second scan, there are very slight changes in the peak positions and decreased peak currents, which can be associated with the slightly reduced diffusion pathway for the intercalating ions. However, the CV curves observed for successive cycles, are almost superimposable, thus indicating that the electrochemical reversibility in the cathode is stable. Importantly, it is also possible that the probable role of proton intercalation can be considered for the layered cathode structure. In general, the electrochemical Zn^{2+}/H^+-intercalation process can be influenced by the multi-step complex reaction during the discharge process [42].

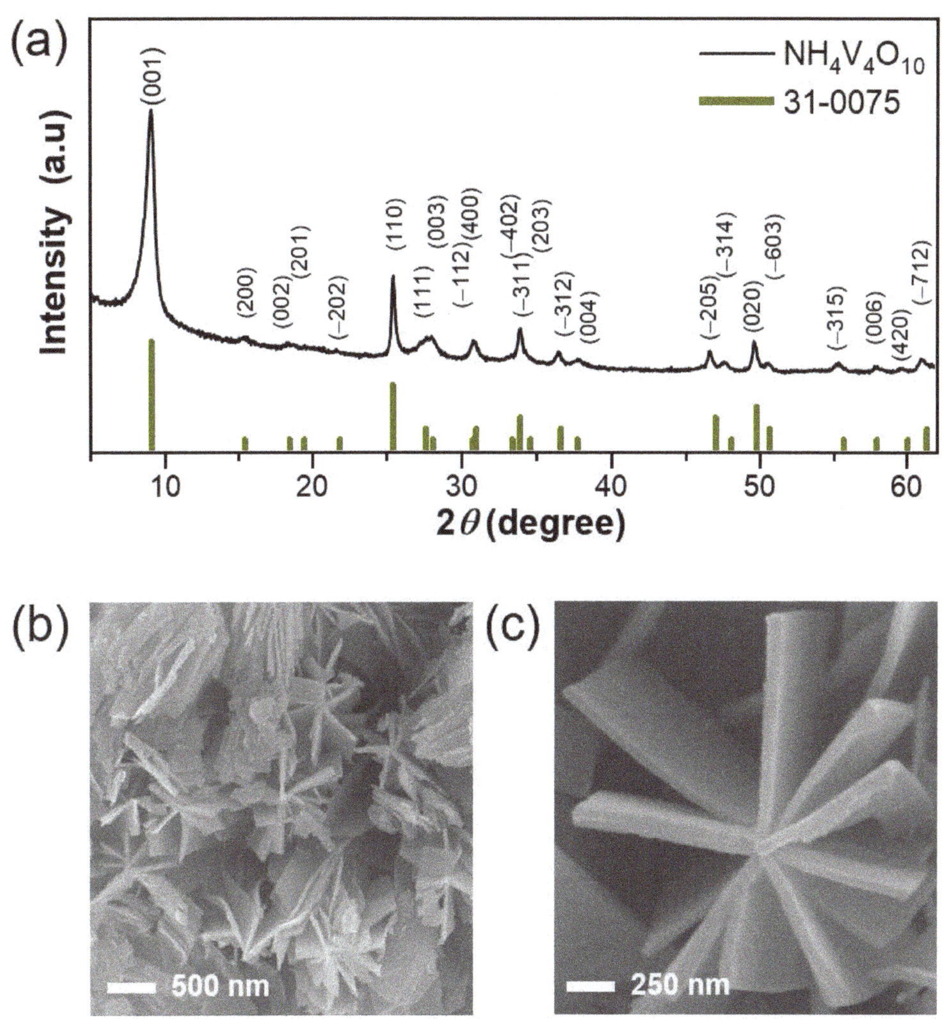

Figure 1. (**a**) Powder XRD pattern of $NH_4V_4O_{10}$ (NHVO) cathode prepared by a microwave reaction, (**b**) Low and (**c**) high magnification FE-SEM images of the NHVO electrode.

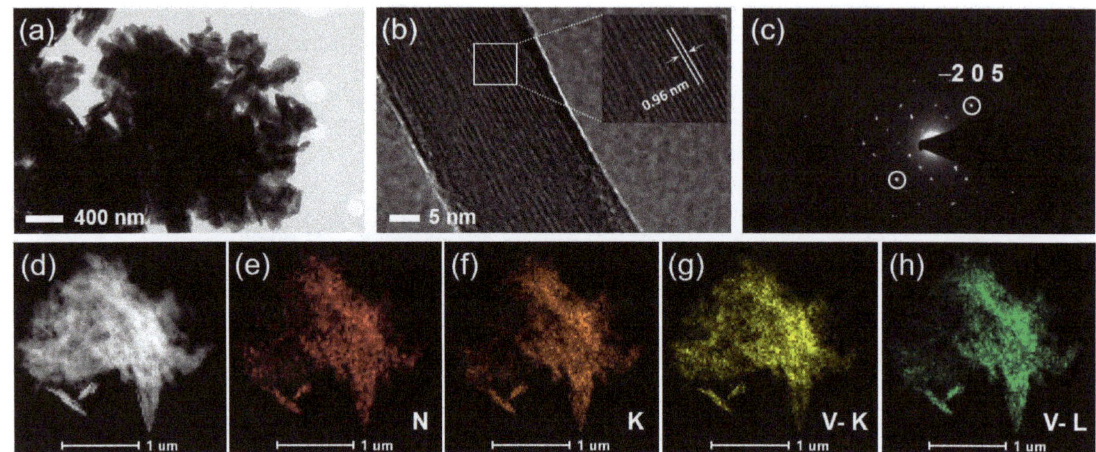

Figure 2. (**a**) Low and (**b**) high resolution TEM images and (**c**) corresponding SAED pattern of the prepared NHVO electrode. (**d–h**) FE-TEM and elemental mapping images recorded from the area shown in (**b**). The colored regions correspond to specific elements.

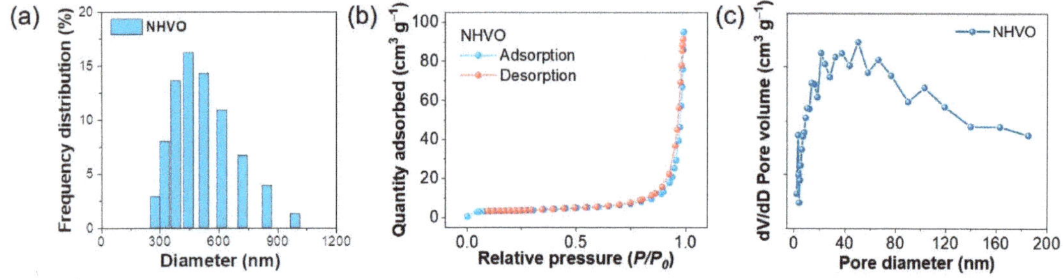

Figure 3. (**a**) DLS result for the prepared NHVO (**b**) N_2 adsorption/desorption plot and (**c**) corresponding pore-size distribution plot for the prepared NHVO sample.

The initial two galvanostatic discharge/charge profiles recorded at a medium current density of 0.25 A g^{-1} within the 1.2–0.2 V potential range for the prepared NHVO cathode using a coin-type test cell are provided in Figure 4b. The entire process of intercalation/de-intercalation is represented by the typical S-shaped sloping trend that is indicative of a single-phase reaction corresponding to the solid-solution behavior; the observation being in congruence with the CV results. The initial discharge capacity observed for the present NHVO cathode is 417 mAh g^{-1}. This is followed by an initial charge capacity of 374 mAh g^{-1} thereby facilitating a 90% Coulombic efficiency by the layered cathode. However, in the second cycle, almost 100% reversible capacity is achieved by the present NHVO cathode (Figure 4b). Upon further cycling, although stable coulombic efficiency is maintained, the capacity fading related to vanadium dissolution seems to be experienced at a moderate current density of 0.25 A g^{-1}, as observed from the gradually decreasing trend of the cycle-life profile shown in Figure 4c. In specific, nearly 67% of the highest initial capacity is sustained after 150 discharge/charge electrochemical cycles. Although higher performances are demonstrated for earlier reported NHVO cathodes, the performances presented here are competitive [32,33]. It is apparent that the advantage presented here is that the phase pure NHVO is successfully prepared under a lower reaction time for useful ARZIB applications.

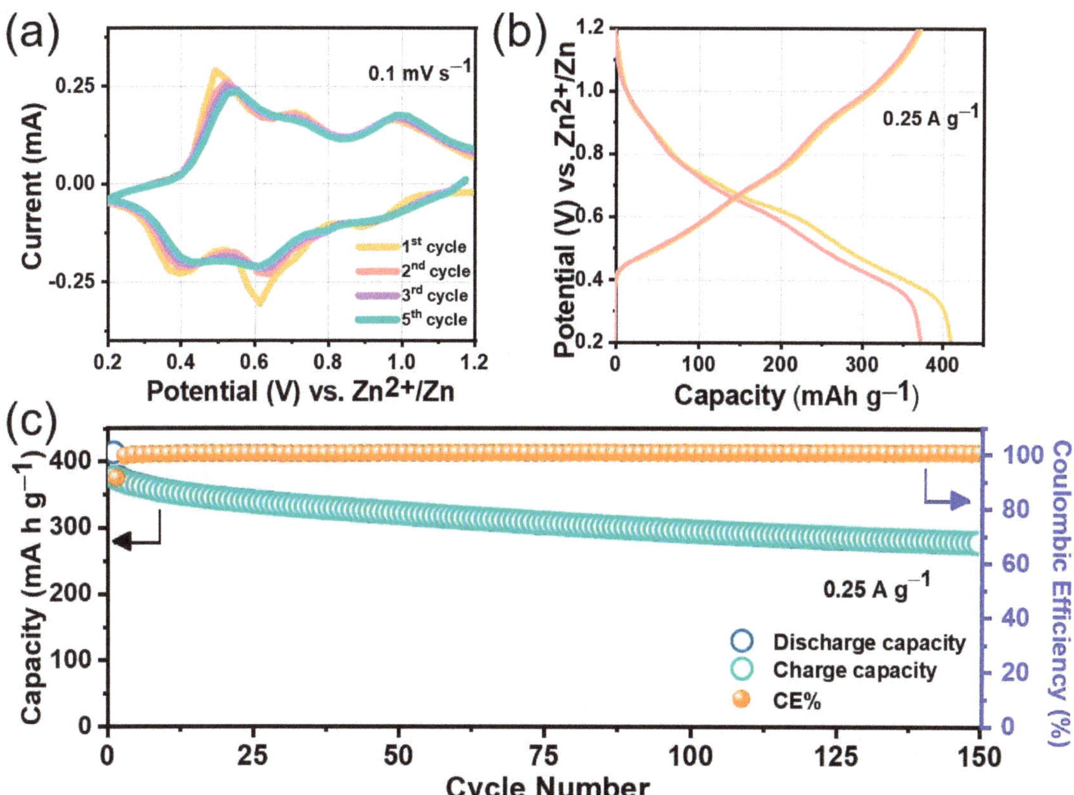

Figure 4. (**a**) The initial five CV profiles of NHVO electrode in ARZIBs, (**b**) Galvanostatic discharge/charge profiles at 250 mA g^{-1} current density for initial two cycles and (**c**) corresponding cyclability configuration of NHVO at the same applied current density.

The prolonged cycling ability of NHVO cathode in Zn(CF$_3$SO$_3$)$_2$ electrolyte at very high current densities was also tested using galvanostatic studies. The resulting long-term cyclability pattern is provided in Figure 5a. To achieve steady capacities, the electrode was cycled for the initial five cycles at 0.2 A g^{-1}. Thereafter, the cycling current density was maintained as high as 2.5 A g^{-1}. Obviously, the NHVO electrode showed 100% coulombic efficiency with a high reversible capacity of almost 128 mAh g^{-1} even after 1500 cycles. Although the capacity declines considerably under long-term cycling, the problem of vanadium dissolution appears to persist and further studies to stabilize the electrode for prolonged cycling are required. Nevertheless, this result indicates that the present layered cathode can consistently demonstrate electrochemical reversibility even at high current rates. For further confirmation on the structural stability of the NHVO cathode under alternate discharge/charge conditions, the rate performance study at progressively increasing current surges between 0.1 and 6.4 A g^{-1} within the voltage window of 1.2–0.2 V in Zn(CF$_3$SO$_3$)$_2$ electrolyte was studied. At each current density, four discharge/charge cycling of the layered cathode was performed to measure the average zinc storage capacities. After the completion of measuring one set of current densities was tested, the next set of testing current densities was applied reversely, i.e., in the progressively decreasing current surges. The sequence of applying two sets of increasing and decreasing current densities alternatively was continued until 90 discharge/charge cycling of the NHVO cathode was performed. The resulting rate performance provided in Figure 5b shows that the NHVO cathode showcased high durability at all applied current rates. In specific, the electrode shows average discharge capacities of 453, 407, 359, 317, 287, 234 and 170 mAh g^{-1} at 0.1, 0.2, 0.4, 0.8, 1.6, 3.2 and 6.4 A g^{-1}, respectively. Upon reversing to progressively decreasing current densities in the next set of rate measurements, there is a very insignificant decrease (<9%) in the average specific capacities compared to those corresponding values registered for the initial set of increasing current densities. More importantly, for the consecutive sets of increasing and decreasing current densities applied, the rate performance response of the NHVO electrode suggests no significant variation in the average specific capacities at each applied current density when compared to the corresponding values recorded in the previous sets. For example, the change in the specific capacities registered at 6.4 A g^{-1} at the 26th (165 mAh g^{-1}) and 68th (151 mAh g^{-1}) discharge cycles, respectively, for the first and second set of varying current density measurements is relatively less (<9%). Finally, the corresponding discharge/charge profiles for the initial set of progressive current densities (Figure 5c) also indicate that the standard S-type trend is maintained except for the loss in capacity due to the usual difference in the applied current rates. Further examination on the structural stability of the prepared electrode was performed via ex situ XRD studies. The XRD patterns recorded after initial discharge, initial charge and after 1500 discharge cycles, respectively, in comparison to the electrode at OCV are presented in Figure 6. The XRD pattern in the initial discharge state shows slight reduction in the peak intensities for all the characteristic planes. While the plane positions are mostly unaltered, the mother plane of (001) slightly shifts towards higher scanning angles after discharge, as observed from the magnified view of the scanning angle around the mother peak in Figure 6. These observations are mostly triggered by the intercalation of Zn^{2+}. Upon charging, all the characteristic planes are observed with increased intensities; albeit the peak intensities are not as high as the parent material at OCV. Further, the (001) mother plane regains its original position. After the 1500th discharge cycle, the mother plane of NHVO disappears (Figure 6) indicating that the layered structure of the active material is disintegrated. While the remaining peaks could be still identified; however, some slight shifts in the respective peak positions are observed. In other words, the reversibility of Zn-intercalation in NHVO is effective during the early cycling and upon prolonged cycling (~ over 1000 cycles), the cathode material loses its layered structure. Overall, these results clearly validate the structural stability of the present NHVO electrode prepared by microwave reaction.

To understand the cathode potential of the present NHVO cathode in the ARZIB system, the energy and power densities were determined based on the cathode mass at different current densities and the results are presented as a Ragone plot in Figure 7a. The Zn//NHVO system demonstrated an energy density of 306 Wh Kg^{-1} for a given power density of 72 W Kg^{-1} at 0.1 A g^{-1} current density. Further, the energy loss at high specific power is also relatively less; at the high specific power density values of 2304 and 4608 W Kg^{-1}, high energy densities of 157 and 107 Wh Kg^{-1} was achieved, as evidenced from Figure 7a. In comparison to some of the vanadate cathodes including ZnV$_2$O$_7$ [22], LiV$_3$O$_8$ [28], Na$_2$V$_6$O$_{16}$·3H$_2$O [43], VS$_2$ [30], Na$_3$V$_2$(PO$_4$)$_3$ [29], and K$_2$V$_6$O$_{16}$·2.7 H$_2$O [39] reported for the ARZIB system, the performance of the present NHVO cathode is remarkable and better suited for plausible real-time applications.

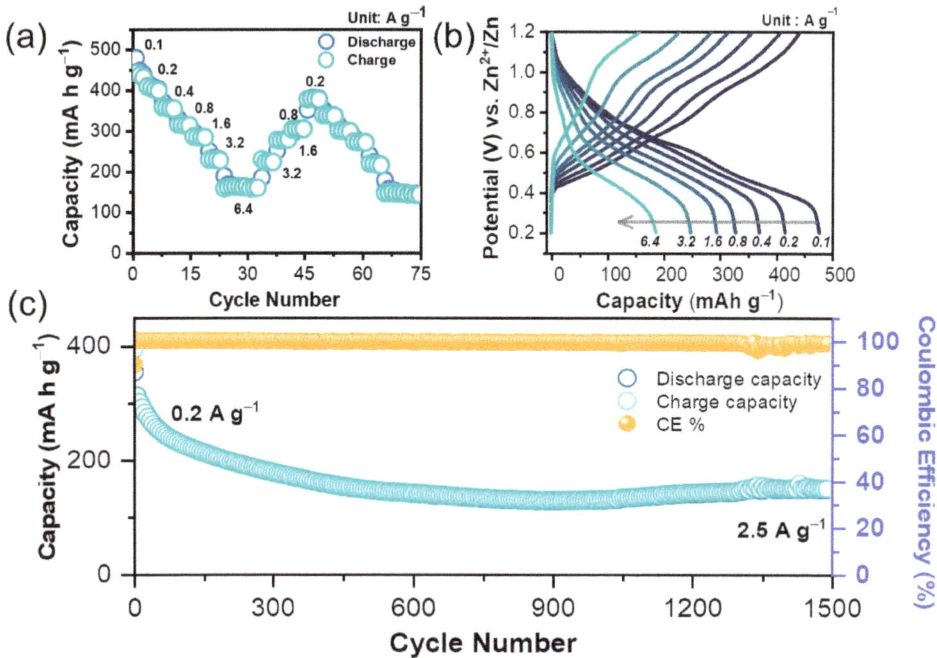

Figure 5. (**a**) Cycle lifespan of the NHVO electrode under prolonged cycling of 1500 cycles at 2.5 A g^{-1} (**b**) Rate performance at different progressively varying (increasing and decreasing alternatively) current densities for the NHVO cathode for ARZIB applications. (**c**) Galvanostatic discharge/charge profiles corresponding to one set of progressively increasing current density from 100 mA g^{-1} to 6.4 A g^{-1}.

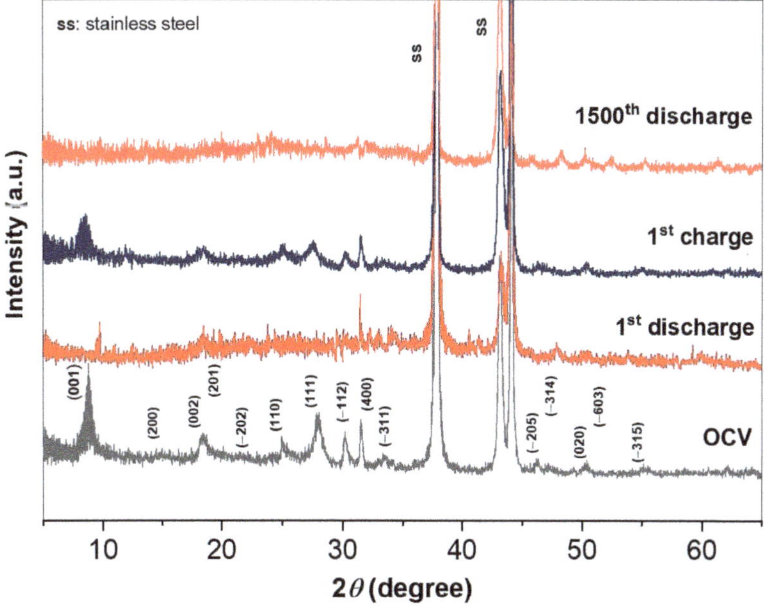

Figure 6. Ex-situ XRD patterns of the NHVO electrode recovered at different conditions.

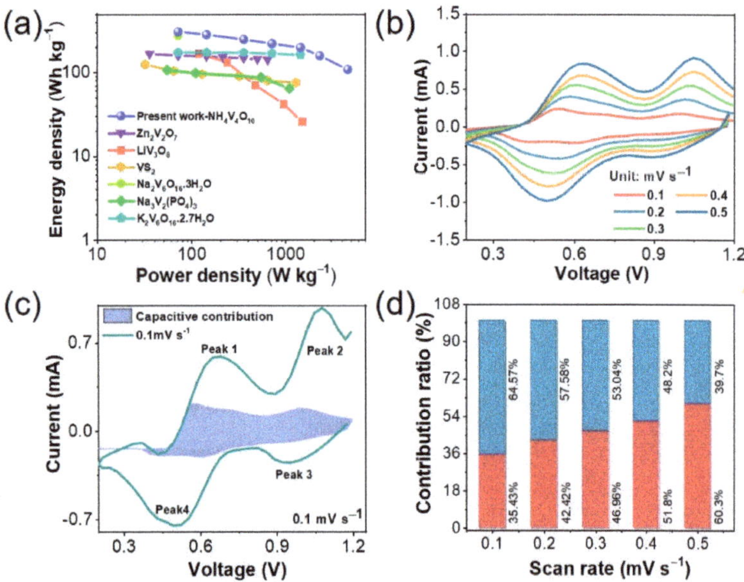

Figure 7. (**a**) Ragone plot for the present NHVO along with reported vanadium-based cathode materials for ARZIBs. (**b**) CV curves at various scan rates between 0.1 and 0.5 mV s^{-1}. (**c**) Surface-controlled capacitive contribution (shaded area) to the overall charge storage at 0.1 mV s^{-1}. (**d**) Ratio of surface-controlled and diffusion-induced contribution to the charge capacity depicted at different scan rates.

The interpretation of the electrochemical reaction kinetics from the CV response at various scan rates ranging between 0.1 and 0.5 mV s^{-1} for the present NHVO cathode was performed and the results are presented in Figure 7b–d. The trend of the CV curves at increasing scan rates is mostly preserved, but with gradual broadening in the shape. Hence, the major redox peaks are slightly shifted towards higher or lower voltage, respectively, as shown in Figure 7b. The total current reflects on the overall specific capacity registered for the NHVO electrode. The individual contribution of the surface-controlled capacitive and diffusion-induced processes, respectively, to the overall specific capacity at a specific potential during electrochemical cycling can be determined using the below relation [44,45]:

$$i = k_1 v + k_2 \, v^{1/2} \tag{1}$$

where, i represents the measured peak current and v refers to the scan rate (mV s^{-1}). $k_1 v$ and $k_2 v^{1/2}$ represent the capacitive and diffusion-controlled process, respectively, for a specific sweep rate. Accordingly, considering the case for the CV curve recorded at 0.1 mV s^{-1} scan rate, the shaded portion corresponding to ~35% of the overall specific-capacity represents the surface-controlled capacitive process (Figure 7c). In other words, this implies that a dominant diffusion-controlled process influences the present NHVO electrode synthesized by the microwave reaction. Nevertheless, the contribution of from the surface and diffusion-controlled process to the overall specific capacity at different scan rates starting from 0.1 to 0.5 mV s^{-1} are determined and presented in Figure 7d. As anticipated, the plot clearly indicates that the surface-related capacitive process gradually increases with increasing scan rates and is predominant at higher scan rates. This implies that the diffusion reactions highly influence the overall specific capacity at low and medium scan rates.

The electrochemical mechanism for layered vanadium oxide cathodes of ARZIBs, in general, can be explained by a multi-stage intercalation mechanism, as observed from the CV and galvanostatic studies. In fact, the initial study on layered vanadium oxide presented the feasibility of reversible intercalation of pure Zn^{2+} via single and two-phase reactions in a $Zn_{0.25}V_2O_5$ cathode [25]. Similarly, our group also reported that a complex mechanism involving more than one phase transition is evident in a LiV_3O_8 cathode upon reaction with zinc [28]. However, in the following studies, the role of proton involvement as a co-intercalation ion to Zn^{2+} ion in the process of intercalation in layered vanadium oxide cathodes including $NaV_3O_8 \cdot 1.5H_2O$ and $Zn_{0.3}V_2O_5 \cdot 1.5H_2O$ was stressed [41,46]. The present study supports this finding that Zn^{2+} and H^+ ions reversibly intercalate into the NHVO host, as observed from the CV and galvanostatic studies. Overall, the plausible electrochemical reaction for the present NHVO cathode in ARZIB system can be expressed by the following equations:

At the cathode:

$$x Zn^{2+} + NH_4V_4O_{10} + 2xe^- + yH^+ + ye^- \leftrightarrow Zn_xH_yNH_4V_4O_{10} \tag{2}$$

At the anode:

$$Zn \leftrightarrow Zn^{2+} + 2e^- \tag{3}$$

Although it has been widely accepted that Zn^{2+} and H^+ ions contribute to the intercalation process, the sequence followed by the guest ions during intercalation is hugely debated. While one group of researchers consider that Zn^{2+} and H^+ ions concomitantly access the cathode host, another group believes that these ions consecutively access into/from the host [47,48]. In this scenario, it is a stiff challenge to identify the exact reaction pathway of the intercalating Zn^{2+} and H^+ ions and requires further studies. Overall, the present study focuses on the presentation of a layered-type $NH_4V_4O_{10}$ cathode with flower-like morphology prepared by a microwave reaction that lasts for a very short duration of ~0.5 h for useful ARZIB applications. This NHVO cathode prepared under short microwave irradiation presents competent or even more impressive electrochemical properties in terms of higher specific capacity and long-term cycling stability than some of the reported counterparts synthesized by the hydrothermal method using a relatively long reaction duration (see Table 1 for comparison) [32,49–52]. Thus, the present microwave NHVO cathode demonstrates notable zinc storage properties and structural stability for electrochemical reversibility under long-term cycling. However, further tuning of the material in terms of improving the performance, particularly during extreme cycling conditions, is required to suit the demands for real-time application of ARZIBs.

Table 1. Comparison of reported ammonium vanadate cathodes and their performance in ARZIBs.

Ammonium Vanadate Based Cathodes	Synthesis Method	Synthesis Time (h)	Maximum Discharge Capacity (mAh g^{-1})/(mA g^{-1})	Rate Capability (mAh g^{-1})/Rate (mA g^{-1})
$NH_4V_4O_{10}$ [32]	Hydrothermal	48	~400/300	~180/10,000
Mo-doped $NH_4V_4O_{10}$ [49]	Hydrothermal	10	~330/100	~150/2000
$NH_4V_4O_{10} \cdot 0.28H_2O$ [50]	Hydrothermal	~	~400/200	~112/10,000
$NH_4V_4O_{10-x} \cdot xH_2O$ [51]	Hydrothermal	~	~410/50	~120/3000
$(NH_4)_xV_2O_5 \cdot nH_2O$ [52]	Hydrothermal	48	~370/100	N/A
$NH_4V_4O_{10}$ [This work]	Microwave	0.5	~450/100	~170/6400

4. Conclusions

In summary, a layered $NH_4V_4O_{10}$ cathode with flower-like particles was synthesized by a microwave method that avoids the undesirable extension of reaction time. The prepared cathode was effectively utilized to develop a high energy ARZIB system in $Zn(CF_3SO_3)_2$ electrolyte medium. The NHVO cathode demonstrated impressive reversible specific capacities under cycling (277 mAh g^{-1} with 75% capacity retention after 150 cycles at 0.25 A g^{-1} current density), rapid Zn-storage properties (128 mAh g^{-1} after 1500 cycles at 2.5 A g^{-1}) with exceptional coulombic efficiencies of almost 100%. Galvanostatic and potentiodynamic measurements indicated the participation of both Zn^{2+} and H$^+$ ions in the electrochemical (de-)intercalation (from)into the layered structure and that the reaction mechanism is reversible. Further, the proposed ARZIB system enclosing the present NHVO electrode exhibits a high energy density of 107 Wh Kg^{-1} under high specific power of 4608 W kg^{-1} based on the cathode mass. These values comparatively better those of commercial Pb-acid (~30 Wh Kg^{-1}) and the Ni-Cd (~50 Wh Kg^{-1}) batteries thereby demonstrating the prospective use of ARZIBs for real-time applications. It is also worth to mention that further improvements in the NHVO cathode can be attempted by tailoring the material using performance-enhancing strategies of forming composites with conductive materials.

Author Contributions: Conceptualization, S.K. (Seokhun Kim) and V.S.; methodology, S.K. (Seokhun Kim) and V.S.; validation, S.K. (Seokhun Kim); investigation, S.K. (Seokhun Kim) and V.S.; resources, S.K. (Sungjin Kim); writing—original draft preparation, S.K. (Sungjin Kim) and V.S.; writing—review and editing, B.S. and V.M.; supervision, J.K..; project administration, J.K. and J.-Y.H.; funding acquisition, J.K. All authors have read and agreed to the published version of the manuscript.

Funding: National Research Foundation of Korea (NRF) grants No. 2020R1A2C3012415 and 2018R1A5A1025224.

Institutional Review Board Statement: Not applicable.

Informed Consent Statement: Not applicable.

Data Availability Statement: The data presented in this study are available on request from the corresponding author.

Acknowledgments: This work was supported by the National Research Foundation of Korea (NRF) grants funded by the Korea government (MSIT) No. 2020R1A2C3012415 and 2018R1A5A1025224.

Conflicts of Interest: The authors declare no conflict of interest.

References

1. Jia, X.; Liu, C.; Neale, Z.G.; Yang, J.; Cao, G. Active Materials for Aqueous Zinc Ion Batteries: Synthesis, Crystal Structure, Morphology, and Electrochemistry. *Chem. Rev.* **2020**, *120*, 7795–7866. [CrossRef] [PubMed]
2. Cai, Y.; Liu, F.; Luo, Z.; Fang, G.; Zhou, J.; Pan, A.; Liang, S. Pilotaxitic Na$_{1.1}$V$_3$O$_{7.9}$ nanoribbons/graphene as high-performance sodium ion battery and aqueous zinc ion battery cathode. *Energy Storage Mater.* **2018**, *13*, 168–174. [CrossRef]
3. Zhang, N.; Cheng, F.; Liu, Y.; Zhao, Q.; Lei, K.; Chen, C.; Liu, X.; Chen, J. Cation-Deficient Spinel ZnMn$_2$O$_4$ Cathode in Zn(CF$_3$SO$_3$)$_2$ Electrolyte for Rechargeable Aqueous Zn-Ion Battery. *J. Am. Chem. Soc.* **2016**, *138*, 12894. [CrossRef]
4. Zhao, H.B.; Hu, C.J.; Cheng, H.W.; Fang, J.H.; Xie, Y.P.; Fang, W.Y.; Doan, T.N.L.; Hoang, T.K.A.; Xu, J.Q.; Chen, P. Novel Rechargeable M$_3$V$_2$(PO$_4$)$_3$//Zinc (M = Li, Na) Hybrid Aqueous Batteries with Excellent Cycling Performance. *Sci. Rep.* **2016**, *6*, 25809. [CrossRef]
5. Hou, Z.; Zhang, X.; Li, X.; Zhu, Y.; Liang, J.; Qian, Y. Surfactant widens the electrochemical window of an aqueous electrolyte for better rechargeable aqueous sodium/zinc battery. *J. Mater. Chem. A* **2017**, *5*, 730–738. [CrossRef]
6. Soundharrajan, V.; Sambandam, B.; Kim, S.; Mathew, V.; Jo, J.; Kim, S.; Lee, J.; Islam, S.; Kim, K.; Sun, Y.-K.; et al. Aqueous Magnesium Zinc Hybrid Battery: An Advanced High Voltage and High Energy MgMn$_2$O$_4$ Cathode. *ACS Energy Lett.* **2018**, *3*, 1998–2004. [CrossRef]
7. Mathew, V.; Sambandam, B.; Kim, S.; Kim, S.; Park, S.; Lee, S.; Alfaruqi, M.H.; Soundharrajan, V.; Islam, S.; Putro, D.Y.; et al. Manganese and Vanadium Oxide Cathodes for Aqueous Rechargeable Zinc-Ion Batteries: A Focused View on Performance, Mechanism, and Developments. *ACS Energy Lett.* **2020**, *5*, 2376–2400. [CrossRef]
8. Soundharrajan, V.; Sambandam, B.; Kim, S.; Islam, S.; Jo, J.; Kim, S.; Mathew, V.; Sun, Y.; Kim, J. The dominant role of Mn^{2+} additive on the electrochemical reaction in ZnMn$_2$O$_4$ cathode for aqueous zinc-ion batteries. *Energy Storage Mater.* **2020**, *28*, 407–417. [CrossRef]
9. Zhang, L.; Chen, L.; Zhou, X.; Liu, Z. Towards High-Voltage Aqueous Metal-Ion Batteries Beyond 1.5 V: The Zinc/Zinc Hexacyanoferrate System. *Adv. Energy Mater.* **2015**, *5*, 1400930. [CrossRef]
10. Chamoun, M.; Brant, W.R.; Tai, C.-W.; Karlsson, G.; Noréus, D. Rechargeability of aqueous sulfate Zn/MnO$_2$ batteries enhanced by accessible Mn^{2+} ions. *Energy Storage Mater.* **2018**, *15*, 351–360. [CrossRef]
11. Alfaruqi, M.H.; Gim, J.; Kim, S.; Song, J.; Jo, J.; Kim, S.; Mathew, V.; Kim, J. Enhanced reversible divalent zinc storage in a structurally stable α-MnO$_2$ nanorod electrode. *J. Power Sources* **2015**, *288*, 320–327. [CrossRef]

12. Bischoff, C.F.; Fitz, O.S.; Burns, J.; Bauer, M.; Gentischer, H.; Birke, K.P.; Henning, H.-M.; Biro, D. Revealing the Local pH Value Changes of Acidic Aqueous Zinc Ion Batteries with a Manganese Dioxide Electrode during Cycling. *J. Electrochem. Soc.* **2020**, *167*, 020545. [CrossRef]

13. Pan, H.; Shao, Y.; Yan, P.; Cheng, Y.; Han, K.S.; Nie, Z.; Wang, C.; Yang, J.; Li, X.; Bhattacharya, P.; et al. Reversible aqueous zinc/manganese oxide energy storage from conversion reactions. *Nat. Energy* **2016**, *1*, 16039. [CrossRef]

14. Jiang, B.; Xu, C.; Wu, C.; Dong, L.; Li, J.; Kang, F. Manganese Sesquioxide as Cathode Material for Multivalent Zinc Ion Battery with High Capacity and Long Cycle Life. *Electrochim. Acta* **2017**, *229*, 422–428. [CrossRef]

15. Islam, S.; Alfaruqi, M.H.; Mathew, V.; Song, J.; Kim, S.; Kim, S.; Jo, J.; Baboo, J.P.; Pham, D.T.; Putro, D.Y.; et al. Facile synthesis and the exploration of the zinc storage mechanism of β-MnO_2 nanorods with exposed (101) planes as a novel cathode material for high performance eco-friendly zinc-ion batteries. *J. Mater. Chem. A* **2017**, *5*, 23299–23309. [CrossRef]

16. Sun, W.; Wang, F.; Hou, S.; Yang, C.; Fan, X.; Ma, Z.; Gao, T.; Han, F.; Hu, R.; Zhu, M.; et al. Zn/MnO_2 Battery Chemistry With H+ and Zn2+ Coinsertion. *J. Am. Chem. Soc.* **2017**, *139*, 9775–9778. [CrossRef]

17. Cheng, Y.; Huang, J.; Li, J.; Cao, L.; Xu, Z.; Wu, J.; Cao, S.; Hu, H. Structure-controlled synthesis and electrochemical properties of $NH_4V_3O_8$ as cathode material for Lithium ion batteries. *Electrochim. Acta* **2016**, *212*, 217–224. [CrossRef]

18. Soundharrajan, V.; Sambandam, B.; Song, J.; Kim, S.; Jo, J.; Duong, P.T.; Kim, S.; Mathew, V.; Kim, J. Facile green synthesis of a $Co_3V_2O_8$ nanoparticle electrode for high energy lithium-ion battery applications. *J. Colloid Interface Sci.* **2017**, *501*, 133–141. [CrossRef]

19. Soundharrajan, V.; Sambandam, B.; Song, J.; Kim, S.; Jo, J.; Pham, D.T.; Kim, S.; Mathew, V.; Kim, J. Bitter gourd-shaped $Ni_3V_2O_8$ anode developed by a one-pot metal-organic framework-combustion technique for advanced Li-ion batteries. *Ceram. Int.* **2017**, *43*, 13224–13232. [CrossRef]

20. Soundharrajan, V.; Sambandam, B.; Song, J.; Kim, S.; Jo, J.; Kim, S.; Lee, S.; Mathew, V.; Kim, J. $Co_3V_2O_8$ Sponge Network Morphology Derived from Metal–Organic Framework as an Excellent Lithium Storage Anode Material. *ACS Appl. Mater. Interfaces* **2016**, *8*, 8546–8553. [CrossRef] [PubMed]

21. Zakharova, G.S.; Ottmann, A.; Ehrstein, B.; Tyutyunnik, A.P.; Zhu, Q.; Lu, S.; Voronin, V.I.; Enyashin, A.N.; Klingeler, R. A new polymorph of $NH_4V_3O_7$: Synthesis, structure, magnetic and electrochemical properties. *Solid State Sci.* **2016**, *61*, 225–231. [CrossRef]

22. Sambandam, B.; Soundharrajan, V.; Kim, S.; Alfaruqi, M.H.; Jo, J.; Kim, S.; Mathew, V.; Sun, Y.-K.; Kim, J. Aqueous rechargeable Zn-ion battery: An imperishable and high-energy $Zn_2V_2O_7$ nanowire cathode through intercalation regulation. *J. Mater. Chem. A* **2018**, *6*, 3850–3856. [CrossRef]

23. Sambandam, B.; Kim, S.; Pham, D.T.; Mathew, V.; Lee, J.; Lee, S.; Soundharrajan, V.; Kim, S.; Alfaruqi, M.H.; Hwang, J.-Y.; et al. Hyper oxidized $V_6O_{13+x}\cdot nH_2O$ layered cathode for aqueous rechargeable Zn battery: Effect on dual carriers transportation and parasitic reactions. *Energy Storage Mater.* **2021**, *35*, 47–61. [CrossRef]

24. Liu, Y.; Wu, X. Review of vanadium-based electrode materials for rechargeable aqueous zinc ion batteries. *J. Energy Chem.* **2021**, *56*, 223–237. [CrossRef]

25. Kundu, D.; Adams, B.D.; Duffort, V.; Vajargah, S.H.; Nazar, L.F. A high-capacity and long-life aqueous rechargeable zinc battery using a metal oxide intercalation cathode. *Nat. Energy* **2016**, *1*, 16119. [CrossRef]

26. He, P.; Quan, Y.; Xu, X.; Yan, M.; Yang, W.; An, Q.; He, L.; Mai, L.Q. High-Performance Aqueous Zinc-Ion Battery Based on Layered $H_2V_3O_8$ Nanowire Cathode. *Small* **2017**, *13*, 1702551. [CrossRef]

27. Xia, C.; Guo, J.; Li, P.; Zhang, X.; Alshareef, H.N. Highly Stable Aqueous Zinc-Ion Storage Using a Layered Calcium Vanadium Oxide Bronze Cathode. *Angew. Chem. Int. Ed.* **2018**, *57*, 3943–3948. [CrossRef]

28. Alfaruqi, M.H.; Mathew, V.; Song, J.; Kim, S.; Islam, S.; Pham, D.T.; Jo, J.; Kim, S.; Baboo, J.P.; Xiu, Z.; et al. Electrochemical Zinc Intercalation in Lithium Vanadium Oxide: A High-Capacity Zinc-Ion Battery Cathode. *Chem. Mater.* **2017**, *29*, 1684–1694. [CrossRef]

29. Li, G.; Yang, Z.; Jiang, Y.; Jin, C.; Huang, W.; Ding, X.; Huang, Y. Towards Polyvalent Ion Batteries: A Zinc-Ion Battery Based on NASICON Structured $Na_3V_2(PO_4)_3$. *Nano Energy* **2016**, *25*, 211. [CrossRef]

30. He, P.; Yan, M.; Zhang, G.; Sun, R.; Chen, L.; An, Q.; Mai, L. Layered VS_2 Nanosheet-Based Aqueous Zn Ion Battery Cathode. *Adv. Energy Mater.* **2017**, *7*, 1601920. [CrossRef]

31. He, P.; Zhang, G.; Liao, X.; Yan, M.; Xu, X.; An, Q.; Liu, J.; Mai, L.Q. Sodium Ion Stabilized Vanadium Oxide Nanowire Cathode for High-Performance Zinc-Ion Batteries. *Adv. Energy Mater.* **2018**, *8*, 1702463. [CrossRef]

32. Tang, B.; Zhou, J.; Fang, G.; Liu, F.; Zhu, C.; Wang, C.; Pan, A.; Liang, S. Engineering the interplanar spacing of ammonium vanadates as a high-performance aqueous zinc-ion battery cathode. *J. Mater. Chem. A* **2019**, *7*, 940–945. [CrossRef]

33. Li, Q.; Rui, X.; Chen, D.; Feng, Y.; Xiao, N.; Gan, L.; Zhang, Q.; Yu, Y.; Huang, S. A High-Capacity Ammonium Vanadate Cathode for Zinc-Ion Battery. *Nano-Micro Lett.* **2020**, *12*, 67. [CrossRef]

34. Yang, G.; Wei, T.; Wang, C. Self-Healing Lamellar Structure Boosts Highly Stable Zinc-Storage Property of Bilayered Vanadium Oxides. *ACS Appl. Mater. Interfaces* **2018**, *10*, 35079–35089. [CrossRef]

35. Qiu, N.; Chen, H.; Yang, Z.; Zhu, Y.; Liu, W.; Wang, Y. Porous hydrated ammonium vanadate as a novel cathode for aqueous rechargeable Zn-ion batteries. *Chem. Commun.* **2020**, *56*, 3785–3788. [CrossRef]

36. Bilecka, I.; Niederberger, M. Microwave chemistry for inorganic nanomaterials synthesis. *Nanoscale* **2010**, *2*, 1358–1374. [CrossRef]

37. Balaji, S.; Mutharasu, D.; Sankara Subramanian, N.; Ramanathan, K. A review on microwave synthesis of electrode materials for lithium-ion batteries. *Ionics* **2009**, *15*, 765. [CrossRef]

38. Haruna, A.B.; Ozoemena, K.I. Effects of microwave irradiation on the electrochemical performance of manganese-based cathode materials for lithium-ion batteries. *Curr. Opin. Electrochem.* **2019**, *18*, 16–23. [CrossRef]

39. Sambandam, B.; Soundharrajan, V.; Kim, S.; Alfaruqi, M.H.; Jo, J.; Kim, S.; Mathew, V.; Sun, Y.; Kim, J. $K_2V_6O_{16} \cdot 2.7H_2O$ nanorod cathode: An advanced intercalation system for high energy aqueous rechargeable Zn-ion batteries. *J. Mater. Chem. A* **2018**, *6*, 15530–15539. [CrossRef]

40. Liu, W.; Dong, L.; Jiang, B.; Huang, Y.; Wang, X.; Xu, C.; Kang, Z.; Mou, J.; Kang, F. Layered vanadium oxides with proton and zinc ion insertion for zinc ion batteries. *Electrochim. Acta* **2019**, *320*, 134565. [CrossRef]

41. Wan, F.; Zhang, L.; Dai, X.; Wang, X.; Niu, Z.; Chen, J. Aqueous rechargeable zinc/sodium vanadate batteries with enhanced performance from simultaneous insertion of dual carriers. *Nat. Commun.* **2018**, *9*, 1656. [CrossRef]

42. Esparcia, E.A.; Chae, M.S.; Ocon, J.D.; Hong, S.-T. Ammonium Vanadium Bronze ($NH_4V_4O_{10}$) as a High-Capacity Cathode Material for Nonaqueous Magnesium-Ion Batteries. *Chem. Mater.* **2018**, *30*, 3690–3696. [CrossRef]

43. Soundharrajan, V.; Sambandam, B.; Kim, S.; Alfaruqi, M.H.; Putro, D.Y.; Jo, J.; Kim, S.; Mathew, V.; Sun, Y.-K.; Kim, J. $Na_2V_6O_{16} \cdot 3H_2O$ Barnesite Nanorod: An Open Door to Display a Stable and High Energy for Aqueous Rechargeable Zn-Ion Batteries as Cathodes. *Nano Lett.* **2018**, *18*, 2402–2410. [CrossRef] [PubMed]

44. Soundharrajan, V.; Sambandam, B.; Alfaruqi, M.H.; Kim, S.; Jo, J.; Kim, S.; Mathew, V.; Sun, Y.; Kim, J. $Na_{2.3}Cu_{1.1}Mn_2O_{7-\delta}$ nanoflakes as enhanced cathode materials for high-energy sodium-ion batteries achieved by a rapid pyrosynthesis approach. *J. Mater. Chem. A* **2020**, *8*, 770–778. [CrossRef]

45. Soundharrajan, V.; Alfaruqi, M.H.; Lee, S.; Sambandam, B.; Kim, S.; Kim, S.; Mathew, V.; Pham, D.T.; Hwang, J.-Y.; Sun, Y.-K.; et al. Multidimensional $Na_4VMn_{0.9}Cu_{0.1}(PO_4)_3/C$ cotton-candy cathode materials for high energy Na-ion batteries. *J. Mater. Chem. A* **2020**, *8*, 12055–12068. [CrossRef]

46. Wang, L.; Huang, K.-W.; Chen, J.; Zheng, J. Ultralong cycle stability of aqueous zinc-ion batteries with zinc vanadium oxide cathodes. *Sci. Adv.* **2019**, *5*, eaax4279. [CrossRef] [PubMed]

47. Kundu, D.; Hosseini Vajargah, S.; Wan, L.; Adams, B.; Prendergast, D.; Nazar, L.F. Aqueous vs. nonaqueous Zn-ion batteries: Consequences of the desolvation penalty at the interface. *Energy Environ. Sci.* **2018**, *11*, 881–892. [CrossRef]

48. Oberholzer, P.; Tervoort, E.; Bouzid, A.; Pasquarello, A.; Kundu, D. Oxide versus Nonoxide Cathode Materials for Aqueous Zn Batteries: An Insight into the Charge Storage Mechanism and Consequences Thereof. *ACS Appl. Mater. Interfaces* **2019**, *11*, 674–682. [CrossRef]

49. Wang, H.; Jing, R.; Shi, J.; Zhang, M.; Jin, S.; Xiong, Z.; Guo, L.; Wang, Q. Mo-doped $NH_4V_4O_{10}$ with enhanced electrochemical performance in aqueous Zn-ion batteries. *J. Alloys Compd.* **2021**, *858*, 158380. [CrossRef]

50. Zhu, T.; Mai, B.; Hu, P.; Liu, Z.; Cai, C.; Wang, X.; Zhou, L. Ammonium Ion and Structural Water Co-Assisted Zn^{2+} Intercalation/De-Intercalation in $NH_4V_4O_{10} \cdot 0.28H_2O$. *Chin. J. Chem.* **2021**. [CrossRef]

51. He, T.; Ye, Y.; Li, H.; Weng, S.; Zhang, Q.; Li, M.; Liu, T.; Cheng, J.; Wang, X.; Lu, J.; et al. Oxygen-deficient ammonium vanadate for flexible aqueous zinc batteries with high energy density and rate capability at $-30\ ^\circ C$. *Mater. Today* **2021**, *43*, 53–61. [CrossRef]

52. Xu, L.; Zhang, Y.; Zheng, J.; Jiang, H.; Hu, T.; Meng, C. Ammonium ion intercalated hydrated vanadium pentoxide for advanced aqueous rechargeable Zn-ion batteries. *Mater. Today Energy* **2020**, *18*, 100509. [CrossRef]

nanomaterials

Review

Recent Advances in Transition Metal Dichalcogenide Cathode Materials for Aqueous Rechargeable Multivalent Metal-Ion Batteries

Vo Pham Hoang Huy, Yong Nam Ahn and Jaehyun Hur *

Department of Chemical and Biological Engineering, Gachon University, Seongnam 13120, Gyeonggi, Korea; vophamhoanghuy@yahoo.com.vn (V.P.H.H.); yahn@gachon.ac.kr (Y.N.A.)
* Correspondence: jhhur@gachon.ac.kr

Abstract: The generation of renewable energy is a promising solution to counter the rapid increase in energy consumption. Nevertheless, the availability of renewable resources (e.g., wind, solar, and tidal) is non-continuous and temporary in nature, posing new demands for the production of next-generation large-scale energy storage devices. Because of their low cost, highly abundant raw materials, high safety, and environmental friendliness, aqueous rechargeable multivalent metal-ion batteries (AMMIBs) have recently garnered immense attention. However, several challenges hamper the development of AMMIBs, including their narrow electrochemical stability, poor ion diffusion kinetics, and electrode instability. Transition metal dichalcogenides (TMDs) have been extensively investigated for applications in energy storage devices because of their distinct chemical and physical properties. The wide interlayer distance of layered TMDs is an appealing property for ion diffusion and intercalation. This review focuses on the most recent advances in TMDs as cathode materials for aqueous rechargeable batteries based on multivalent charge carriers (Zn^{2+}, Mg^{2+}, and Al^{3+}). Through this review, the key aspects of TMD materials for high-performance AMMIBs are highlighted. Furthermore, additional suggestions and strategies for the development of improved TMDs are discussed to inspire new research directions.

Keywords: transition metal dichalcogenide; aqueous multivalent metal-ion batteries; zinc-ion batteries; magnesium-ion batteries; aluminum-ion batteries

Citation: Hoang Huy, V.P.; Ahn, Y.N.; Hur, J. Recent Advances in Transition Metal Dichalcogenide Cathode Materials for Aqueous Rechargeable Multivalent Metal-Ion Batteries. *Nanomaterials* **2021**, *11*, 1517. https://doi.org/10.3390/nano11061517

Academic Editor: Sergio Brutti

Received: 20 May 2021
Accepted: 7 June 2021
Published: 8 June 2021

Publisher's Note: MDPI stays neutral with regard to jurisdictional claims in published maps and institutional affiliations.

1. Introduction

Nowadays, owing to their depleting sources and adverse effects on the environment, traditional fossil fuels have been replaced by various renewable sources, such as sunlight, tides, and wind. However, the non-continuous supply of these renewable resources has sparked the need to develop safe and low-cost energy storage devices, especially for large-scale grid applications [1–3]. Lithium-ion batteries (LIBs) have been extensively investigated as an energy resource solution for grid applications and as the primary source of electrical machines owing to their high energy/power density and durability [4–16]. Nevertheless, the low availability of lithium, its growing cost, and safety issues hamper the application of these batteries in the production of commercial storage devices. Other monovalent metal-ion batteries, namely sodium-ion batteries (SIBs) and potassium-ion batteries (KIBs) have also been widely researched owing to their similar chemical properties to those of LIBs and the relatively higher abundance of Na and K [17–21]. Nevertheless, like LIBs, SIBs and KIBs use highly toxic electrolytes, which raise safety concerns and increase the fabrication costs.

Compared to batteries using organic electrolytes, aqueous rechargeable batteries (ARBs) possess tremendous competitive potential for energy storage devices because of their outstanding advantages such as: (i) high safety due to the absence of harmful solvents with high volatility; (ii) low cost and environmental friendliness; (iii) high ionic conductivity of aqueous electrolytes compared to that of organic electrolytes (1000 times higher); and (iv) the absence of a solid electrolyte interphase (SEI) layer during the electrochemical reaction [22–24]. Therefore, ARBs can potentially meet the demands of high energy density, high elemental abundance, and better safety features [25,26]. In particular, the use of aqueous multivalent metal-ion batteries (AMMIBs) is an ingenious solution for meeting the rapidly increasing demand for high-performance and cost-effective energy storage devices. According to the data in Table 1, multivalent metal anode has high volumetric energy density and theoretical capacity due to the multi-electron transfer capability. Consequently, various efforts have been made to study the various types of AMMIBs, such as zinc-ion batteries (ZIBs) [27–30], magnesium-ion batteries (MIBs) [31], aluminum-ion batteries (AIBs) [32], and calcium-ion batteries (CIBs) [33].

Transition metal dichalcogenides (TMDs) have attracted significant attention as potential materials in diverse applications pertaining to energy storage [34–38]. TMDs with the general formula MX_2 (M: transition metal, X: chalcogen) are often considered as inorganic analogues of graphite. Every single TMD layer is formed by three atoms (X-M-X) with an M layer sandwiched by two X atomic layers. The weak interlayer van der Waals (vdW) force facilitates the insertion of ions [39–42]. This property is especially adequate for AMMIB electrodes that utilize large hydrated metal ions. For example, molybdenum disulfide (MoS_2) with an interlayer distance of 0.62 nm is considered as a representative member of the TMD family and can easily accommodate hydrated Zn^{2+} ions in its framework. Nevertheless, numerous studies have demonstrated the difficulty of Zn^{2+} storage in bulk MoS_2 [24]. This indicates the incompatibility of MoS_2 as a ZIB cathode despite its sufficient interlayer distance. Consequently, the simple adoption of TMDs as potential cathodes is not a great strategy for realizing high-performance ZIBs.

Table 1. Comparison of multivalent and monovalent metal-ions [23,30,33,43–46].

Charge Carrier Ions	Zn^{2+}	Mg^{2+}	Al^{3+}	Ca^{2+}	Li^+	Na^+	K^+
Atomic mass (g mol^{-1})	65.41	24.31	26.98	40.08	6.94	22.99	39.1
Density (at 20 °C) (g cm^{-3})	7.11	1.74	2.7	1.55	0.53	0.97	0.89
Crystal structure	Hexagonal	Hexagonal	face-centered cubic	face-centered cubic	body-centered cubic	body-centered cubic	body-centered cubic
Abundance [a]	25	8	3	5	33	6	7
Metal cost (USD kg^{-1})	2.2	2.2	1.9	2.28	19.2	3.1	13.1
Ionic radius (Å)	0.75	0.72	0.53	1.00	0.76	1.02	1.38
Hydrated ionic radius (Å)	4.3	4.28	4.75	4.12	3.82	3.58	3.31
Redox potential vs. SHE	−0.763	−2.356	−1.676	−2.84	−3.04	−2.713	−2.924
Gravimetric specific capacity of metal anode (mAh g^{-1})	820	2206	2980	1337	3860	1166	685
Volumetric specific capacity (mAh cm^{-1})	5855	3834	8046	2072	2061	1129	610

[a] Ranking on the basis of the abundance of all the elements on earth.

Recently, there have been many reviews focusing on the TMD materials [47,48] and other cathode materials [49] for battery applications. While most previous reviews on TMD materials for battery applications covered various rechargeable batteries regardless of the electrolyte type [47,48], the reviews for the cathode materials dealt with a wide range of cathode candidate materials beyond TMDs [49]. The main viewpoint in this review is centered on the TMD materials as a cathode and multivalent metal ions dissolved in water as a electrolyte, respectively, which is distinctively different from most previous review articles. Figure 1a,b highlight some recent reviews and the numbers of publications on TMDs over the last 10 years [36,40,41,47,48,50–59]. The number of publications on various TMD materials has been steadily increased from the year of 2010. In particular, the increase in publication number has been more radical since the year of 2015. In the perspectives of applications, TMDs have been widely used in energy, sensors, optoelectronic devices, piezoelectric devices, and biosensors (Figure 1c). Considering the high interest in energy storage using TMD materials, the reviews focusing on the AMMIB cathode materials have been rare. To stimulate new research strategies in this direction, this review provides an overview of TMDs as cathode materials in AMMIB. This review summarizes the current developments of TMD cathodes in various AMMIBs and discusses the main challenges as well as the advantages of employing TMDs in AMMIBs. Some modification strategies for TMDs have also been discussed to enhance multivalent-ion storage in aqueous batteries. Finally, the future perspective and outlook toward next-generation AMMIB cathode research will be discussed, which would be beneficial for the rational design of TMD-based materials for AMMIBs.

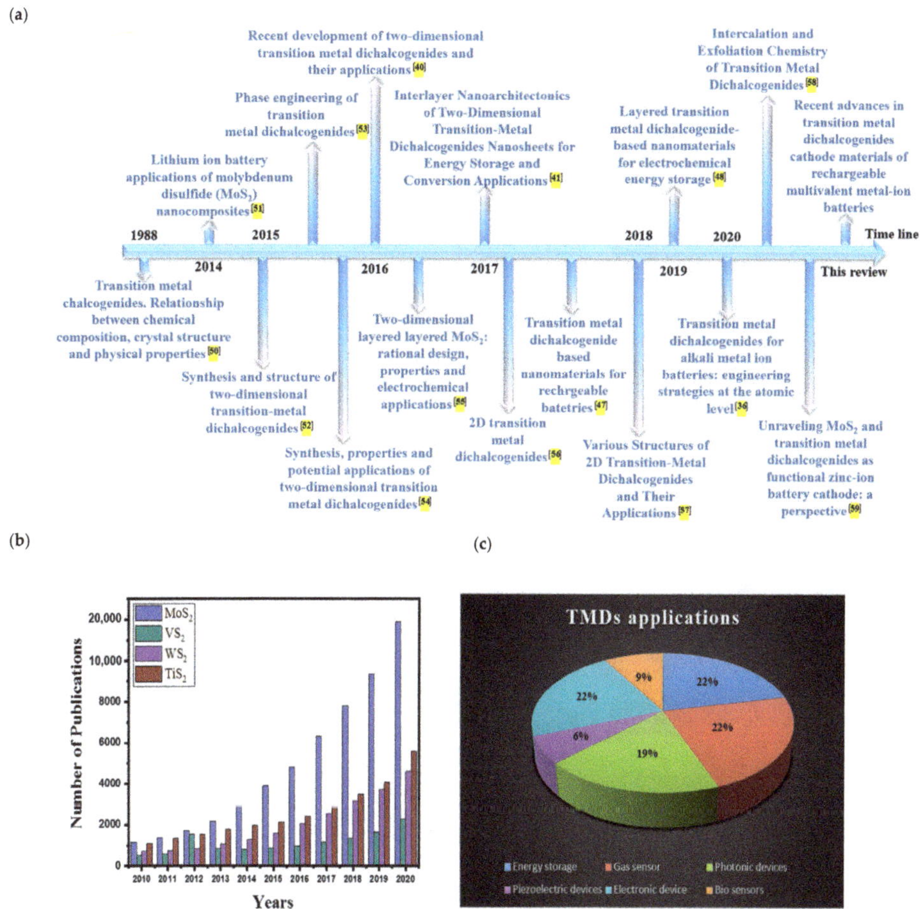

Figure 1. (**a**) Brief summary of recent reviews on transition metal dichalcogenides (TMD) materials, (**b**) Year-wise publication plot for TMD materials including MoS_2, VS_2, WS_2, and TiS_2 in the period of 2010–2020. (searched by Google Scholar, 2 June 2021), (**c**) The applications of TMDs in the period of 2010–2020.

2. Brief Introduction of Transition Metal Dichalcogenides (TMDs)

2.1. Concept and Principle of TMDs

The chemical formula of TMDs is MX_2, where M is the transition metal from groups 4–10 in the periodic table and X is the chalcogen, as illustrated in Figure 2a. In general, TMD materials with transition metals from groups 4–7 have a layered structure, whereas some transition metals from groups 8–10, such as pyrite, have a non-layered structure [39]. The atoms in layered MX_2 are arranged as polytypes with a transition metal atom M surrounded by six chalcogen X atoms (Figure 2b). In a TMD monolayer (basal plane), strong covalent bonds between the transition metal and chalcogen result in the formation of stacking polytypes (stacking order) and polymorphs (metal coordination geometry). As shown in Figure 3a, the TMD structures have typical configurations such as 1T, 2H, and 3R, which indicate one (1), two (2), and three (3) layers per stacked cell unit in tetragonal (T), hexagonal (H), and rhombohedral (R) phases, respectively [53,57,60,61]. For example, MoS_2 has all the three polytypes with a regular layered structure consisting of chalcogen atoms surrounding Mo transition metal atoms (as shown in Figure 3b) [62]. It is well-known that 1T-MoS_2 is a metastable metal phase, whereas 2H-MoS_2 and 3R-MoS_2 are semiconductor phases with thermodynamic stability. In particular, 1T- and 2H-MoS_2 are different in terms of the horizontal movement of one of the two sulfur planes. The intercalation of lithium ions (Li^+ ions) can induce the 2H-to-1T phase transition [63].

While the electronic structure of graphene is based on the hybridization of s and p orbitals, the electronic properties of TMDs are determined by the electrons filled in the d orbitals of the transition metal. Although graphene and TMDs exhibit structural similarities, the electrical properties of TMDs are determined by the number of electrons in their non-bonding d orbitals, as well as the geometrical coordination of the transition metal atoms [64]. The degree of electron filling in the d orbital significantly affects the electrical properties of TMDs, where a partially filled d orbital imparts metallic properties, whereas a completely filled d orbital leads to a semiconducting behavior [39]. As a result, transition metal atoms have a greater influence on the electronic structure of TMDs than chalcogen atoms.

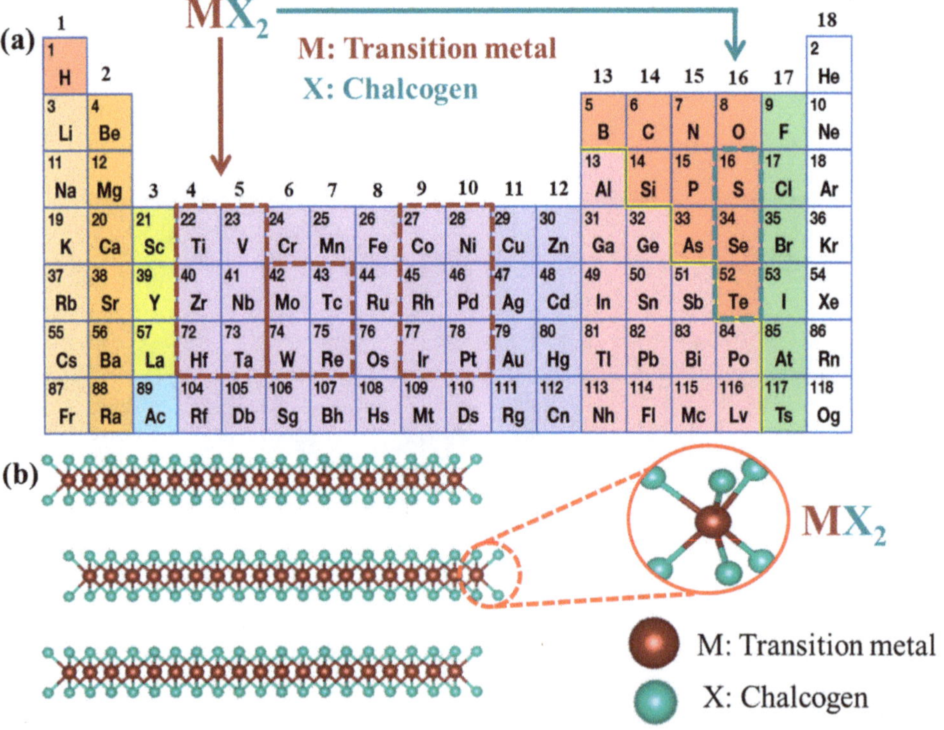

Figure 2. (**a**) TMDs with the MX_2 structure consisting of M from the 16 transition metals indicated by the red dotted box and X from the three halogen elements indicated by the green dotted box, (**b**) layered structure of MX_2.

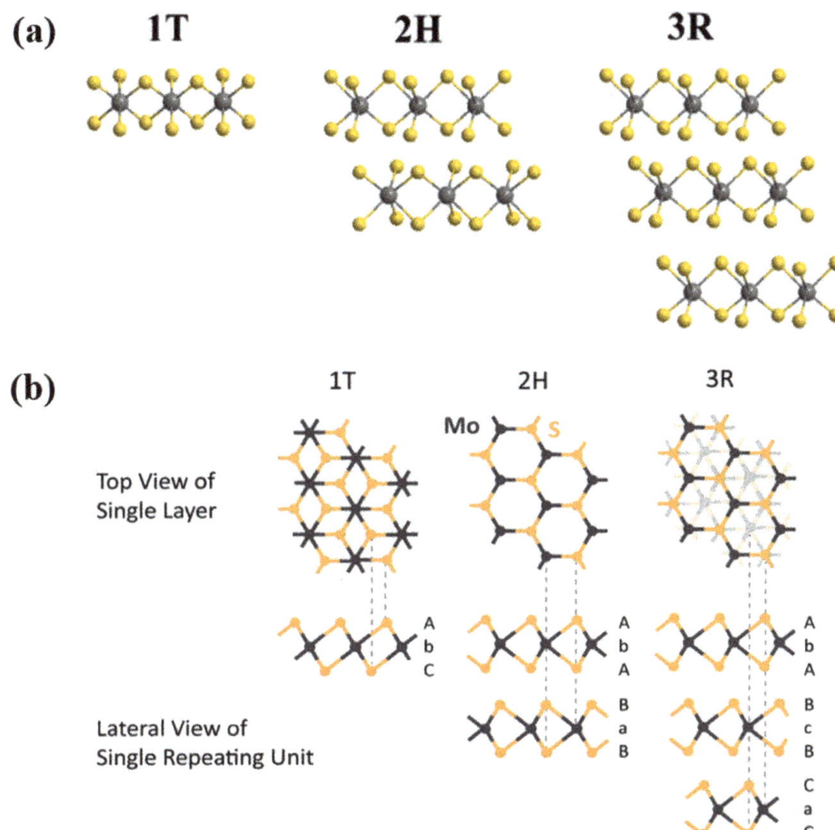

Figure 3. (a) Polytype structure of TMDs (1T, 2H, and 3R). Reprinted with permission from Coogan et al. [61] Copyright 2021, Royal Society of Chemistry. (b) Polytype structure of MoS$_2$. Reprinted with permission from Song et al. [62] Copyright 2015, Royal Society of Chemistry.

2.2. Advantages and Challenges of TMDs

Inspired by the great success of graphene, many two-dimensional (2D) materials with unique physical and chemical properties have recently gained significant attention [65–68]. In particular, layered TMDs have shown immense prospects for implementation in energy storage, catalysis, photonics, etc. [42,69,70]. Because of their large surface-to-volume ratio, which allows significantly increased interaction between the active material and electrolyte, graphene-like 2D TMDs are highly advantageous for battery applications [71,72]. Owing to the weak van der Waals (vdW) force, the ions can diffuse rapidly through the interlayer gap of MX$_2$ layer. The large interlayer distance between the MX$_2$ layer, in particular, makes it possible to accommodate multivalent ions such as Zn^{2+}, Mg^{2+}, Al^{3+}, and Ca^{2+} [51,72,73]. The interlayer distance and bandgaps of various TMDs commonly used in AMMIBs are shown in Figure 4.

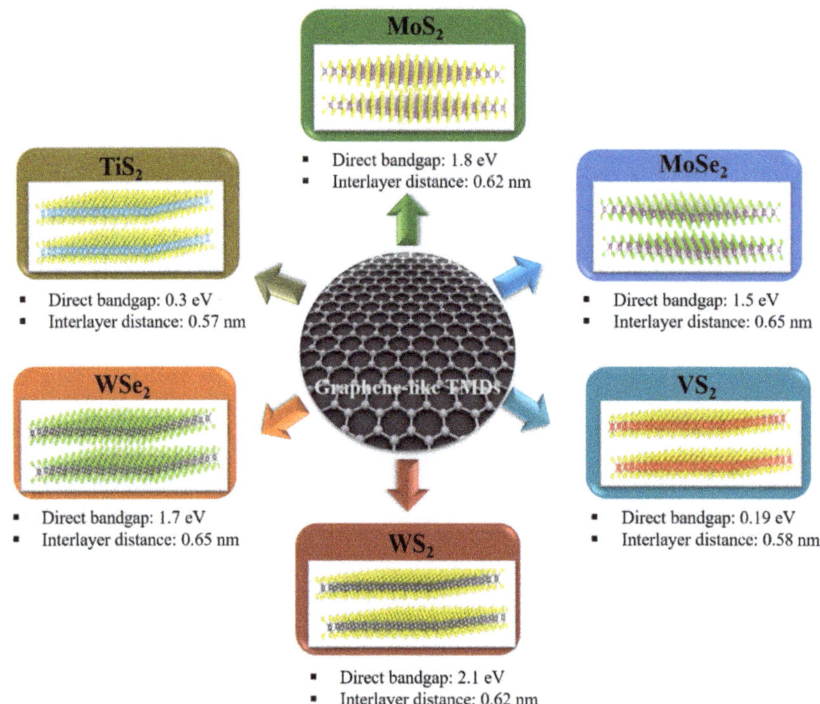

Figure 4. Direct bandgap and interlayer distance of various types of TMD.

Despite these benefits, the poor rate performance and cyclic stability of TMDs due to their low ionic conductance and large volume expansion limit their application as cathode materials in AMMIBs. Furthermore, the irreversible side effects during the charging and discharging processes decrease the coulombic efficiency of TMDs [59]. These disadvantages severely limit the application of TMDs as cathode materials for AMMIBs. For example, MoS_2 is considered as a suitable cathode for Zn^{2+} storage owing to its large interlayer distance (0.62 nm) compared to its much smaller size (0.15 nm). Nevertheless, as reported by Liu et al., bulk MoS_2 delivers a specific capacity of only 40 mAh g^{-1} because of the lack of clear redox peaks during electrochemical reactions [24]. This clearly shows that the interlayer distance of MoS_2 is not the only factor that guarantees the high performance of ZIBs. The more important issue is the efficient ion adsorption on the electrode, which influences the electromigration characteristics of multivalent ions (M^{z+}) in the aqueous state. Hence, the ionic radius in the hydrated state (i.e., $M^{z+}(H_2O)_n$) is more important than the general ionic radius (i.e., M^{z+}). In practice, the radius of hydrated Zn^{2+} ions (0.404–0.430 nm) can obstruct their absorption into the minimum interlayer distances that the host TMD material can accommodate [74]. Another important issue affecting the performance of TMD cathode materials is the variation in their structural features during the electrochemical reaction. Typically, the conductivity retained by the 1T MoS_2 phase is 10^7 times higher than that retained by the 2H MoS_2 phase. In addition, while the 1T MoS_2 phase is hydrophilic, the 2H MoS_2 phase is hydrophobic. These different features of 1T and 2H MoS_2 affect the adsorption and diffusion of multivalent ions in water, which eventually results in their distinct electrochemical performance in rechargeable aqueous ZIBs [75]. In addition to these fundamental aspects, the following section discusses the recent advancements of TMD cathode materials for AMMIBs.

3. TMD Cathode Materials for Multivalent Metal-Ion Batteries (MMIBs)

3.1. Zinc-Ion Batteries (ZIBs)

As can be observed from Table 1, among the different types of rechargeable metal-ion batteries, ZIBs possess the following appealing characteristics: (i) high abundance and low cost of zinc, facilitating their mass production, (ii) high redox potential (-0.763V vs. standard hydrogen electrode (SHE)), (iii) high theoretical volumetric energy density of zinc anodes (5855 mAh cm^{-1}), and (iv) high safety. These features make zinc ions suitable for use in both aqueous and non-aqueous batteries [23,24,76]. Recently, aqueous ZIBs (AZIBs) have gained considerable attention because of their safety, low cost, and environmental friendliness. In particular, the ionic conductance of Zn^{2+} in water (1000 mS cm^{-1}) is much higher than that in organic solvents (1–10 mS cm^{-1}). Nevertheless, Zn^{2+} ions easily coordinate with six water molecules, making the diffusion of Zn^{2+} into the electrode more difficult. As a result, since Liu et al. [24] first reported the poor diffusion of Zn^{2+} ions into bulk MoS_2, there have been a limited number of studies on this material with satisfactory results. For instance, Liang et al. proposed an increase in the interlayer distance and hydrophilicity of MoS_2 via oxygen incorporation using a hydrothermal method to enhance the inherent diffusion of Zn^{2+} into the MoS_2 layers [77]. The pristine MoS_2 layers were activated by adding oxygen (replacing sulfur with oxygen) because of its smaller atomic radius (48 pm) than that of S (88 pm). The vdW interaction between the interlayers was reduced by the formation of the Mo–O bond; thus, the interlayer distance widened from 6.2 to 9.5 Å (Figure 5a). As a result, the presence of a small amount of oxygen (5%) in the layered MoS_2 increased the Zn^{2+} diffusion coefficient by three orders of magnitude. Consequently, the activated MoS_2 had a high Zn^{2+} storage capacity of 232 mAh g^{-1} (increased by approximately 10 times after the oxygen activation), and showed pronounced redox peaks as compared to the unmodified counterpart (Figure 5b,c). Li et al. also reported the expansion of the interlayer distance through its vertical alignment on a carbon fiber fabric by hydrothermal reaction (Figure 5d) [78]. This novel structure made the assembly suitable for Zn^{2+} diffusion because of the following reasons: (i) The interfacial contact between MoS_2 and the electrolyte was improved through the 3D network of the carbon fibers, which facilitated the formation of Zn^{2+} diffusion pathways. (ii) The extended distance between the layers reduced the ionic diffusion resistance, thereby enhancing the reaction kinetics and reducing the energy barrier for Zn^{2+} diffusion. (iii) The carbon fibers derived from glucose facilitated the interaction between MoS_2 and the carbon-based substrate through a good conducting network. As such, the expanded MoS_2 delivered a discharge capacity of 202.6 mAh g^{-1} and excellent cycling performance (capacity retention of 98.6% after 600 cycles) along with a high rate capability (Figure 5e). They proposed the following Zn^{2+} storage mechanism: (1) at the cathode: $xZn^{2+} + x2e^- + MoS_2 \leftrightarrow Zn_xMoS_2$; (2) at the anode: $xZn - x2e^- \leftrightarrow xZn$.

On a different note, Liu et al. reported that the proportion of the 1T MoS_2 phase affects the performance of ZIBs [75]. The 1T MoS_2 phase content was controlled via a hydrothermal reaction over the temperature range of 140–220 °C. At 160 °C, the percentage of 1T MoS_2 was as high as 70%, which reduced significantly to 0.4% at 220 °C. The 1T MoS_2 phase showed a significantly lower energy barrier for Zn^{2+} than the 2H MoS_2 phase, further facilitating the diffusion of Zn^{2+}. Consequently, the MoS_2 with a high 1T phase content exhibited excellent electrochemical performance with a high capacity retention of 98.1% and a coulombic efficiency of ~100% after 400 cycles. These results suggest that the 1T MoS_2 phase can effectively accelerate charge transfer because of its lower diffusion energy barrier for Zn^{2+} than that of the 2H MoS_2 phase for the same interlayer distance. This study demonstrated the importance of phase control for MoS_2 which can influence the electron density in TMDs, eventually affecting the battery performance. Understanding the exact mechanism and the conditions regarding how different phases can trigger the phase transition can provide new opportunities to realize high-performance TMD electrodes in the future.

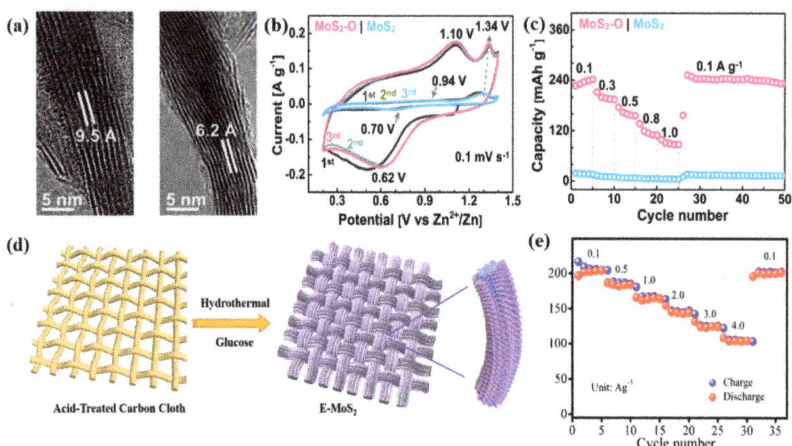

Figure 5. (**a**) Transmission electron microscopy (TEM) image of MoS$_2$ (left:MoS$_2$ with oxygen incorporation, right: bulk MoS$_2$). (**b**) cyclic voltammetry curves of MoS$_2$-O (pink) and bulk MoS$_2$ (light blue) at a scan rate of 0.1 mV s^{-1}. (**c**) Rate capability of MoS$_2$-O and bulk MoS$_2$ at various current densities. Reprinted with permission from Liang et al. [77] Copyright 2019, American Chemical Society. (**d**) Illustration for the preparation of E-MoS$_2$. (**e**) Rate capability of E-MoS$_2$ at various current densities. Reprinted with permission from Li et al. [78] Copyright 2018, Elsevier B.V.

Xu et al. applied defect engineering to prepare defect-rich MoS$_2$ nanosheets via a facile hydrothermal process accompanied by heat treatment [79]. The introduction of defects into nanomaterials can be promising strategies because they can store more foreign ions and enhance the electrochemical performance of these batteries. From many theoretical predictions, researchers have found that defects can increase the active sites in the electrode. The presence of defects increases the surface energy of the electrode and provides new active sites, which promotes the ion adsorption and increase the capacity. Numerous edge sites and vacancies were created in these nanosheets in a controlled manner. Indeed, these defects facilitated the diffusion of Zn^{2+} and significantly improved the reversibility of the activated MoS$_2$ as compared to its pristine counterpart. Additionally, the defects in MoS$_2$ could increase its interaction with Zn^{2+}, and at the same time, could extend its interlayer distance. Figure 6a shows the (100) plane stacking fault and disordered atomic disposition on the surface of the material. The inefficient plane stacking resulted in a significant increase in the edge spacing and a slightly increased interlayer distance (0.686 nm). In addition, a disturbance of the atoms caused cracking of the planes, leading to the formation of additional edges. The defect-rich MoS$_2$ accelerated the diffusion of Zn^{2+} owing to the formation of new transport pathways at the numerous edges and defects. As a result, a satisfactory discharge capacity of 102.4 mAh g^{-1} was obtained after 600 cycles at 500 mA g^{-1}, demonstrating the promising energy storage ability of MoS$_2$ (Figure 6b). Yang et al. reported novel MoS$_2$ nanosheets with a porous tubular structure prepared via template-assisted thermal decomposition [80]. As can be observed from the transmission electron microscopy (TEM) image (Figure 6c), the MoS$_2$ nanotube exhibited a layered structure with a interlayer distance of 0.65 nm. The tubular MoS$_2$ showed the following merits for Zn^{2+} diffusion: (i) the formation of voids in the structure of the MoS$_2$ nanosheets (the ratio of Mo and S atoms of 1:1.4) improved its ability to accommodate Zn^{2+}; (ii) the tubular structure of the MoS$_2$ nanosheets enhanced their electrolyte uptake and ion diffusion, which facilitated electron transport during the charge/discharge process. Thus, the novel MoS$_2$ nanosheets exhibited a good discharge capacity of 146.2 mAh g^{-1} at 200 mA g^{-1} and excellent cycling performance with 74.0% capacity retention after 800 cycles.

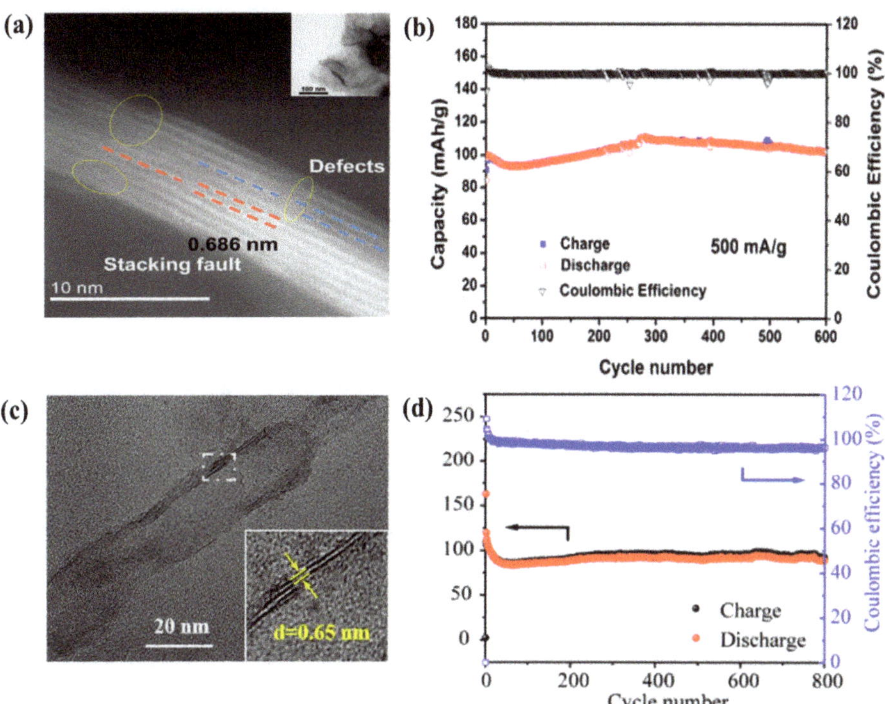

Figure 6. (**a**) TEM image of defect-rich MoS$_2$ nanosheets, (**b**) cyclic performance of defect-rich MoS$_2$ nanosheets at 200 mA g^{-1}. Reprinted with permission from Xu et al. [79] Copyright 2018, Elsevier B.V. (**c**) TEM image of tubular MoS$_2$, (**d**) cyclic performance of tubular MoS$_2$ at 500 mA g^{-1}. Reprinted with permission from Yang et al. [80] Copyright 2020, ESG Publication.

In addition to MoS$_2$, vanadium-based TMDs have also received considerable attention as promising materials for ZIB cathodes. For instance, He et al. prepared vanadium disulfide (VS$_2$) nanosheets via a facile hydrothermal reaction for the first time [81]. The layered VS$_2$ with high specific capacity and cyclic stability showed a huge potential for application in ZIBs. As shown in Figure 7a, the interlayer distance of the prepared VS$_2$ was as high as 0.576 nm, which facilitated the diffusion of Zn^{2+}. The as-prepared VS$_2$ exhibited a high discharge capacity of 190.3 mAh g^{-1} at 50 mA g^{-1} and showed good cyclic stability (retained 98.0% of the initial capacity after 200 cycles) (Figure 7b). They proposed the occurrence of the following electrochemical reactions: (1) at the cathode, VS$_2$ + 0.09Zn^{2+} + 0.18e$^-$ ↔ Zn$_{0.09}$VS$_2$, and Zn$_{0.09}$VS$_2$ + 0.14Zn^{2+} + 0.28e$^-$ ↔ Zn$_{0.23}$VS$_2$; (2) at the anode: Zn^{2+} + 2e$^-$ ↔ Zn.

Jiao et al. prepared a hierarchical 1T VS$_2$ directly on a stainless steel mesh substrate (VS$_2$@SS) without inactive materials such as binder and conductive additives to increase the active material content [82]. The open-flower structure of VS$_2$ improved the interaction between the active material and electrolyte, which contributed to the favorable ion and electron transport. The free-standing VS$_2$@SS electrodes showed several outstanding advantages: the VS$_2$ flower structure was suitable for accommodating volume expansion, thereby reducing the chances of electrode degradation. In addition, the layered structure of the VS$_2$ flower (interlayer spacing of 0.58 nm) was favorable for ion transport because it interacted well with the electrolyte and shortened the Zn^{2+} diffusion length. In general, the maximum electrochemical performance is achieved when the loading amount of the active material is small (below 3 mg cm^{-2}). However, in this work, even when the active material loading was 4–5 mg cm^{-2}, the VS$_2$@SS electrode showed high discharge capacity (198 mAh g^{-1}

at 50 mA g^{-1}) and excellent cycling performance over 2000 cycles at 2000 mA g^{-1}. With the increase in the mass loading, the thickness of electrode increases. An excessively high electrode thickness is likely to be fractured or even delaminated from the current collector during electrochemical reactions, resulting in the cycling instability. Moreover, the presence of binder and conductive additives in the conventional slurry-coated method reduces the gravimetric/volumetric energy density of the electrode. Therefore, the novel structure of VS$_2$@SS without binder and conductive additives could resolve this issue. The good cyclic stability of VS$_2$@SS was still achieved (90% retention for 600 cycles) even when the active material loading was increased to be 11 mg cm^{-2} in this study.

Wu et al. synthesized ultrathin vanadium diselenide (VSe$_2$) nanosheets via a wet-chemical route [83]. The aqueous ZIB exhibited excellent rate capability with the discharge capacities of 131.8, 114.6, 105.2, 93.9, and 79.5 mAh g^{-1} at the current densities of 100, 200, 500, 1000, and 2000 mA g^{-1}, respectively. When the current density returned to 100 mA g^{-1}, the recovered capacity was 118.4 mAh g^{-1} (Figure 7c). In addition, the VSe$_2$ nanosheets showed good specific capacity (131.8 mAh g^{-1} at 100 mA g^{-1}) and cycling performance (retained capacity of 80.8% after 500 cycles) (Figure 7d). The excellent performance of the VSe$_2$ nanosheet cathode could be attributed to the following reasons: (i) reversible intercalation/de-intercalation of Zn^{2+}; (ii) rapid Zn^{2+} diffusion dynamics in the ultrathin 2D structures; (iii) metallic features of VSe$_2$, which promoted its Zn^{2+} storage kinetics; and (iv) structural robustness during long-term cycling. The possible Zn^{2+} storage mechanism is as follows: (1) at the cathode: VSe$_2$ + 0.23Zn^{2+} + 0.46e$^-$ $\leftrightarrow$ Zn$_{0.23}$VSe$_2$, and Zn$_{0.23}$VSe$_2$ + 0.17Zn^{2+} + 0.34e$^-$ $\leftrightarrow$ Zn$_{0.4}$VSe$_2$; (2) at the anode: 0.4 Zn $\leftrightarrow$ 0.4Zn^{2+} + 0.8e$^-$.

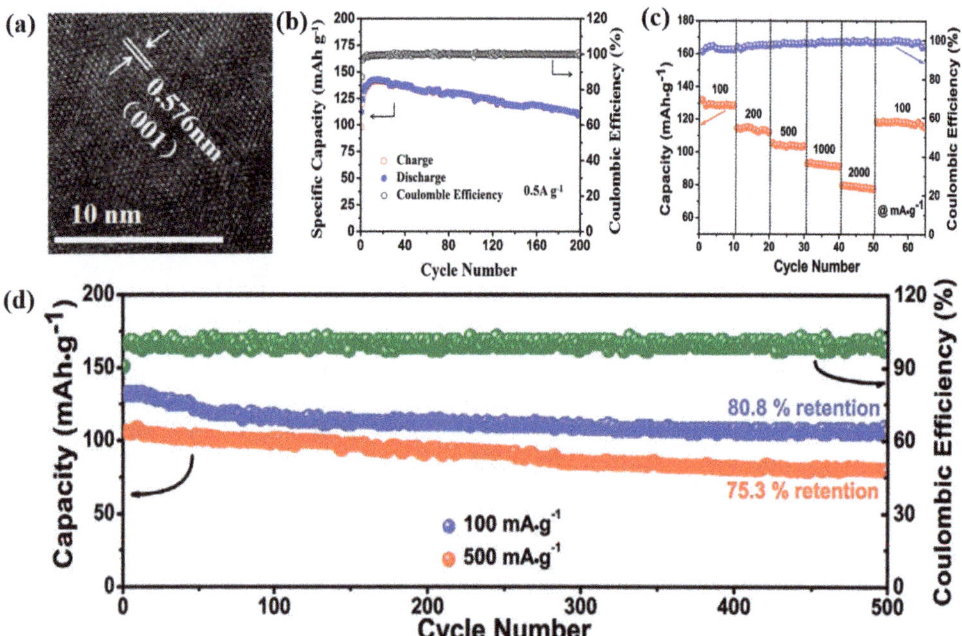

Figure 7. (**a**) TEM image of VS$_2$ nanosheets, (**b**) cyclic performance of VS$_2$ nanosheets at 500 mA g^{-1}. Reprinted with permission from He et al. [81] Copyright 2017,WILEY-VCH Verlag GmbH and Co. KGaA, Weinheim. (**c**) Rate capability of VS$_4$, (**d**) cyclic performance of VSe$_2$ nanosheets at 100 and 500 mA g^{-1}. Reprinted with permission from Wu et al. [83] Copyright 2020, Wiley-VCH GmbH.

Zhu et al. prepared a chain crystal framework of vanadium tetrasulfide (VS$_4$) via a hydrothermal method as a cathode material for ZIBs [84]. With a loosely stacked structure formed by the atomic chains of VS_4 bonded by a weak vdW force, VS_4 was considered as a potential cathode for ZIBs. The Zn^{2+} reaction mechanism was explained as follows: (1) at the cathode, $VS_4 + 0.49Zn^{2+} + 0.98e^- \leftrightarrow Zn_{0.49}VS_4$, and $Zn_{0.49}VS_4 + 0.54Zn^{2+} + 1.08e^- \leftrightarrow Zn_{1.03}VS_4$; (2) at the anode: $Zn \leftrightarrow Zn^{2+} + 2e^-$. As a result, the maximum capacity reached 110 mAh g^{-1}, and high capacity was obtained even after 500 cycles, which is favorable for practical applications. The electrochemical performances of the recently reported TMD cathodes for ZIBs are summarized in Table 2.

Table 2. Electrochemical performance of TMDs as zinc-ion batteries (ZIB) cathodes.

TMDs	Interlayer Spacing of Activated (Å)	Specific Capacity (mAh g^{-1})	Capacity Retention (%)	Cycle	Current Density (mA g^{-1})	Voltage Ranges (V)	Comments (Main Findings)	Ref.
MoS_2	-	18.0	-	50	50	0.1–1.9	Study of zinc ion storage in pristine TMDs	[24]
WS_2	-	22.0	-	50	50	0.1–1.9	Study of zinc ion storage in pristine TMDs	[24]
MoS_2-O	9.5	232	-	20	100	0.2–1.4	Reduction in intercalation energy barrier by oxygen incorporation	[77]
E-MoS_2	7.0	202.6	98.6	600	100	0.3–1.5	A novel structure of MoS_2	[78]
1T-MoS_2	6.8	-	98.4	400	100	0.25–1.25	Effect of different phase contents on the distinct performance	[75]
Defect-rich MoS_2	6.86	88.6	87.8	1000	100	0.25–1.25	Development of defect rich MoS_2 nanosheets	[79]
Tubular MoS_2	6.5	146.2	74	800	200	0.25–1.25	A novel structure of MoS_2	[80]
VS_2	5.76	110.9	98	200	500	0.4–1.0	Storage mechanism of Zn/VS_2	[81]
VS_2@SS	5.8	198	80	2000	2000	0.4–1.0	Binder-free hierarchical VS_2@SS electrode	[82]
VSe_2	6.1	131.8	80.8	500	100	0.25–1.50	Zinc-ion transport behavior in VSe_2 nanosheets	[83]
VS_4	5.83	110	85	500	2500	0.2–1.6	Energy storage mechanism of VS_4	[84]

3.2. Magnesium-Ion Batteries (MIBs)

Over the past few years, MIBs have been extensively studied owing to their environmentally friendliness, high abundance of Mg, high energy density, low reduction potential, and virtually non-dendrite formation [85]. The energy storage mechanism of MIBs is similar to that of ZIBs: intercalation and conversion processes. Although the ionic radius (0.72Å) and hydrated ionic radius of Mg^{2+} (4.28 Å) are similar to those of Li^+ (ionic radius of 0.76 Å and hydrated ionic radius of 4.3 Å of Li^+), most cathode materials for LIBs are not favorable for MIBs owing to the high charge density of Mg^{2+} [86]. Cathode materials for MIBs should exhibit high reversible capacities under adequate operating voltages. The kinetics of Mg^{2+} diffusion through the cathode material is usually slow because of the high valence of Mg^{2+} ions, redistribution of divalent cations in the host material, and strong ionic interactions. Various strategies have been reported to overcome these limitations.

Yang et al. carried out first-principle studies and proposed MoS_2 with a zigzag structure as a favorable cathode for rechargeable MIBs [87]. This study showed the detailed diffusion pathways for Mg^{2+} in the MoS_2 nanoribbons. The specific diffusion path of Mg^{2+} was shown to be through a nearest-neighbor H site in a zigzag manner between two adjacent T sites in MoS_2. This pathway in the MoS_2 nanoribbons was different from the conventional pathways of graphene nanoribbons, where Mg^{2+} ions undergo hopping between the intralayer hollow sites across the C–C bridge. The maximum theoretical capacity of a MoS_2 nanoribbon (width of 5 nm) that can accommodate six Mg atoms was predicted to be 232.2 mAh g^{-1}. Liang et al. synthesized highly exfoliated, graphene-like MoS_2 (G-MoS_2) as a cathode material for MIBs through a solvothermal process [88]. The synthesis of G-MoS$_2$ was from solvothermal reaction between MoO$_3$ and thioacetamide with pyridine as a solvent. The nature of few layer MoS$_2$ was confirmed by XRD which showed significantly reduced intensity of (002) peak for G-MoS$_2$ compared with bulk-MoS$_2$ (B-MoS$_2$). In addition, the interlayer distance of the prepared G-MoS$_2$ was 0.65–0.70 nm, which is larger than that of B-MoS$_2$ (0.63 nm) (Figure 8a). Apart from cathode side, the anode part was also studied; the ultra-small Mg nanoparticles (N-Mg, diameter of ~2.5 nm) was prepared by ionic liquid-assisted chemical reduction and compared with bulk Mg (B-Mg). The full cell performance of four different cells (G-MoS$_2$/N-Mg, G-MoS$_2$/B-Mg, B-MoS$_2$/N-Mg, and B-MoS$_2$/B-Mg) was systematically compared (Figure 8b). The results showed significantly improved cyclic performance for G-MoS$_2$/N-Mg cell in terms of specific capacity and cyclic stability. The measured Mg storage capacity of MoS$_2$ was estimated to be 170 mAh g^{-1} after 50 cycles. It was noted that this capacity was 24% lower than its theoretical capacity (232.2 mAh g^{-1}), which probably originated in the partial restacking of the MoS$_2$ layers and the intercalation of Mg only on one side of the MoS$_2$ layer. The high performance of G-MoS$_2$/N-Mg cell was attributed to the reduction in the passive film on the surface of N-Mg that increases the diffusion coefficient and increased available intercalation sites in g-MoS$_2$. Liu et al. fabricated MoS$_2$/C microspheres with a sandwich structure via a hydrothermal process and heat treatment [89]. The carbon formed by the hydrothermal glucose carbonization improved the ionic conductance and extended the interlayer spacing, facilitating the diffusion of Mg^{2+} and making the process reversible. In addition, the graphene-like MoS$_2$/C structure promoted the access of Mg^{2+} ions to the electrolyte, facilitating Mg^{2+} transport. As a result, the sandwiched MoS$_2$/C delivered a discharge capacity of 118.8 mAh g^{-1} after 20 cycles at 50 mA g^{-1} and exhibited excellent cycling stability as compared to bulk MoS$_2$. Truong et al. reported the rapid exfoliation of MoS$_2$ and molybdenum diselenide (MoSe$_2$) nanosheets using the supercritical fluid (SCF) method [90]. SCF exfoliation is a facile and fast process used to produce high-quality TMD nanosheets (few-layer). They applied this process to MoS$_2$ and MoSe$_2$ and obtained few-layer (1–10 layers) nanosheets with hexagonal structures (2H stacking sequence). The structure was precisely observed using atomic-resolution high-angle annular dark-field imaging. When used as MIB electrodes, the MoS$_2$ and MoSe$_2$ nanosheets delivered the specific capacities of 81 mAh g^{-1} after 10 cycles and 55 mAh g^{-1} after five cycles at 0.02 A g^{-1}, respectively, which are superior to those of pristine MoS$_2$ and MoSe$_2$.

Mao et al. studied layered MX$_2$ (M = Ti, V; X = O, S, Se) as a model to examine how different chalcogen species affect the Mg intercalation dynamics of MIBs [91]. TiSe$_2$ showed the best electrochemical performance among all the samples investigated owing to the following factors: (i) the interlayer spacing of TiSe$_2$ was greater than the size of Mg^{2+}; (ii) the vdW interaction between the basal planes of TiSe$_2$ was very weak, which facilitated the diffusion of Mg^{2+} into the layers; (iii) the electronic conductivity of TiSe$_2$ was higher than those of its counterparts. In addition, the authors proposed further modification methods to improve the intercalation kinetics of the compound, including: (i) using anions with less electronegativity, (ii) diminishing the electrostatic interaction between Mg^{2+} and the host (e.g., incorporation of monovalent anions on the host materials); (iii) developing an open-tunnel structure for the MX$_2$ to redistribute the charge and electric conductivity. This report provides guidelines for choosing and designing high-performance cathodes with rapid dynamics for MIBs. Gu et al. developed a MIB with a micrometer-sized TiSe$_2$ cathode operated at room temperature [92]. They found that the crucial factor that improves the reversible Mg^{2+}-intercalation/deintercalation is charge displacement in the metal-binding units, which is induced by the strong d-p orbital hybridization in TiSe$_2$. In the case of selenides, the highly overlapped d and p orbitals promote d-p orbital hybridization because of the larger 4p-orbital dimensions of selenides than those of oxides or sulfides. Furthermore, 2D ion-conducting channels in the gap between the basal planes of TiSe$_2$ reduce the coulombic repulsion among the Mg ions. This study has demonstrated the importance of d-p orbital hybridization between transition metal atom and the chalcogen atom in TMDs. The strong d-p orbital hybridization induced by their close energy levels might be one of the key factors, which improves the reversible intercalation/deintercalation of Mg^{2+}. It will be interesting whether such concept can be applicable to other types of multivalent ions (Zn^{2+} and Al^{3+}) for the TMD electrodes. Xu et al. employed a simple one-step hydrothermal method to synthesize a flower-like tungsten diselenide (WSe_2) nanosheet [93]. Figure 8c shows the lattice spacings of 0.68 and 0.28 nm, which correspond to the d-spacing of the (002) and (100) crystal plane of WSe_2, respectively. There have been only a limited number of studies utilizing WSe_2 for MIBs due to its low conductivity and structural instability that leads to the rapid degradation during electrochemical reactions. However, in this work, they overcame these by developing a novel structure (i.e., orderly flower structure) that effectively increased the contact area between electrode and electrolyte and allowed abundant ion channels through highly connected three-dimensional nanostructure. The WSe_2 cathode showed a high reversible capacity of 265 mAh g^{-1}, excellent cycling performance with a capacity retention of 90% after 100 cycles at 50 mA g^{-1}, and excellent rate capability with 70% capacity retention even at 500 mA g^{-1} (Figure 8d). The following Mg^{2+} storage mechanism was proposed: (1) at the cathode: $6WSe_2 + 4Mg^{2+} + 8e^- \leftrightarrow Mg_4W_6Se_{12}$; (2) at the anode: $4Mg \leftrightarrow 4Mg^{2+} + 8e^-$. The hydrothermal method is one of the common processes for the synthesis of novel TMD structures. The property of TMD is strongly related to its phase, size, morphology, and crystallinity. These features can be controlled by rational design and careful tuning of the hydrothermal process conditions. Therefore, it is important to understand the general mechanism of the hydrothermal growth of TMD (e.g., how parameters such as the precursor, substrate, additive, temperature, reaction time, and solvent affect the growth of TMD materials). The electrochemical performances of the recently reported MD-based MIB cathodes are summarized in Table 3. Based on Table 3, it can be clearly seen that the new structure design of TMD is one of the important strategies to create high-performance cathodes in MIBs. To prevent the restacking of TMD nanosheets, one of the most effective approaches is to prepare layered TMD nanomaterials with a hierarchical structure. In this hierarchical structure, the high surface-to-volume ratio offers abundant electrochemical active sites for ion storage, shortens the diffusion distance of Mg^{2+} ions, and alleviates the volume change of layered TMD nanomaterials by the presence of large voids during the repetitive charge/discharge processes. In this context, the hierarchical WSe_2 nanoflower has been demonstrated as a promising material in terms of the specific capacity and capacity retention (Table 3) [93]. Moreover, the carbon coating of layered TMD nanomaterials is another effective strategy because it can give high electrical conductivity and good elasticity for TMD materials.

Table 3. Electrochemical performance of TMD cathodes for magnesium-ion batteries (MIBs).

TMDs	Interlayer Spacing of Activated (Å)	Specific Capacity (mAh g^{-1})	Capacity Retention (%)	Cycle	Current Density (mA g^{-1})	Voltage Range (V)	Comments (Main Findings)	Ref.
Zigzag MoS_2	-	170	-	-	-	-	Study of Mg adsorption sites with DFT calculations	[87]
G-WS_2	0.7	161.5	95	50	20	0.5–3.0	A novel structure of G-MoS_2 cathode and ultra-small Mg nanoparticle anode	[88]
MoS_2/C	1.07	118.8	-	50	50	0–2.4	A novel structure of MoS_2/C cathode and AZ31 Mg alloy anode	[89]
Bulk MoS_2	-	81	68.7	10	20	0.2–2.2	Exfoliation of TMD into high-quality nanosheets	[90]
Bulk $MoSe_2$	-	55	73.3	5	20	0.2–2.2	Exfoliation of TMD into high-quality nanosheets	[90]
TiS_2	-	80	63.0	40	5	0.5–2.0	Kinetic study of Mg^{2+} migration in layered TMD	[91]
$TiSe_2$	-	86	74.8	40	5	0.5–2.0	Kinetic study of Mg^{2+} migration in layered TMD	[91]
1T-$TiSe_2$	-	108	-	40	-	0.25–1.8	Study of phase effect in $TiSe_2$ on the battery performance	[92]
WSe_2	6.8	239	91.9	100	50	0–2.5	A novel WSe_2 structure	[93]

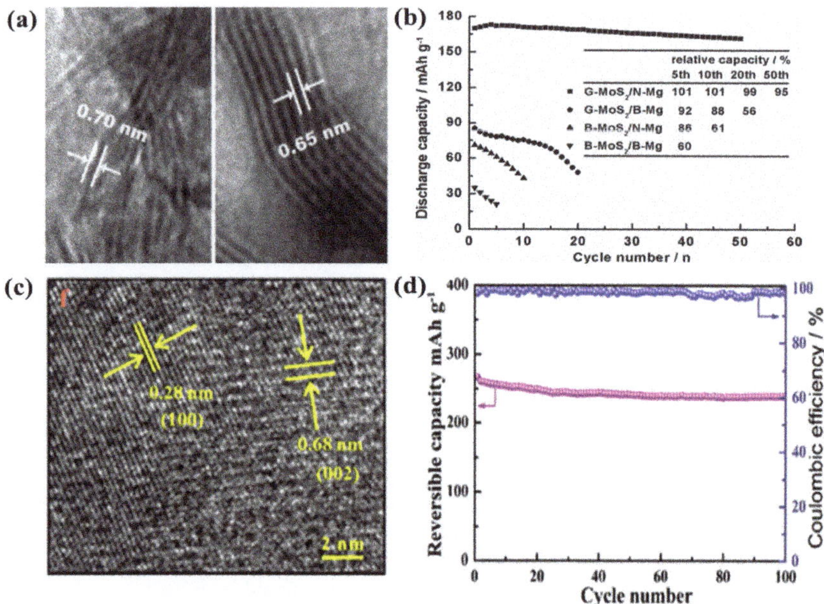

Figure 8. (**a**) TEM image of G-MoS$_2$ nanosheet, (**b**) cyclic performance of G-MoS$_2$ and B-MoS$_2$ with Mg nanoparticle (N-Mg) and bulk Mg (B-Mg) anodes at 20 mA g^{-1}. Reprinted with permission from Liang et al. [88] Copyright 2011, WILEY-VCH Verlag GmbH and Co. KGaA, Weinheim.(**c**) TEM image of WSe$_2$ nanosheets, (**d**) cyclic performance of WSe$_2$ nanosheets at 50 mA g^{-1}. Reprinted with permission from Xu et al. [93] Copyright 2020, Elsevier Inc.

3.3. Aluminum-Ion Batteries (AIBs)

The concept of rechargeable AIBs was first discovered in 1970 [94]. AIBs offer many advantages such as low cost, high safety, and high electrochemical energy storage. Nevertheless, AIBs suffer from many challenges, including low ionic diffusion [95], material disintegration [96], poor durability [97–103], and the formation of passive oxide layers [46]. The energy storage mechanism of AIBs is similar to that of the other AMMIBs (the conversion and intercalation reactions); however, the hydrated ionic radius of Al^{3+} (4.75 Å) is larger than those of other multivalent charge carriers (Zn^{2+}, Mg^{2+}, and Ca^{2+}). Thus, it is difficult to select a suitable cathode material for AIBs.

Fan et al. prepared interlayer-expanded MoS$_2$ nanosheets on graphene foam via a hydrothermal process as a cathode material for AIBs [104]. They expanded the interlayer distance of the MoS$_2$ nanosheets to 1.0 nm (Figure 9a), which improved the structural stability of the nanosheets and prevented large volume changes due to the facile intercalation of Al^{3+}. Therefore, the diffusion barrier of Al^{3+} and ion trapping were greatly reduced, thus increasing the number of ion storage locations. The amorphous carbon used in this study was formed by heat treatment at 450 °C, which enhanced the interaction between the MoS$_2$ sheets and graphene foam, resulting in the formation of a highly conductive three-dimensional (3D) structure. As a result, the MoS$_2$ nanosheets exhibited a specific capacity of 105 mAh g^{-1} after 20 cycles, and the capacity decreased gradually to 87.6 mAh g^{-1} after 120 cycles at 200 mA g^{-1} (Figure 9b). Li et al. prepared MoS$_2$ microspheres as a cathode material for AIBs via a simple hydrothermal process [105]. They analyzed the Al^{3+} intercalation and deintercalation sites in the MoS$_2$ microspheres as the S–Mo–S bonding (A1) and vdW gap (A2) sites. During the charge insertion/deinsertion process, the A1 sites tended to lose their capacity and underwent a phase transition, whereas the A2 sites were more stable for the (de)intercalation of the Al^{3+} ions. They described the process of (de) intercalation of Al^{3+} as follows: (i) the discharge process: Al^{3+} initially intercalated at the A1 sites, then continued to intercalate at the A2 sites; (ii) charge process: Al^{3+} first deintercalated from the A2 sites and finally from the A1 sites of the cathode. The high electrostatic interaction between Al^{3+} and S^{2-} anionic network in S-Mo-S (A1 site) led to the electrochemical polarization of MoS_2, which was not favorable for the reversible intercalation/deintercalation of Al^{3+}. This electrostatic interaction causes loss in capacity and phase transition due to high electrochemical polarization of MoS_2. On the other hand, Al^{3+} ions adsorbed at A2 site (constructed by the van der Walls force with a small electrostatic effect) tend to undergo more reversible intercalation/deintercalation. As a result, the AIB exhibited the excellent discharge capacities of 253.6 and 66.7 mA h g^{-1} at 20 and 40 mA g^{-1}, respectively, after 100 cycles. Similar results were reported by Divya et al. [106] Al^{3+} intercalated into two vacant sites, M1 and M2, where M1 is the X site in X-Mo-X atoms (X: S, Se, and SSe) and M2 represents the interlayer distance of MoX_2 (Figure 9c). The phase transformation after the first few intercalations/deintercalations led to the accommodation of more Al^{3+} ions, resulting in a higher capacity than that of the original MoS_2. On the other hand, this 2H-to-1T structural change led to potential deformation and disorder, resulting in an unstable capacity depending on the X atoms in MoX_2. As a result, while $MoSe_2$ showed a high coulombic efficiency of 95% at the current density of 100 mA g^{-1}, the coulombic efficiencies of MoS_2 and MoSSe were much lower (Figure 9d).

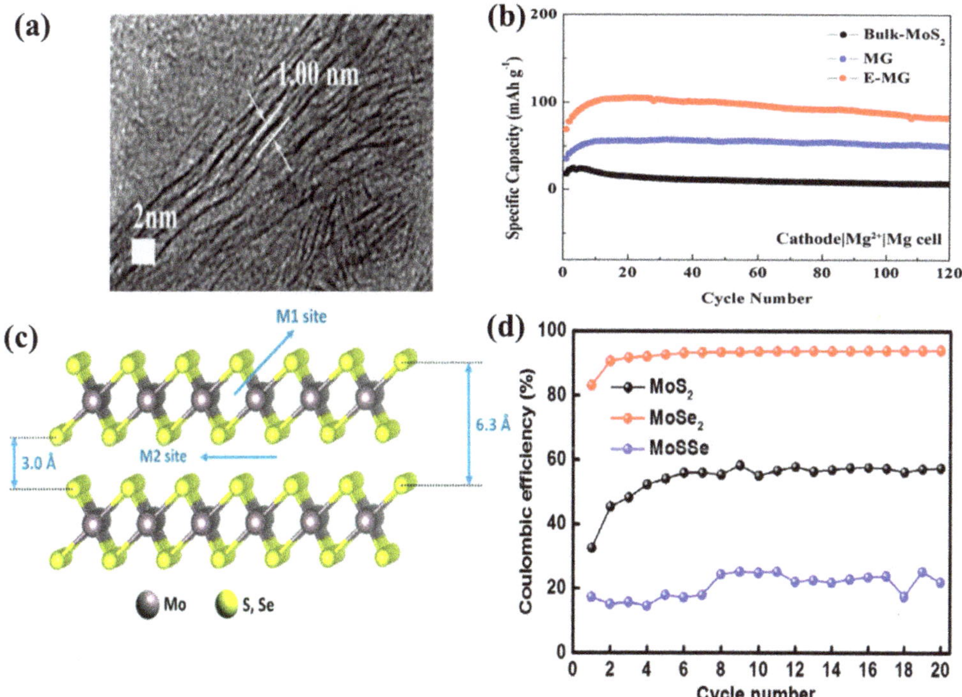

Figure 9. (a) TEM image of freestanding MoS_2-graphene foam composite with glucose (E-MG), (b) cyclic performance of bulk-MoS_2, MoS_2-graphene foam without glucose (MG), and E-MG at 20 mA g^{-1}. Reprinted with permission from Fan et al. [104] Copyright 2017, WILEY-VCH Verlag GmbH and Co. KGaA, Weinheim. (c) Schematic of the MoX_2 structure (X: S, Se) with M1 and M2 site, (d) coulombic efficiency of MoX_2 at 100 mA g^{-1}. Reprinted with permission from Divya et al. [106] Copyright 2020, WILEY-VCH Verlag GmbH and Co. KGaA, Weinheim.

Wu et al. prepared VS_2 nanosheets as a highly-efficient cathode material for AIBs by physically depositing a coating on graphene (G-VS_2) [107]. Owing to the synergistic effect of the modified layered VS_2 and graphene (layered spacing of 5.75 Å), the AIB with G-VS_2 showed significantly improved electrochemical performance as compared to that with pristine VS_2. Owing to its highly stable framework, G-VS_2 provided a good support for ion diffusion and improved electron transport properties. As a result, G-VS_2 delivered a discharge capacity of 50 mAh g^{-1} at 100 mA g^{-1} with a coulombic efficiency of ~100% after 50 cycles (vs. 22 mAh g^{-1} of pristine VS_2 under the same conditions). The Al^{3+} storage mechanism of G-VS_2 is as follows: (1) at the cathode: $VS_2 + xAl^{3+} + 3xe^- \leftrightarrow Al_xVS_2$; (2) at the anode: $Al + AlCl_4^- \leftrightarrow Al_2Cl_7^-$.

Geng et al. investigated the insertion/extraction behavior of Al^{3+} in layered titanium disulfide (TiS_2) and spinel-based cubic $Cu_{0.31}Ti_2S_4$ in AIBs [108]. In their work, the Al^{3+} ions occupied mainly the octahedral sites in the layered TiS_2 owing to the less adaptive nature of Al^{3+} ions in the layered TiS_2. The major obstacle for the facile intercalation/deintercalation of Al^{3+} ions was associated with the Al^{3+} diffusion coefficient, as revealed by galvanostatic intermittent titration analysis results. The strong coulombic interaction between the Al^{3+} ions and anionic sulfide sites increased the energy barrier for Al^{3+} diffusion. This effect was more pronounced in the spinel-based cubic $Cu_{0.31}Ti_2S_4$ than in the layered TiS_2. As a result, the discharge capacity of the layered TiS_2 at 50 °C was considerably higher than that at room temperature. The capacity of the layered TiS_2 increased gradually after the first few cycles and stabilized at approximately 70 mAh g^{-1}, whereas the discharge capacity of the cubic $Cu_{0.31}Ti_2S_4$ at 50 °C was only approximately 25 mAh g^{-1} (Figure 10a,b). Geng et al. also prepared Mo_6S_8 particles using a precipitation procedure [109]. In this study, they not only demonstrated the good performance of Mo_6S_8, but elucidated the two distinct Al adsorption sites in chevrel phase Mo_6S_8. From Figure 10c, the larger site (Al_1) can be visualized as a cubic center of a hexahedron with eight Mo_6S_8 units as the vertices, whereas the smaller site (Al_2) can be visualized as a face centered hexahedron. Al^{3+} ions could occupy the Al_1 position more easily than the Al_2 position because of the strong electrostatic interaction between the Al^{3+} ions at the Al_1 position. Al^{3+} ions could occupy the Al_1 position more easily than the Al_2 position because of the strong electrostatic interaction between the Al^{3+} ions at the Al_1 position. Although the number of available sites for the Al intercalation was six on the hexahedron, the actual number of Al filled in the sites was only two ($Al_2Mo_6S_8$) due to the strong electrostatic force from Al cation with three positive charges. The Al^{3+} storage mechanism of the Mo_6S_8 particles was described as follows: (1) at the cathode: $8[Al_2Cl_7]^- + 6e^- + Mo_6S_8 \leftrightarrow Al_2Mo_6S_8 + 14[AlCl_4]^-$; (2) at the anode: $Al + 7[AlCl_4]^- \leftrightarrow 4[Al_2Cl_7]^- + 3e^-$. The capacity of Mo_6S_8 was rapidly stabilized after the first cycle and maintained at 70 mAh g^{-1} after 50 cycles (Figure 10d). The electrochemical performances of the recently reported MD-based AIBs cathodes are summarized in Table 4. At present, it is still challenging to select suitable cathode materials for AIBs because the hydrated ionic radius of Al^{3+} (4.75 Å) is larger than those of other multivalent charge carriers. Therefore, most researches regarding AIBs with TMDs are only in the immature stage of understanding the reversible intercalation and extraction of Al in various TMDs materials. Based on Table 4, MoS_2 and G-VS_2 have been demonstrated as impressive candidates for AIB cathodes in terms of specific capacity and capacity retention. Similar to other types of AMMIBs (ZIBs and MIBs), the control of MoS_2 nanostructure (MoS_2 on graphene foam [104] or MoS_2 microsphere [105]) is efficient strategy in realizing good performance in AIBs. As for VS_2, it possesses the potential due to its high theoretical capacity based on the multiple chemical oxidation states in vanadium and sulfide atom. Besides, the interlayer space of VS_2 is 5.76 Å, which is large enough to enable the facile intercalation/de-intercalation of Al^{3+} ions. Furthermore, the strategy of VS_2 hybridization with graphene (G-VS_2) is one of the promising approaches to improve the performance because it can facilitate the Al^{3+} ion transport and reduce the restacking in VS_2 cathode materials [107].

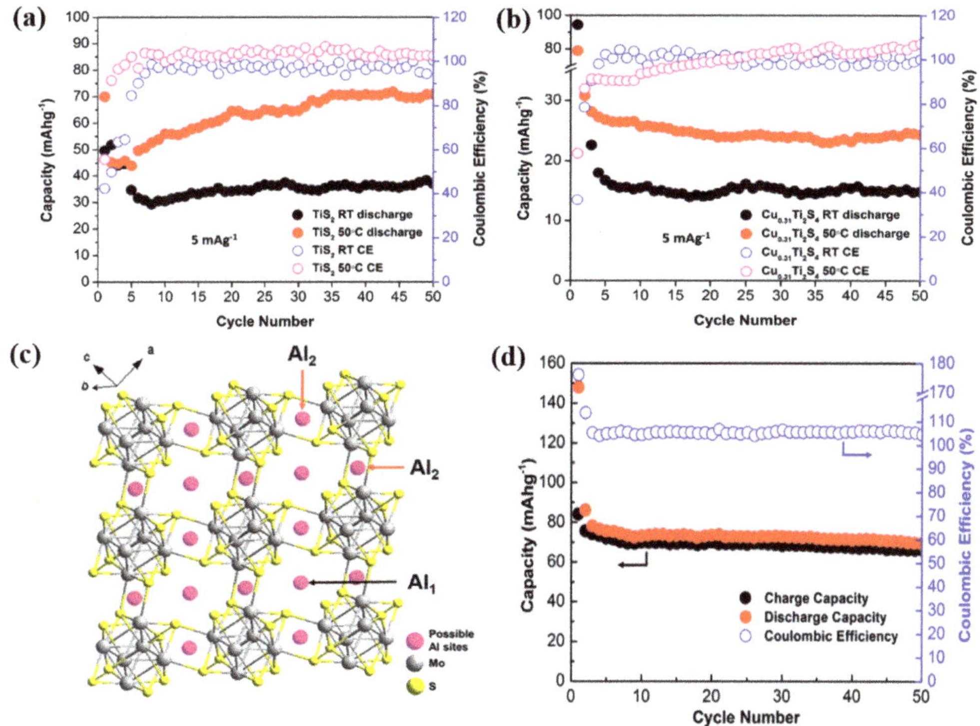

Figure 10. Cyclic performance of (**a**) TiS$_2$ and (**b**) Cu$_{0.31}$Ti$_2$S$_4$ at room temperature and 50 °C at 5 mA g^{-1}. Reprinted with permission from Geng et al. [108] Copyright 2017, American Chemical Society. (**c**) Schematic structure of Al insertion sites in Mo$_6$S$_8$, (**d**) cyclic performance of Mo$_6$S$_8$. Reprinted with permission from Geng et al. [109] Copyright 2015, American Chemical Society.

Table 4. Electrochemical performance of TMD cathodes for aluminum-ion batteries (AIBs).

TMDs	Interlayer Spacing of Activated (Å)	Specific Capacity (mAh g^{-1})	Capacity Retention (%)	Cycle	Current Density (mA g^{-1})	Voltage Range (V)	Comments (Main Findings)	Ref.
TiS$_2$	-	70	-	50	5	0.2–1.3	Reversible insertion and extraction of Al in TiS$_2$	[108]
MoS$_2$ (E-MG)	1.0	87.6	-	120	20	0.01–2.0	A novel structure of MoS$_2$ (E-MG)	[104]
MoS$_2$	6.2	66.7	-	100	40	0.5–2.0	Phase transition mechanism during the charge-discharge process in MoS$_2$	[105]
MoS$_2$	6.3	30	-	-	100	0.01–2.5	Intercalation mechanism of Al^{3+} into MoS$_2$	[106]
G-VS$_2$	5.75	50	33.6	50	100	0.3–1.7	Intercalation mechanism of Al^{3+} into G-VS$_2$	[107]
Mo$_6$S$_8$	-	70	-	50	-	0.1–1.2	Reversible intercalation and extraction of Al^{3+} in Mo$_6$S$_8$	[109]

4. Modification Strategies for TMDs toward High-Performance Aqueous Multivalent Metal-Ion Batteries (AMMIBs)

The results discussed thus far emphasize the need to activate TMD cathodes to achieve good electrochemical performance for AMMIB applications. Although TMD materials have large interlayer spacing, their high energy barrier of multivalent ions limits their application as high-performance cathodes. Many modification strategies have been proposed to create structures that can realize the high electrochemical performance of TMD materials.

In this section, we have summarized the modification strategies used to improve the structural stability of TMD materials in order to realize their AMMIB applications. The main modification strategies for TMDs include: (1) interlayer modification, (2) defect modification, (3) hybridization, and (4) phase modification.

4.1. Interlayer Modification

The metal ion (M^{n+}) intercalation channel in a cathodc crystal significantly affects its electrochemical behavior. Therefore, the adjustment of cation channels and coordination in the interlayer not only encourages the intercalation/deintercalation of M^{n+}, but also enhances the electrochemical performance of the TMD cathode. The introduction of intercalants into 2D layered TMD materials is a viable strategy for enhancing their electrochemical performance [110].

Small guest intercalants (SGIs) such as ions, small molecules, and polymers can alter the interlayer spacing of TMDs [58,111–114]. Nevertheless, the following factors need to be considered for the incorporation of these small intercalants into the interlayer spacing of TMDs: (i) SGIs can occupy the active site and block the cationic pathway. This interaction may deteriorate the diffusion kinetics of the cations. Thus, a rigorous management of the quantity, geometry, functional structure, and interaction sites in the interlayer spacing should be taken into account to ensure effective cation diffusion; (ii) the integrity of SGIs in the TMD framework is crucial for the long lifespan of the electrode. Small molecules and low-valence ions are more susceptible to deintercalation from the lattices than high-molecular-weight polymers or high-valence ions; (iii) the introduction of SGIs with unnecessary mass and volume leads to a decrease in the theoretical power and energy densities of the electrode. In general, the interlayer modification of cathodic materials should be carried out carefully to ensure cationic intercalation and deintercalation without negatively affecting the electrochemical properties of the electrode [115].

Chemical vapor transport, electrochemical treatment, ion exchange, and oxide-reduction methods have been used to intercalate SGIs into TMDs [77,112]. The insertion of SGIs can modify the electron filling state in the orbital and Fermi stages of TMDs, often resulting in unusual properties such as superconductivity, charge density waves, and Peierls instability (characteristic phenomenon of one-dimensional electron-lattice systems). The increased interlayer distance is extremely effective in lowering the M^{n+} intercalation energy barrier. In view of these benefits, there have been several reports on using various SGIs to modify the interlayer spacing of TMDs. The benefits and drawbacks of intercalation strategies (intercalation of hydrophilic and carbon species and exfoliation technique) are discussed in the following sections.

4.1.1. Intercalation of Hydrophilic Species into TMDs

The introduction of an oxygen atom into MoS_2 has been suggested as an effective strategy for lowering the energy barrier of Zn^{2+}, enhancing the intrinsic diffusion and facilitating structural changes in layered MoS_2 (Figure 11). Because the interlayer distance of pristine MoS_2 (3.1 Å) is too small, the insertion of large Zn^{2+} hydrate molecules (5.5 Å) is difficult. According to density functional theory (DFT) calculations, a 66 kcal mol^{-1} energy per coordination (Zn–O) bond is required to break a hydrated Zn^{II}–OH$_2$ bonds to form Zn^{II}–S bonds, which is a large energetic penalty (Figure 11a). However, the energetic requirement can be significantly reduced when the interlayer distance is increased by introducing OH functional groups to form Zn–OH bonds without breaking the water molecules complexed with Zn^{2+} ion. The interlayer distance of MoS_2-O can be increased to accommodate hydrated Zn^{2+} molecules (five water molecules complexed with Zn^{2+} to form a solvated shell, as shown in Figure 11b). In addition, MoS_2 possesses hydrophobic properties, which make the intercalation process challenging owing to the weak interaction between Zn-H$_2$O and S. To resolve this, Liang et al. used oxygen atoms to replace the sulfur atom in MoS_2, which improved the hydrophilic properties of MoS_2 and enhanced the interaction of Zn-H$_2$O-O in it (Figure 11c). Figure 11d shows the effective energy required depending on the hydration level of Zn^{2+} cations for intercalation into MoS_2 and MoS_2-O. An increase in the interlayer distance of MoS_2 favors a reduction in the number of $Zn–H_2O$ bonds, resulting in a reduced energy barrier. The interlayer modification method used by Liang et al. can be utilized to improve the ion storage efficiency of TMDs, creating new trends for the development of advanced materials for energy storage devices.

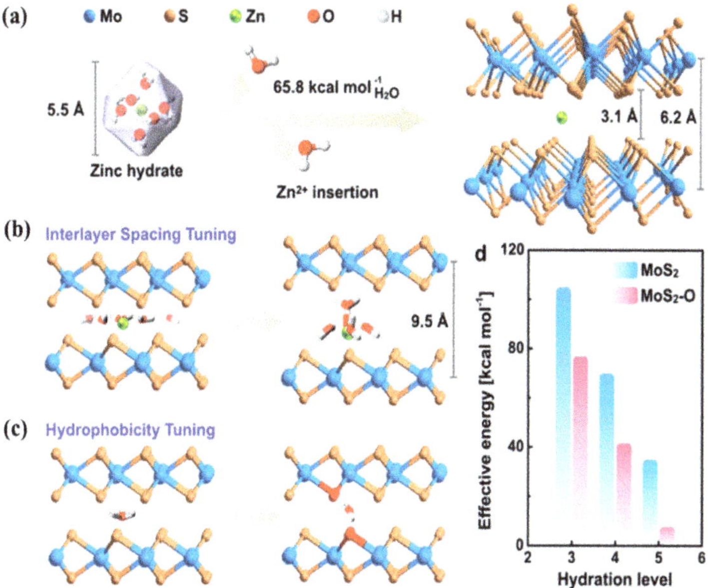

Figure 11. (**a**) Schematic showing the difficulty in the intercalation of Zn hydrates into bulk MoS_2 owing to the large energy barrier between the layers, (**b**) the expanded interlayer distance that supports the diffusion of Zn^{2+}, (**c**) hydrophobicity control by the $Zn–H_2O–O$ interaction, (**d**) theoretical energy barrier between MoS_2 and MoS_2–O depending on the hydration level of Zn^{2+}. Reprinted with permission from Liang et al. [77] Copyright 2019, American Chemical Society.

4.1.2. Exfoliation

Liquid-phase exfoliation is a direct chemical method, which has been demonstrated to be an appropriate method for the production of industrial-scale nanosheets [116–121]. Exfoliation under slight sonification [118] or high-shear mixing [121] in stable organic solvents or water-based surfactants can be utilized to prepare TMD nanosheets. Sonification-assisted liquid-phase exfoliation in organic solvents is a facile procedure for the synthesis of TMDs with few layers. The energy required for the exfoliation process is reduced owing to the strong interaction of the solvent with the bulk materials and the solvation process. The solvents used in the exfoliation method should have high solvation power. However, the limitations of this method include sensitivity to the ambient environment, structural deformation, and changes in the electronic properties.

Recently, the SCF method has been widely used to prepare high-quality graphene and inorganic nanosheets for industrial applications [122–124]. The main advantage of this method is that it can disperse the products without changing their original nature [122]. The SCF method was recently utilized to exfoliate MoS_2 into few-layer nanosheets. Solvents used in the SCF method should have low surface tension and high diffusion coefficient, serving as a good medium to exfoliate TMDs. Truong et al. used a SCF to create few-layer (1–10) MoS_2 and $MoSe_2$ nanosheets. They demonstrated that the exfoliated nanosheets of both MoS_2 and $MoSe_2$ could retain their 2H stacking sequence during the exfoliation process (Figure 12a), [90]. Yang et al. systematically investigated the absorption of Mg on zigzag MoS_2 nanoribbons. The results indicated that ionic bonds were formed predominantly through interactions between the guest Mg atoms and MoS_2 nanoribbon substrate, while some covalent hybridizations still existed simultaneously to some extent. The T position at the edge of the nanoribbon was the most stable for the absorption of the guest Mg atoms. The Mg diffusion pathway was identified as two adjacent T and H positions on the zigzag MoS_2 nanoribbons (Figure 12b). As a result, the MoS_2 nanoribbons exhibited a maximum theoretical capacity of 223.2 mAh g^{-1} [87].

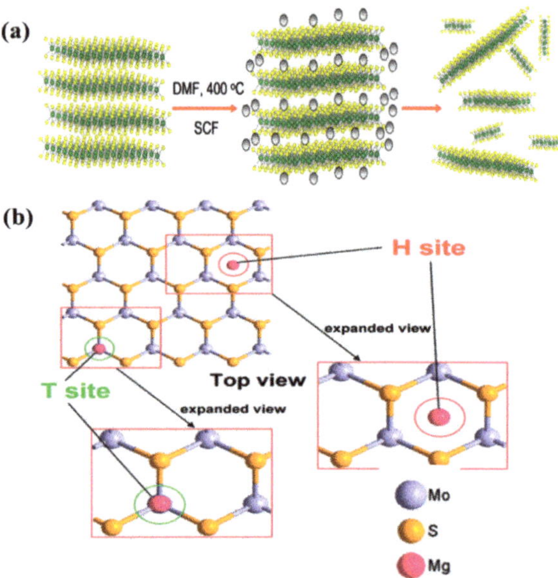

Figure 12. (**a**) Schematic diagram of the supercritical fluid (SCF) procedure to synthesize TMDs. Reprinted with permission from Truong et al. [90] Copyright 2017, American Chemical Society.(**b**) Illustration of two Mg adsorption positions (H and T sites) on MoS_2 nanoribbon:. Reprinted with permission from Yang et al. [87] Copyright 2012, American Chemical Society.

4.2. Defect Modification

In recent years, the defect technique has been widely used as an effective method to modify the surface properties and electronic structure of electrodes. Defect modification of the crystal structure proceeds according to the second law of thermodynamics [125–127]. Point defects, which play an important role in defect modification, can be divided into two categories: intrinsic and non-intrinsic defects [128–130]. The thermal vibration of the atomic lattice is responsible for the generation of internal defects, which do not affect the overall composition of the crystal. Schottky and Frenkel defects are examples of intrinsic defects [131,132]. Schottky defects are formed from lattice vacancies generated by the thermal vibration of atoms or ions in the crystal structure at the initial lattice position, whereas Frenkel defects are formed by the intercalation of atoms or ions into the lattice sites. Non-intrinsic defects are due to impurity atoms or ions being embedded into the lattice, and hence are also known as heteroatomic defects [133]. Point defects cause lattice distortion by disturbing the surrounding atoms, which can affect the electronic structure, chemical properties, or ionic conductance of the material, thereby regulating its electrochemical properties. The defects can be formed ranging from faults on an atomic scale that are inherent to crystallographic structures to larger defects that are introduced during fabrication process. Generally, it is difficult to control the defects caused by the fabrication process because they are non-uniform and unpredictable, leading to a serious failure during fabrication. The misapplication of the manufacturing process or lack of control at any stage may introduce defects and residual stresses that can affect the performance of the materials, making it susceptible to failure. These types of defects can include holes, cracks, segregation, inclusions, surface marks, notches and other undesirable or unintentional property changes within the material [134].

Defect sites can be the active, storage, and absorption sites that can immobilize the foreign ions during the reaction process. Defect modification affects the (de)intercalation of metal ions in layered materials, causing a decrease in the stress and electrostatic force between the adjacent layers, which reduces the diffusion energy barrier and promotes ion dissolution as well as charge transfer during metal ion intercalation. The main effects of defect modification in a material are as follows: (i) the ionic charge can be redistributed through defect engineering, thereby promoting the process of ion diffusion and electron transfer; (ii) the ion storage can be increased through the defect sites, leading to an increase in the capacity of the material; (iii) defect engineering also contributes to the enhanced dynamics in electrochemical reactions and phase transitions owing to the formation of multiple active sites in TMDs; (iv) defect engineering enhances the structural stability of TMD materials, making them resistant to structural damage caused by the insertion/extraction of foreign ions. As a result, the defect modification of TMDs may be a feasible approach to improve their electrochemical performance for application as ZIB cathodes. For instance, to facilitate Zn^{2+} intercalation in MoS_2, Xu et al. used defect engineering by controlling its multiple edge and void sites [79]. The defects so formed included exposed edge sites and defects within the basal planes of MoS_2. The defects formed were S vacancies combined with a disturbed atomic arrangement, resulting in the cracking of the basal plane and the formation of complementary edges or boundaries. These defects generated a large number of active sites in the electrode, which improved its electrochemical performance in the ZIB. Similar defect strategies have been applied in manganese oxide cathode materials (MnO_2). Xiong et al. [135] studied the use of oxygen deficient σ-MnO_2 as a cathode material for ZIB. Gibbs free energy of Zn^{2+} adsorption in the vicinity of oxygen deficient region can be reduced to a thermoneutral value ($\approx$0.05 eV) by generating oxygen vacancies (OV) in the MnO_2 lattice. This suggests that Zn^{2+} adsorption/desorption process on oxygen-deficient MnO_2 is more reversible as compared to pristine MnO_2. In addition, because fewer electrons are needed for Zn–O bonding in oxygen-deficient MnO_2, more valence electrons can be contributed to the delocalized electron cloud of the material, which promotes the attainable capacity. As a result, the stable Zn/oxygen-deficient MnO_2 battery was able to deliver one of the highest capacities of 345 mAh g^{-1} reported for a birnessite MnO_2 system.

4.3. Hybridization with Carbon

Hybridization of two or more materials is a technique that combines the advantages of all the components to create novel materials with improved functionalities and electrochemical performance [136,137]. Layered TMD nanomaterials are often hybridized with other materials such as carbon, metal sulfides, metal oxides, and conductive polymers. Among these, carbon-based materials are the most commonly used materials to form hybrids with TMDs [138]. The high electrical conductivity of carbon materials accelerates the electron transport in TMDs and reduces the diffusion energy barrier, thereby improving their electrochemical performance.

In the TMDs hybridized with carbon, the carbon can be a supporting material, coated material, and intercalated material depending on the role and its length scale. Supporting carbon means the carbon that acts as a matrix for TMD materials. It is generally used for a carbon in a more or less macroscopic scale. Carbon coating generally means the carbon coated on the TMD surface. Intercalated carbon means the carbon that has been intercalated into layered TMDs in much smaller length scale (less than 1 nm). For example, glucose was converted to a polysaccharide through hydrothermal synthesis with abundant hydroxyl functional groups (Figure 13a) [139,140]. After converting it to the graphene foam as a carbon support, MoS_2 was synthesized on this substrate to form a freestanding MoS_2/graphene foam. The carbon coating of TMD nanoplatelets with other hybrid materials provides a large surface area and effectively inhibits of the restacking of the nanoplatelets [141–143]. Carbon coatings often form a graphene-like framework that improves the electrochemical performance of the cathode in ZIBs [78], MIBs [89,104], and AIBs [107]. Besides, the coated carbon enhances the ionic conductance of the TMD, which facilitates the diffusion of M^{n+} through the 3D foam structure (Figure 13b). Intercalated carbon has also been explored as an effective material for the interlayer expansion of TMDs (Figure 13c). Li et al. demonstrated the increase in MoS_2 interlayer distance from 0.62 to 0.70 nm after carbon intercalation.

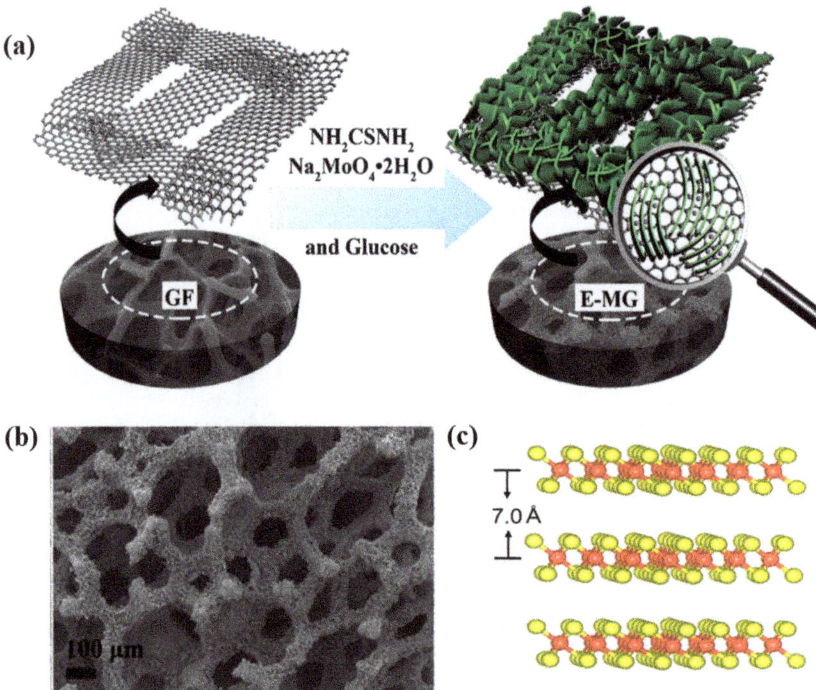

Figure 13. (**a**) Schematic for the synthesis of E-MG. (**b**) Scanning electron microscopy (SEM) image of free-standing MoS_2/graphene foam. Reprinted with permission from Fan et al. [104] Copyright 2017, WILEY-VCH Verlag GmbH and Co. KGaA, Weinheim. (**c**) The atomic structure with the expanded interlayer distance of MoS_2. Reprinted with permission from Li et al. [78] Copyright 2018, Elsevier B.V.

Over the past few decades, various methods have been proposed to control the morphology and structure of layered TMD hybrid materials. Hydrothermal/solvothermal synthesis methods are cost-effective and offer large product yields. The chemical vapor deposition method is a popular bottom-up method for developing high-quality layered TMD nanomaterials on many substrates. Exfoliation is a top-down method for preparing TMD nanosheets from bulk TMDs. As hybrids of layered TMDs with other materials show the merits of all the constituents, they are considered as potential candidates for application in energy storage devices [68,69,144]. Thus, the use of hybrid TMDs is an efficient approach for designing high-performance AMMIBs.

4.4. Phase Modification

The electronic structures of graphene and silicon are determined by the hybridization of their s and p orbitals, whereas the electronic structure of TMD layers is mainly affected by the d-orbital filling of the transition metal. The electron density of the d-orbitals in the transition metal affects the phase state of TMDs, and the electron filling of the d-orbitals causes phase modification in TMDs. Phase modification is often used to alter the electronic properties of a material. In general, phase modification in TMDs occurs incompletely, thereby resulting in the formation of mixed 1T and 2H phases.

As a representative TMD, MoS_2 possesses different structural stages depending on the metal coordination geometry, properties, and stacking order. For instance, bulk MoS_2 has two different phases: the 2H phase with a triangular prismatic coordinated geometry (bandgap of 1.3 eV) and the 1T phase with an octahedral coordination between Mo and S atoms (metallic properties), as shown in Figure 14a [63,145,146]. The ionic conductivity of the 1T MoS_2 phase is much larger than that of the 2H MoS_2 phase [147–149]. The different phases of MoS_2 affect its electrochemical performance. For example, the phase engineering of MoS_2 has been demonstrated to be an effective strategy for modifying its catalytic capacity [150–153]. Because of the different atomic and electronic structures of the hydrophobic 2H and hydrophilic 1T phases, the different concentrations of the two phases in MoS_2 can significantly affect the adsorption sites and diffusion pathways for Zn^{2+} ions in rechargeable aqueous ZIBs (Figure 14b). Liu et al. proposed an effective method to produce MoS_2 nanosheets with different phase contents through phase modification [75]. The MoS_2 nanosheets with a high 1T phase content (~70%) showed better specific capacity and cycling performance in ZIBs than those with the 2H phase. In addition, as the distance between the interlayers increased, the diffusion energy barrier of Zn also decreased (Figure 14c).

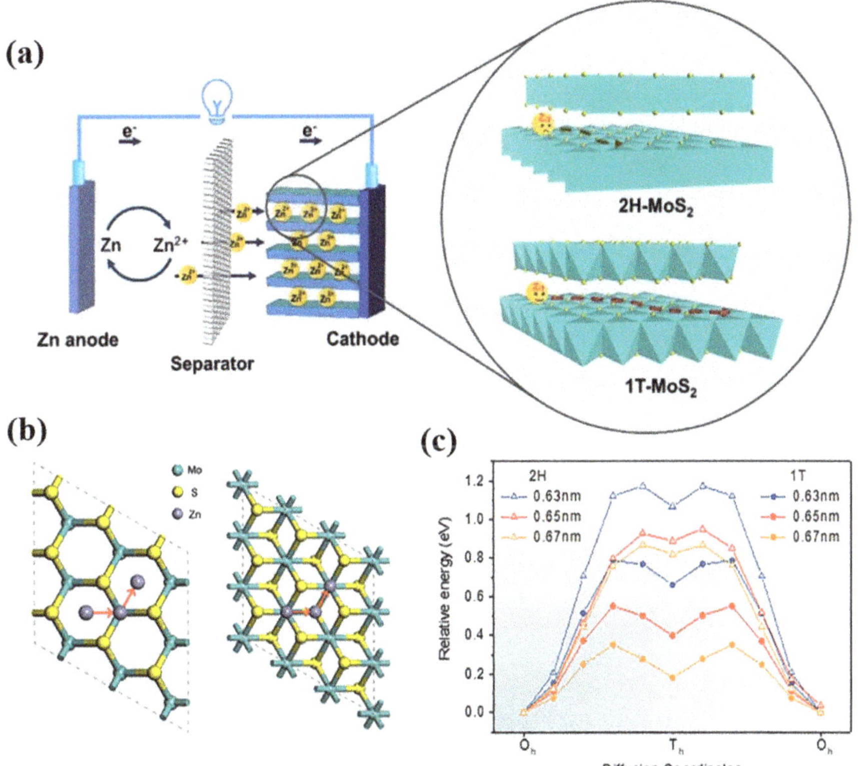

Figure 14. (**a**) Schematic of a rechargeable MoS_2 cathode with different phases (1T- and 2H-MoS_2). (**b**) The adsorption sites and diffusion pathway of Zn^{2+} (left:2H phase MoS_2 and right: 1T phase MoS_2). (**c**) Calculation of Zn^{2+} diffusion energy barrier on the1T and 2H phases. Reprinted with permission from Liu et al. [75] Copyright 2020, Elsevier B.V.

5. Conclusion and Outlook

In this review, the potential of layered TMDs for application as AMMIB cathodes is discussed. TMDs have many advantages over other cathode materials: (i) it is facile to control the interlayer spacing; (ii) ion transport is fast benefited from layer structure; (iii) they are abundant in nature and environmentally friendly; (iv) they can be utilized easily for flexible battery applications owing to its mechanical flexibility; (v) they can be produced on a large-scale using exfoliation of bulk TMDs (top-down approach) and sol-gel synthesis (bottom-up approach). However, there are still challenges that should be overcome when using TMD as a cathode material in AMMIBs: (i) the electronic conductivities of most TMDs are not sufficiently high due to the nature of the semiconductor with a certain bandgap; (ii) a highly dispersed TMDs can be restacked during repeated charge/discharge processes due to interfacial instability among TMDs with few layers; (iii) the increase in surface area of TMDs tends to produce various byproducts (e.g., gel-like polymer) from electrolyte decomposition, leading to the irreversibility of the electrochemical reaction. Despite their sufficient interlayer spacing as compared to the size of M^{n+}, TMDs suffer from ionic diffusion in aqueous environments because of the hydration effect, which increases the hydrodynamic size of M^{n+}.

Therefore, special efforts should be made to overcome these obstacles. Various strategies such as (1) intercalation modification, (2) defect modification, (3) hybridization, and (4) phase modification have been proposed to reduce the ion diffusion energy barrier, increase the interlayer spacing, improve the ion absorption, increase the electronic conductivity, and increase the hydrophilicity of TMD electrodes as discussed earlier. In addition, doping of appropriate materials in TMDs can increase the electronic conductivity especially for the wide bandgap TMDs. Moreover, as predicted from many recent theoretical studies using DFT calculations, the TMD heterostructure (e.g., $MoS_2/MoSe_2$, MoS_2/WS_2, etc.) can retain much higher theoretical capacity than single component TMDs. Toward this direction, a facile and reliable synthesis of TMD heterostructure is proposed as a promising opportunity.

Although great progress has been achieved in the development of TMD-based cathode materials for AMMIBs, extensive efforts are still required to design novel TMD cathode materials for future energy storage devices. It is still challenging to synthesize MX_2 nanostructures with controllable size, interlayer distance, number of layers, phases, composition, and amounts of intercalated foreign species through facile, large-scale, and green synthesis methods. The correlation between the interlayer spacing of TMDs and the hydrodynamic size of ionic species is one of the basic factors to be considered for realizing high-performance AMMIBs. In addition, understanding the adsorption and diffusion mechanisms, origin of possible instability during intercalation/deintercalation, and the role of structural features in TMDs are all important aspects for designing novel TMD materials that can create new opportunities in AMMIBs.

To meet these expectations, a systematic and comprehensive study of characterization techniques will be beneficial. Various in situ analysis techniques will be helpful in monitoring the structural changes in MX_2 during the intercalation of foreign species. Real-time measurements can provide important dynamic information regarding the origin of interlayer expansion. Density functional theory calculation is another useful method for understanding the role of interlayer engineering by quantitatively predicting the dependence of material properties on the interlayer spacing and intercalated species. The strategies and results summarized in this review will provide a guideline for realizing the potential applications of TMDs in future AMMIBs.

Author Contributions: Conceptualization, J.H. and V.P.H.H.; validation, Y.N.A.; investigation, V.P.H.H.; writing—original draft preparation, V.P.H.H.; writing—review and editing, Y.N.A. and J.H.; supervision, J.H.; funding acquisition, J.H. All authors have read and agreed to the published version of the manuscript.

Funding: This work was supported by the Korea Institute of Energy Technology Evaluation and Planning (KETEP) and the Ministry of Trade, Industry and Energy (MOTIE) of the Republic of Korea (No. 20194030202290) and the Gachon University research fund of 2019 (GCU-2019-0359).

Conflicts of Interest: The authors declare no conflict of interest.

References

1. Jiao, Y.; Kang, L.; Berry-Gair, J.; McColl, K.; Li, J.; Dong, H.; Jiang, H.; Wang, R.; Corà, F.; Brett, D.J.L.; et al. Enabling stable MnO$_2$ matrix for aqueous zinc-ion battery cathodes. *J. Mater. Chem. A* **2020**, *8*, 22075–22082. [CrossRef]
2. Pan, H.; Shao, Y.; Yan, P.; Cheng, Y.; Han, K.S.; Nie, Z.; Wang, C.; Yang, J.; Li, X.; Bhattacharya, P.; et al. Reversible aqueous zinc/manganese oxide energy storage from conversion reactions. *Nat. Energy* **2016**, *1*, 16039. [CrossRef]
3. Zampardi, G.; La Mantia, F. Prussian blue analogues as aqueous Zn-ion batteries electrodes: Current challenges and future perspectives. *Curr. Opin. Electrochem.* **2020**, *21*, 84–92. [CrossRef]
4. Scrosati, B.; Garche, J. Lithium batteries: Status, prospects and future. *J. Power Sources* **2010**, *195*, 2419–2430. [CrossRef]
5. Goodenough, J.B.; Park, K.-S. The Li-Ion Rechargeable Battery: A Perspective. *J. Am. Chem. Soc.* **2013**, *135*, 1167–1176. [CrossRef]
6. Winter, M.; Barnett, B.; Xu, K. Before Li Ion Batteries. *Chem. Rev.* **2018**, *118*, 11433–11456. [CrossRef] [PubMed]
7. Mun, Y.S.; Pham, T.N.; Hoang Bui, V.K.; Tanaji, S.T.; Lee, H.U.; Lee, G.-W.; Choi, J.S.; Kim, I.T.; Lee, Y.-C. Tin oxide evolution by heat-treatment with tin-aminoclay (SnAC) under argon condition for lithium-ion battery (LIB) anode applications. *J. Power Sources* **2019**, *437*, 226946. [CrossRef]
8. Nguyen, T.L.; Park, D.; Kim, I.T. Fe(x)Sn(y)O(z) Composites as Anode Materials for Lithium-Ion Storage. *J. Nanosci. Nanotechnol.* **2019**, *19*, 6636–6640. [CrossRef] [PubMed]
9. Nguyen, T.L.; Kim, J.H.; Kim, I.T. Electrochemical Performance of Sn/SnO/Ni$_3$Sn Composite Anodes for Lithium-Ion Batteries. *J. Nanosci. Nanotechnol.* **2019**, *19*, 1001–1005. [CrossRef] [PubMed]
10. Pham, T.N.; Tanaji, S.T.; Choi, J.-S.; Lee, H.U.; Kim, I.T.; Lee, Y.-C. Preparation of Sn-aminoclay (SnAC)-templated Fe$_3$O$_4$ nanoparticles as an anode material for lithium-ion batteries. *RSC Adv.* **2019**, *9*, 10536–10545. [CrossRef]
11. Nguyen, T.P.; Kim, I.T. Ag Nanoparticle-Decorated MoS(2) Nanosheets for Enhancing Electrochemical Performance in Lithium Storage. *Nanomaterials* **2021**, *11*, 626. [CrossRef]
12. Nguyen, T.P.; Kim, I.T. Self-Assembled Few-Layered MoS(2) on SnO(2) Anode for Enhancing Lithium-Ion Storage. *Nanomaterials* **2020**, *10*, 2558. [CrossRef]
13. Preman, A.N.; Lee, H.; Yoo, J.; Kim, I.T.; Saito, T.; Ahn, S.-k. Progress of 3D network binders in silicon anodes for lithium ion batteries. *J. Mater. Chem. A* **2020**, *8*, 25548–25570. [CrossRef]
14. Nguyen, T.P.; Kim, I.T. W2C/WS2 Alloy Nanoflowers as Anode Materials for Lithium-Ion Storage. *Nanomaterials* **2020**, *10*, 1336. [CrossRef] [PubMed]
15. Kim, W.S.; Vo, T.N.; Kim, I.T. GeTe-TiC-C Composite Anodes for Li-Ion Storage. *Materials* **2020**, *13*, 4222. [CrossRef] [PubMed]
16. Vo, T.N.; Kim, D.S.; Mun, Y.S.; Lee, H.J.; Ahn, S.-K.; Kim, I.T. Fast charging sodium-ion batteries based on Te-P-C composites and insights to low-frequency limits of four common equivalent impedance circuits. *Chem. Eng. J.* **2020**, *398*, 125703. [CrossRef]
17. Yabuuchi, N.; Kubota, K.; Dahbi, M.; Komaba, S. Research Development on Sodium-Ion Batteries. *Chem. Rev.* **2014**, *114*, 11636–11682. [CrossRef]
18. Pramudita, J.C.; Sehrawat, D.; Goonetilleke, D.; Sharma, N. An Initial Review of the Status of Electrode Materials for Potassium-Ion Batteries. *Adv. Energy Mater.* **2017**, *7*, 1602911. [CrossRef]
19. Wu, Z.; Xie, J.; Xu, Z.J.; Zhang, S.; Zhang, Q. Recent progress in metal–organic polymers as promising electrodes for lithium/sodium rechargeable batteries. *J. Mater. Chem. A* **2019**, *7*, 4259–4290. [CrossRef]
20. Wang, S.; Sun, C.; Wang, N.; Zhang, Q. Ni- and/or Mn-based layered transition metal oxides as cathode materials for sodium ion batteries: Status, challenges and countermeasures. *J. Mater. Chem. A* **2019**, *7*, 10138–10158. [CrossRef]
21. Xiong, W.; Huang, W.; Zhang, M.; Hu, P.; Cui, H.; Zhang, Q. Pillar[5]quinone–Carbon Nanocomposites as High-Capacity Cathodes for Sodium-Ion Batteries. *Chem. Mater.* **2019**, *31*, 8069–8075. [CrossRef]
22. Ponrouch, A.; Bitenc, J.; Dominko, R.; Lindahl, N.; Johansson, P.; Palacin, M.R. Multivalent rechargeable batteries. *Energy Storage Mater.* **2019**, *20*, 253–262. [CrossRef]
23. Song, M.; Tan, H.; Chao, D.; Fan, H.J. Recent Advances in Zn-Ion Batteries. *Adv. Funct. Mater.* **2018**, *28*, 1802564. [CrossRef]
24. Liu, W.; Hao, J.; Xu, C.; Mou, J.; Dong, L.; Jiang, F.; Kang, Z.; Wu, J.; Jiang, B.; Kang, F. Investigation of zinc ion storage of transition metal oxides, sulfides, and borides in zinc ion battery systems. *Chem. Commun.* **2017**, *53*, 6872–6874. [CrossRef]
25. Wang, H.; Yu, D.; Kuang, C.; Cheng, L.; Li, W.; Feng, X.; Zhang, Z.; Zhang, X.; Zhang, Y. Alkali Metal Anodes for Rechargeable Batteries. *Chem* **2019**, *5*, 313–338. [CrossRef]
26. Stoddart, A. Alkali metal batteries: Preventing failure. *Nat. Rev. Mater.* **2018**, *3*, 18021. [CrossRef]
27. Li, H.; Ma, L.; Han, C.; Wang, Z.; Liu, Z.; Tang, Z.; Zhi, C. Advanced rechargeable zinc-based batteries: Recent progress and future perspectives. *Nano Energy* **2019**, *62*, 550–587. [CrossRef]
28. Fang, G.; Zhou, J.; Pan, A.; Liang, S. Recent Advances in Aqueous Zinc-Ion Batteries. *ACS Energy Lett.* **2018**, *3*, 2480–2501. [CrossRef]
29. Konarov, A.; Voronina, N.; Jo, J.H.; Bakenov, Z.; Sun, Y.-K.; Myung, S.-T. Present and Future Perspective on Electrode Materials for Rechargeable Zinc-Ion Batteries. *ACS Energy Lett.* **2018**, *3*, 2620–2640. [CrossRef]

30. Selvakumaran, D.; Pan, A.; Liang, S.; Cao, G. A review on recent developments and challenges of cathode materials for rechargeable aqueous Zn-ion batteries. *J. Mater. Chem. A* **2019**, *7*, 18209–18236. [CrossRef]

31. Muldoon, J.; Bucur, C.B.; Gregory, T. Quest for Nonaqueous Multivalent Secondary Batteries: Magnesium and Beyond. *Chem. Rev.* **2014**, *114*, 11683–11720. [CrossRef]

32. Cai, T.; Zhao, L.; Hu, H.; Li, T.; Li, X.; Guo, S.; Li, Y.; Xue, Q.; Xing, W.; Yan, Z.; et al. Stable $CoSe_2$/carbon nanodice@reduced graphene oxide composites for high-performance rechargeable aluminum-ion batteries. *Energy Environ. Sci.* **2018**, *11*, 2341–2347. [CrossRef]

33. Gummow, R.J.; Vamvounis, G.; Kannan, M.B.; He, Y. Calcium-Ion Batteries: Current State-of-the-Art and Future Perspectives. *Adv. Mater.* **2018**, *30*, 1801702. [CrossRef]

34. Yu, X.; Prévot, M.S.; Guijarro, N.; Sivula, K. Self-assembled 2D WSe_2 thin films for photoelectrochemical hydrogen production. *Nat. Commun.* **2015**, *6*, 7596. [CrossRef] [PubMed]

35. Wu, W.; Wang, L.; Li, Y.; Zhang, F.; Lin, L.; Niu, S.; Chenet, D.; Zhang, X.; Hao, Y.; Heinz, T.F.; et al. Piezoelectricity of single-atomic-layer MoS_2 for energy conversion and piezotronics. *Nature* **2014**, *514*, 470–474. [CrossRef] [PubMed]

36. Chen, B.; Chao, D.; Liu, E.; Jaroniec, M.; Zhao, N.; Qiao, S.-Z. Transition metal dichalcogenides for alkali metal ion batteries: Engineering strategies at the atomic level. *Energy Environ. Sci.* **2020**, *13*, 1096–1131. [CrossRef]

37. Yuan, H.; Kong, L.; Li, T.; Zhang, Q. A review of transition metal chalcogenide/graphene nanocomposites for energy storage and conversion. *Chin. Chem. Lett.* **2017**, *28*, 2180–2194. [CrossRef]

38. Voiry, D.; Yang, J.; Chhowalla, M. Recent Strategies for Improving the Catalytic Activity of 2D TMD Nanosheets toward the Hydrogen Evolution Reaction. *Adv. Mater.* **2016**, *28*, 6197–6206. [CrossRef]

39. Chhowalla, M.; Shin, H.S.; Eda, G.; Li, L.-J.; Loh, K.P.; Zhang, H. The chemistry of two-dimensional layered transition metal dichalcogenide nanosheets. *Nat. Chem.* **2013**, *5*, 263–275. [CrossRef]

40. Choi, W.; Choudhary, N.; Han, G.H.; Park, J.; Akinwande, D.; Lee, Y.H. Recent development of two-dimensional transition metal dichalcogenides and their applications. *Mater. Today* **2017**, *20*, 116–130. [CrossRef]

41. Xu, J.; Zhang, J.; Zhang, W.; Lee, C.-S. Interlayer Nanoarchitectonics of Two-Dimensional Transition-Metal Dichalcogenides Nanosheets for Energy Storage and Conversion Applications. *Adv. Energy Mater.* **2017**, *7*, 1700571. [CrossRef]

42. Huang, X.; Zeng, Z.; Zhang, H. Metal dichalcogenide nanosheets: Preparation, properties and applications. *Chem. Soc. Rev.* **2013**, *42*, 1934–1946. [CrossRef] [PubMed]

43. Zeng, X.; Hao, J.; Wang, Z.; Mao, J.; Guo, Z. Recent progress and perspectives on aqueous Zn-based rechargeable batteries with mild aqueous electrolytes. *Energy Storage Mater.* **2019**, *20*, 410–437. [CrossRef]

44. Huang, S.; Du, P.; Min, C.; Liao, Y.; Sun, H.; Jiang, Y. Poly(1-amino-5-chloroanthraquinone): Highly Selective and Ultrasensitive Fluorescent Chemosensor For Ferric Ion. *J. Fluoresc.* **2013**, *23*, 621–627. [CrossRef] [PubMed]

45. Liu, T.; Cheng, X.; Yu, H.; Zhu, H.; Peng, N.; Zheng, R.; Zhang, J.; Shui, M.; Cui, Y.; Shu, J. An overview and future perspectives of aqueous rechargeable polyvalent ion batteries. *Energy Storage Mater.* **2019**, *18*, 68–91. [CrossRef]

46. Li, Q.; Bjerrum, N.J. Aluminum as anode for energy storage and conversion: A review. *J. Power Sources* **2002**, *110*, 1–10. [CrossRef]

47. Wu, S.; Du, Y.; Sun, S. Transition metal dichalcogenide based nanomaterials for rechargeable batteries. *Chem. Eng. J.* **2017**, *307*, 189–207. [CrossRef]

48. Yun, Q.; Li, L.; Hu, Z.; Lu, Q.; Chen, B.; Zhang, H. Layered Transition Metal Dichalcogenide-Based Nanomaterials for Electrochemical Energy Storage. *Adv. Mater.* **2020**, *32*, 1903826. [CrossRef]

49. Rashad, M.; Asif, M.; Wang, Y.; He, Z.; Ahmed, I. Recent advances in electrolytes and cathode materials for magnesium and hybrid-ion batteries. *Energy Storage Mater.* **2020**, *25*, 342–375. [CrossRef]

50. Jellinek, F. Transition metal chalcogenides. relationship between chemical composition, crystal structure and physical properties. *React. Solids* **1988**, *5*, 323–339. [CrossRef]

51. Stephenson, T.; Li, Z.; Olsen, B.; Mitlin, D. Lithium ion battery applications of molybdenum disulfide (MoS_2) nanocomposites. *Energy Environ. Sci.* **2014**, *7*, 209–231. [CrossRef]

52. Shi, Y.; Zhang, H.; Chang, W.-H.; Shin, H.; Li, L. Synthesis and structure of two-dimensional transition-metal dichalcogenides. *MRS Bull.* **2015**, *40*, 566–576. [CrossRef]

53. Voiry, D.; Mohite, A.; Chhowalla, M. Phase engineering of transition metal dichalcogenides. *Chem. Soc. Rev.* **2015**, *44*, 2702–2712. [CrossRef]

54. Han, S.A.; Bhatia, R.; Kim, S.-W. Synthesis, properties and potential applications of two-dimensional transition metal dichalcogenides. *Nano Converg.* **2015**, *2*, 17. [CrossRef]

55. Zhang, G.; Liu, H.; Qu, J.; Li, J. Two-dimensional layered MoS_2: Rational design, properties and electrochemical applications. *Energy Environ. Sci.* **2016**, *9*, 1190–1209. [CrossRef]

56. Manzeli, S.; Ovchinnikov, D.; Pasquier, D.; Yazyev, O.V.; Kis, A. 2D transition metal dichalcogenides. *Nat. Rev. Mater.* **2017**, *2*, 17033. [CrossRef]

57. Wei, Z.; Li, B.; Xia, C.; Cui, Y.; He, J.; Xia, J.-B.; Li, J. Various Structures of 2D Transition-Metal Dichalcogenides and Their Applications. *Small Methods* **2018**, *2*, 1800094. [CrossRef]

58. Zhang, Q.; Mei, L.; Cao, X.; Tang, Y.; Zeng, Z. Intercalation and exfoliation chemistries of transition metal dichalcogenides. *J. Mater. Chem. A* **2020**, *8*, 15417–15444. [CrossRef]

59. Lee, W.S.V.; Xiong, T.; Wang, X.; Xue, J. Unraveling MoS_2 and Transition Metal Dichalcogenides as Functional Zinc-Ion Battery Cathode: A Perspective. *Small Methods* **2021**, *5*, 2000815. [CrossRef]

60. Wang, Q.H.; Kalantar-Zadeh, K.; Kis, A.; Coleman, J.N.; Strano, M.S. Electronics and optoelectronics of two-dimensional transition metal dichalcogenides. *Nat. Nanotechnol.* **2012**, *7*, 699–712. [CrossRef] [PubMed]

61. Coogan, Á.; Gun'ko, Y.K. Solution-based "bottom-up" synthesis of group VI transition metal dichalcogenides and their applications. *Mater. Adv.* **2021**, *2*, 146–164. [CrossRef]

62. Song, I.; Park, C.; Choi, H.C. Synthesis and properties of molybdenum disulphide: From bulk to atomic layers. *RSC Adv.* **2015**, *5*, 7495–7514. [CrossRef]

63. Eda, G.; Yamaguchi, H.; Voiry, D.; Fujita, T.; Chen, M.; Chhowalla, M. Photoluminescence from Chemically Exfoliated MoS_2. *Nano Lett.* **2011**, *11*, 5111–5116. [CrossRef] [PubMed]

64. Schmidt, H.; Giustiniano, F.; Eda, G. Electronic transport properties of transition metal dichalcogenide field-effect devices: Surface and interface effects. *Chem. Soc. Rev.* **2015**, *44*, 7715–7736. [CrossRef]

65. Novoselov, K.S.; Geim, A.K.; Morozov, S.V.; Jiang, D.; Zhang, Y.; Dubonos, S.V.; Grigorieva, I.V.; Firsov, A.A. Electric Field Effect in Atomically Thin Carbon Films. *Science* **2004**, *306*, 666–669. [CrossRef] [PubMed]

66. Huang, X.; Yin, Z.; Wu, S.; Qi, X.; He, Q.; Zhang, Q.; Yan, Q.; Boey, F.; Zhang, H. Graphene-Based Materials: Synthesis, Characterization, Properties, and Applications. *Small* **2011**, *7*, 1876–1902. [CrossRef]

67. Zhang, H. Ultrathin Two-Dimensional Nanomaterials. *ACS Nano* **2015**, *9*, 9451–9469. [CrossRef]

68. Tan, C.; Cao, X.; Wu, X.-J.; He, Q.; Yang, J.; Zhang, X.; Chen, J.; Zhao, W.; Han, S.; Nam, G.-H.; et al. Recent Advances in Ultrathin Two-Dimensional Nanomaterials. *Chem. Rev.* **2017**, *117*, 6225–6331. [CrossRef]

69. Tan, C.; Zhang, H. Two-dimensional transition metal dichalcogenide nanosheet-based composites. *Chem. Soc. Rev.* **2015**, *44*, 2713–2731. [CrossRef]

70. Lu, Q.; Yu, Y.; Ma, Q.; Chen, B.; Zhang, H. 2D Transition-Metal-Dichalcogenide-Nanosheet-Based Composites for Photocatalytic and Electrocatalytic Hydrogen Evolution Reactions. *Adv. Mater.* **2016**, *28*, 1917–1933. [CrossRef] [PubMed]

71. Lin, L.; Lei, W.; Zhang, S.; Liu, Y.; Wallace, G.G.; Chen, J. Two-dimensional transition metal dichalcogenides in supercapacitors and secondary batteries. *Energy Storage Mater.* **2019**, *19*, 408–423. [CrossRef]

72. Yu, X.; Yun, S.; Yeon, J.S.; Bhattacharya, P.; Wang, L.; Lee, S.W.; Hu, X.; Park, H.S. Emergent Pseudocapacitance of 2D Nanomaterials. *Adv. Energy Mater.* **2018**, *8*, 1702930. [CrossRef]

73. Chen, J.; Luo, Y.; Zhang, W.; Qiao, Y.; Cao, X.; Xie, X.; Zhou, H.; Pan, A.; Liang, S. Tuning Interface Bridging between $MoSe_2$ and Three-Dimensional Carbon Framework by Incorporation of MoC Intermediate to Boost Lithium Storage Capability. *Nano-Micro Lett.* **2020**, *12*, 171. [CrossRef]

74. Tansel, B. Significance of thermodynamic and physical characteristics on permeation of ions during membrane separation: Hydrated radius, hydration free energy and viscous effects. *Sep. Purif. Technol.* **2012**, *86*, 119–126. [CrossRef]

75. Liu, J.; Xu, P.; Liang, J.; Liu, H.; Peng, W.; Li, Y.; Zhang, F.; Fan, X. Boosting aqueous zinc-ion storage in MoS_2 via controllable phase. *Chem. Eng. J.* **2020**, *389*, 124405. [CrossRef]

76. Ming, J.; Guo, J.; Xia, C.; Wang, W.; Alshareef, H.N. Zinc-ion batteries: Materials, mechanisms, and applications. *Mater. Sci. Eng. R Rep.* **2019**, *135*, 58–84. [CrossRef]

77. Liang, H.; Cao, Z.; Ming, F.; Zhang, W.; Anjum, D.H.; Cui, Y.; Cavallo, L.; Alshareef, H.N. Aqueous Zinc-Ion Storage in MoS_2 by Tuning the Intercalation Energy. *Nano Lett.* **2019**, *19*, 3199–3206. [CrossRef] [PubMed]

78. Li, H.; Yang, Q.; Mo, F.; Liang, G.; Liu, Z.; Tang, Z.; Ma, L.; Liu, J.; Shi, Z.; Zhi, C. MoS_2 nanosheets with expanded interlayer spacing for rechargeable aqueous Zn-ion batteries. *Energy Storage Mater.* **2019**, *19*, 94–101. [CrossRef]

79. Xu, W.; Sun, C.; Zhao, K.; Cheng, X.; Rawal, S.; Xu, Y.; Wang, Y. Defect engineering activating (Boosting) zinc storage capacity of MoS_2. *Energy Storage Mater.* **2019**, *16*, 527–534. [CrossRef]

80. Yang, Y. Synthesis and Properties of Halloysite Templated Tubular MoS_2 as Cathode Material for Rechargeable Aqueous Zn-ion Batteries. *Int. J. Electrochem. Sci.* **2020**, *15*, 6052–6059. [CrossRef]

81. He, P.; Yan, M.; Zhang, G.; Sun, R.; Chen, L.; An, Q.; Mai, L. Layered VS_2 Nanosheet-Based Aqueous Zn Ion Battery Cathode. *Adv. Energy Mater.* **2017**, *7*, 1601920. [CrossRef]

82. Jiao, T.; Yang, Q.; Wu, S.; Wang, Z.; Chen, D.; Shen, D.; Liu, B.; Cheng, J.; Li, H.; Ma, L.; et al. Binder-free hierarchical VS_2 electrodes for high-performance aqueous Zn ion batteries towards commercial level mass loading. *J. Mater. Chem. A* **2019**, *7*, 16330–16338. [CrossRef]

83. Wu, Z.; Lu, C.; Wang, Y.; Zhang, L.; Jiang, L.; Tian, W.; Cai, C.; Gu, Q.; Sun, Z.; Hu, L. Ultrathin VSe_2 Nanosheets with Fast Ion Diffusion and Robust Structural Stability for Rechargeable Zinc-Ion Battery Cathode. *Small* **2020**, *16*, 2000698. [CrossRef] [PubMed]

84. Zhu, Q.; Xiao, Q.; Zhang, B.; Yan, Z.; Liu, X.; Chen, S.; Ren, Z.; Yu, Y. VS_4 with a chain crystal structure used as an intercalation cathode for aqueous Zn-ion batteries. *J. Mater. Chem. A* **2020**, *8*, 10761–10766. [CrossRef]

85. Aurbach, D.; Lu, Z.; Schechter, A.; Gofer, Y.; Gizbar, H.; Turgeman, R.; Cohen, Y.; Moshkovich, M.; Levi, E. Prototype systems for rechargeable magnesium batteries. *Nature* **2000**, *407*, 724–727. [CrossRef] [PubMed]

86. Liu, Y.; He, G.; Jiang, H.; Parkin, I.P.; Shearing, P.R.; Brett, D.J.L. Cathode Design for Aqueous Rechargeable Multivalent Ion Batteries: Challenges and Opportunities. *Adv. Funct. Mater.* **2021**, *31*, 2010445. [CrossRef]

87. Yang, S.; Li, D.; Zhang, T.; Tao, Z.; Chen, J. First-Principles Study of Zigzag MoS_2 Nanoribbon as a Promising Cathode Material for Rechargeable Mg Batteries. *J. Phys. Chem. C* **2012**, *116*, 1307–1312. [CrossRef]
88. Liang, Y.; Feng, R.; Yang, S.; Ma, H.; Liang, J.; Chen, J. Rechargeable Mg Batteries with Graphene-like MoS_2 Cathode and Ultrasmall Mg Nanoparticle Anode. *Adv. Mater.* **2011**, *23*, 640–643. [CrossRef]
89. Liu, Y.; Jiao, L.; Wu, Q.; Du, J.; Zhao, Y.; Si, Y.; Wang, Y.; Yuan, H. Sandwich-structured graphene-like MoS_2/C microspheres for rechargeable Mg batteries. *J. Mater. Chem. A* **2013**, *1*, 5822–5826. [CrossRef]
90. Truong, Q.D.; Kempaiah Devaraju, M.; Nakayasu, Y.; Tamura, N.; Sasaki, Y.; Tomai, T.; Honma, I. Exfoliated MoS_2 and $MoSe_2$ Nanosheets by a Supercritical Fluid Process for a Hybrid Mg–Li-Ion Battery. *ACS Omega* **2017**, *2*, 2360–2367. [CrossRef]
91. Mao, M.; Ji, X.; Hou, S.; Gao, T.; Wang, F.; Chen, L.; Fan, X.; Chen, J.; Ma, J.; Wang, C. Tuning Anionic Chemistry to Improve Kinetics of Mg Intercalation. *Chem. Mater.* **2019**, *31*, 3183–3191. [CrossRef]
92. Gu, Y.; Katsura, Y.; Yoshino, T.; Takagi, H.; Taniguchi, K. Rechargeable magnesium-ion battery based on a $TiSe_2$-cathode with d-p orbital hybridized electronic structure. *Sci. Rep.* **2015**, *5*, 12486. [CrossRef]
93. Xu, J.; Wei, Z.; Zhang, S.; Wang, X.; Wang, Y.; He, M.; Huang, K. Hierarchical WSe_2 nanoflower as a cathode material for rechargeable Mg-ion batteries. *J. Colloid Interface Sci.* **2021**, *588*, 378–383. [CrossRef]
94. Guduru, R.K.; Icaza, J.C. A Brief Review on Multivalent Intercalation Batteries with Aqueous Electrolytes. *Nanomaterials* **2016**, *6*, 41. [CrossRef] [PubMed]
95. Liu, S.; Hu, J.J.; Yan, N.F.; Pan, G.L.; Li, G.R.; Gao, X.P. Aluminum storage behavior of anatase TiO_2 nanotube arrays in aqueous solution for aluminum ion batteries. *Energy Environ. Sci.* **2012**, *5*, 9743–9746. [CrossRef]
96. He, S.; Wang, J.; Zhang, X.; Chen, J.; Wang, Z.; Yang, T.; Liu, Z.; Liang, Y.; Wang, B.; Liu, S.; et al. A High-Energy Aqueous Aluminum-Manganese Battery. *Adv. Funct. Mater.* **2019**, *29*, 1905228. [CrossRef]
97. Lahan, H.; Boruah, R.; Hazarika, A.; Das, S.K. Anatase TiO_2 as an Anode Material for Rechargeable Aqueous Aluminum-Ion Batteries: Remarkable Graphene Induced Aluminum Ion Storage Phenomenon. *J. Phys. Chem. C* **2017**, *121*, 26241–26249. [CrossRef]
98. He, Y.J.; Peng, J.F.; Chu, W.; Li, Y.Z.; Tong, D.G. Retracted Article: Black mesoporous anatase TiO_2 nanoleaves: A high capacity and high rate anode for aqueous Al-ion batteries. *J. Mater. Chem. A* **2014**, *2*, 1721–1731. [CrossRef]
99. Liu, S.; Pan, G.L.; Li, G.R.; Gao, X.P. Copper hexacyanoferrate nanoparticles as cathode material for aqueous Al-ion batteries. *J. Mater. Chem. A* **2015**, *3*, 959–962. [CrossRef]
100. Zhou, A.; Jiang, L.; Yue, J.; Tong, Y.; Zhang, Q.; Lin, Z.; Liu, B.; Wu, C.; Suo, L.; Hu, Y.-S.; et al. Water-in-Salt Electrolyte Promotes High-Capacity $FeFe(CN)_6$ Cathode for Aqueous Al-Ion Battery. *ACS Appl. Mater. Interfaces* **2019**, *11*, 41356–41362. [CrossRef]
101. Zhao, Q.; Zachman, M.J.; Al Sadat, W.I.; Zheng, J.; Kourkoutis, L.F.; Archer, L. Solid electrolyte interphases for high-energy aqueous aluminum electrochemical cells. *Sci. Adv.* **2018**, *4*, eaau8131. [CrossRef]
102. Wu, C.; Gu, S.; Zhang, Q.; Bai, Y.; Li, M.; Yuan, Y.; Wang, H.; Liu, X.; Yuan, Y.; Zhu, N.; et al. Electrochemically activated spinel manganese oxide for rechargeable aqueous aluminum battery. *Nat. Commun.* **2019**, *10*, 73. [CrossRef]
103. Pan, W.; Wang, Y.; Zhang, Y.; Kwok, H.Y.H.; Wu, M.; Zhao, X.; Leung, D.Y.C. A low-cost and dendrite-free rechargeable aluminium-ion battery with superior performance. *J. Mater. Chem. A* **2019**, *7*, 17420–17425. [CrossRef]
104. Fan, X.; Gaddam, R.R.; Kumar, N.A.; Zhao, X.S. A Hybrid Mg^{2+}/Li^+ Battery Based on Interlayer-Expanded MoS_2/Graphene Cathode. *Adv. Energy Mater.* **2017**, *7*, 1700317. [CrossRef]
105. Li, Z.; Niu, B.; Liu, J.; Li, J.; Kang, F. Rechargeable Aluminum-Ion Battery Based on MoS_2 Microsphere Cathode. *ACS Appl. Mater. Interfaces* **2018**, *10*, 9451–9459. [CrossRef]
106. Divya, S.; Johnston, J.H.; Nann, T. Molybdenum Dichalcogenide Cathodes for Aluminum-Ion Batteries. *Energy Technol.* **2020**, *8*, 2000038. [CrossRef]
107. Wu, L.; Sun, R.; Xiong, F.; Pei, C.; Han, K.; Peng, C.; Fan, Y.; Yang, W.; An, Q.; Mai, L. A rechargeable aluminum-ion battery based on a VS_2 nanosheet cathode. *Phys. Chem. Chem. Phys.* **2018**, *20*, 22563–22568. [CrossRef] [PubMed]
108. Geng, L.; Scheifers, J.P.; Fu, C.; Zhang, J.; Fokwa, B.P.T.; Guo, J. Titanium Sulfides as Intercalation-Type Cathode Materials for Rechargeable Aluminum Batteries. *ACS Appl. Mater. Interfaces* **2017**, *9*, 21251–21257. [CrossRef] [PubMed]
109. Geng, L.; Lv, G.; Xing, X.; Guo, J. Reversible Electrochemical Intercalation of Aluminum in Mo_6S_8. *Chem. Mater.* **2015**, *27*, 4926–4929. [CrossRef]
110. Kundu, D.; Adams, B.D.; Duffort, V.; Vajargah, S.H.; Nazar, L.F. A high-capacity and long-life aqueous rechargeable zinc battery using a metal oxide intercalation cathode. *Nat. Energy* **2016**, *1*, 16119. [CrossRef]
111. Bin, D.; Huo, W.; Yuan, Y.; Huang, J.; Liu, Y.; Zhang, Y.; Dong, F.; Wang, Y.; Xia, Y. Organic-Inorganic-Induced Polymer Intercalation into Layered Composites for Aqueous Zinc-Ion Battery. *Chem* **2020**, *6*, 968–984. [CrossRef]
112. Jung, Y.; Zhou, Y.; Cha, J.J. Intercalation in two-dimensional transition metal chalcogenides. *Inorg. Chem. Front.* **2016**, *3*, 452–463. [CrossRef]
113. Zhang, J.; Yang, A.; Wu, X.; van de Groep, J.; Tang, P.; Li, S.; Liu, B.; Shi, F.; Wan, J.; Li, Q.; et al. Reversible and selective ion intercalation through the top surface of few-layer MoS_2. *Nat. Commun.* **2018**, *9*, 5289. [CrossRef] [PubMed]
114. Gu, J.; Zhu, Q.; Shi, Y.; Chen, H.; Zhang, D.; Du, Z.; Yang, S. Single Zinc Atoms Immobilized on MXene (Ti_3C_2Clx) Layers toward Dendrite-Free Lithium Metal Anodes. *ACS Nano* **2020**, *14*, 891–898. [CrossRef] [PubMed]
115. Shi, H.-Y.; Sun, X. Interlayer Engineering of Layered Cathode Materials for Advanced Zn Storage. *Chem* **2020**, *6*, 817–819. [CrossRef]

116. Winchester, A.; Ghosh, S.; Feng, S.; Elias, A.L.; Mallouk, T.; Terrones, M.; Talapatra, S. Electrochemical Characterization of Liquid Phase Exfoliated Two-Dimensional Layers of Molybdenum Disulfide. *ACS Appl. Mater. Interfaces* **2014**, *6*, 2125–2130. [CrossRef]
117. Bang, G.S.; Nam, K.W.; Kim, J.Y.; Shin, J.; Choi, J.W.; Choi, S.-Y. Effective Liquid-Phase Exfoliation and Sodium Ion Battery Application of MoS$_2$ Nanosheets. *ACS Appl. Mater. Interfaces* **2014**, *6*, 7084–7089. [CrossRef]
118. Coleman, J.N.; Lotya, M.; O'Neill, A.; Bergin, S.D.; King, P.J.; Khan, U.; Young, K.; Gaucher, A.; De, S.; Smith, R.J.; et al. Two-Dimensional Nanosheets Produced by Liquid Exfoliation of Layered Materials. *Science* **2011**, *331*, 568. [CrossRef]
119. Smith, R.J.; King, P.J.; Lotya, M.; Wirtz, C.; Khan, U.; De, S.; O'Neill, A.; Duesberg, G.S.; Grunlan, J.C.; Moriarty, G.; et al. Large-Scale Exfoliation of Inorganic Layered Compounds in Aqueous Surfactant Solutions. *Adv. Mater.* **2011**, *23*, 3944–3948. [CrossRef] [PubMed]
120. Shmeliov, A.; Shannon, M.; Wang, P.; Kim, J.S.; Okunishi, E.; Nellist, P.D.; Dolui, K.; Sanvito, S.; Nicolosi, V. Unusual Stacking Variations in Liquid-Phase Exfoliated Transition Metal Dichalcogenides. *ACS Nano* **2014**, *8*, 3690–3699. [CrossRef]
121. Paton, K.R.; Varrla, E.; Backes, C.; Smith, R.J.; Khan, U.; O'Neill, A.; Boland, C.; Lotya, M.; Istrate, O.M.; King, P.; et al. Scalable production of large quantities of defect-free few-layer graphene by shear exfoliation in liquids. *Nat. Mater.* **2014**, *13*, 624–630. [CrossRef] [PubMed]
122. Rangappa, D.; Sone, K.; Wang, M.; Gautam, U.K.; Golberg, D.; Itoh, H.; Ichihara, M.; Honma, I. Rapid and Direct Conversion of Graphite Crystals into High-Yielding, Good-Quality Graphene by Supercritical Fluid Exfoliation. *Chem. A Eur. J.* **2010**, *16*, 6488–6494. [CrossRef] [PubMed]
123. Rangappa, D.; Murukanahally, K.D.; Tomai, T.; Unemoto, A.; Honma, I. Ultrathin Nanosheets of Li$_2$MSiO$_4$ (M = Fe, Mn) as High-Capacity Li-Ion Battery Electrode. *Nano Lett.* **2012**, *12*, 1146–1151. [CrossRef]
124. Truong, Q.D.; Devaraju, M.K.; Honma, I. Benzylamine-directed growth of olivine-type LiMPO$_4$ nanoplates by a supercritical ethanol process for lithium-ion batteries. *J. Mater. Chem. A* **2014**, *2*, 17400–17407. [CrossRef]
125. Huang, Y.; Wang, Y.; Tang, C.; Wang, J.; Zhang, Q.; Wang, Y.; Zhang, J. Atomic Modulation and Structure Design of Carbons for Bifunctional Electrocatalysis in Metal–Air Batteries. *Adv. Mater.* **2019**, *31*, 1803800. [CrossRef] [PubMed]
126. Uchaker, E.; Cao, G. The Role of Intentionally Introduced Defects on Electrode Materials for Alkali-Ion Batteries. *Chem. Asian J.* **2015**, *10*, 1608–1617. [CrossRef] [PubMed]
127. Zhang, Y.; Guo, L.; Tao, L.; Lu, Y.; Wang, S. Defect-Based Single-Atom Electrocatalysts. *Small Methods* **2019**, *3*, 1800406. [CrossRef]
128. Banhart, F.; Kotakoski, J.; Krasheninnikov, A.V. Structural Defects in Graphene. *ACS Nano* **2011**, *5*, 26–41. [CrossRef]
129. Fang, Z.; Bueken, B.; De Vos, D.E.; Fischer, R.A. Defect-Engineered Metal–Organic Frameworks. *Angew. Chem. Int. Ed.* **2015**, *54*, 7234–7254. [CrossRef]
130. Lin, Z.; Carvalho, B.R.; Kahn, E.; Lv, R.; Rao, R.; Terrones, H.; Pimenta, M.A.; Terrones, M. Defect engineering of two-dimensional transition metal dichalcogenides. *2D Mater.* **2016**, *3*, 022002. [CrossRef]
131. Matsunaga, K.; Tanaka, T.; Yamamoto, T.; Ikuhara, Y. First-principles calculations of intrinsic defects in Al$_2$O$_3$. *Phys. Rev. B* **2003**, *68*, 085110. [CrossRef]
132. Meggiolaro, D.; Mosconi, E.; De Angelis, F. Formation of Surface Defects Dominates Ion Migration in Lead-Halide Perovskites. *ACS Energy Lett.* **2019**, *4*, 779–785. [CrossRef]
133. Paier, J.; Penschke, C.; Sauer, J. Oxygen Defects and Surface Chemistry of Ceria: Quantum Chemical Studies Compared to Experiment. *Chem. Rev.* **2013**, *113*, 3949–3985. [CrossRef]
134. Wilby, A.; Neale, D. Defects Introduced into Metals during Fabrication and Service. 2011. Available online: http://www.eolss.net/Sample-Chapters/C05/E6-36-04-01.pdf (accessed on 1 March 2012).
135. Xiong, T.; Yu, Z.G.; Wu, H.; Du, Y.; Xie, Q.; Chen, J.; Zhang, Y.-W.; Pennycook, S.J.; Lee, W.S.V.; Xue, J. Defect Engineering of Oxygen-Deficient Manganese Oxide to Achieve High-Performing Aqueous Zinc Ion Battery. *Adv. Energy Mater.* **2019**, *9*, 1803815. [CrossRef]
136. Huang, X.; Qi, X.; Boey, F.; Zhang, H. Graphene-based composites. *Chem. Soc. Rev.* **2012**, *41*, 666–686. [CrossRef] [PubMed]
137. Wang, Y.; Zhang, X.; Luo, Z.; Huang, X.; Tan, C.; Li, H.; Zheng, B.; Li, B.; Huang, Y.; Yang, J.; et al. Liquid-phase growth of platinum nanoparticles on molybdenum trioxide nanosheets: An enhanced catalyst with intrinsic peroxidase-like catalytic activity. *Nanoscale* **2014**, *6*, 12340–12344. [CrossRef]
138. Deng, Z.; Jiang, H.; Li, C. 2D Metal Chalcogenides Incorporated into Carbon and their Assembly for Energy Storage Applications. *Small* **2018**, *14*, 1800148. [CrossRef]
139. Lou, X.W.; Li, C.M.; Archer, L.A. Designed Synthesis of Coaxial SnO$_2$@carbon Hollow Nanospheres for Highly Reversible Lithium Storage. *Adv. Mater.* **2009**, *21*, 2536–2539. [CrossRef]
140. Qi, X.; Liu, N.; Lian, Y. Carbonaceous microspheres prepared by hydrothermal carbonization of glucose for direct use in catalytic dehydration of fructose. *RSC Adv.* **2015**, *5*, 17526–17531. [CrossRef]
141. Chen, Y.; Song, B.; Tang, X.; Lu, L.; Xue, J. Ultrasmall Fe$_3$O$_4$ Nanoparticle/MoS$_2$ Nanosheet Composites with Superior Performances for Lithium Ion Batteries. *Small* **2014**, *10*, 1536–1543. [CrossRef]
142. Gong, Y.; Yang, S.; Zhan, L.; Ma, L.; Vajtai, R.; Ajayan, P.M. A Bottom-Up Approach to Build 3D Architectures from Nanosheets for Superior Lithium Storage. *Adv. Funct. Mater.* **2014**, *24*, 125–130. [CrossRef]
143. Zhu, C.; Mu, X.; van Aken, P.A.; Yu, Y.; Maier, J. Single-Layered Ultrasmall Nanoplates of MoS$_2$ Embedded in Carbon Nanofibers with Excellent Electrochemical Performance for Lithium and Sodium Storage. *Angew. Chem. Int. Ed.* **2014**, *53*, 2152–2156. [CrossRef]

144. Yang, Y.; Wang, S.; Zhang, J.; Li, H.; Tang, Z.; Wang, X. Nanosheet-assembled $MoSe_2$ and S-doped $MoSe_{2-x}$ nanostructures for superior lithium storage properties and hydrogen evolution reactions. *Inorg. Chem. Front.* **2015**, *2*, 931–937. [CrossRef]
145. Fan, X.; Xu, P.; Li, Y.C.; Zhou, D.; Sun, Y.; Nguyen, M.A.T.; Terrones, M.; Mallouk, T.E. Controlled Exfoliation of MoS_2 Crystals into Trilayer Nanosheets. *J. Am. Chem. Soc.* **2016**, *138*, 5143–5149. [CrossRef] [PubMed]
146. Tang, Q.; Jiang, D.-E. Mechanism of Hydrogen Evolution Reaction on $1T$-MoS_2 from First Principles. *ACS Catal.* **2016**, *6*, 4953–4961. [CrossRef]
147. Lei, Z.; Zhan, J.; Tang, L.; Zhang, Y.; Wang, Y. Recent Development of Metallic (1T) Phase of Molybdenum Disulfide for Energy Conversion and Storage. *Adv. Energy Mater.* **2018**, *8*, 1703482. [CrossRef]
148. Wang, R.; Yu, Y.; Zhou, S.; Li, H.; Wong, H.; Luo, Z.; Gan, L.; Zhai, T. Strategies on Phase Control in Transition Metal Dichalcogenides. *Adv. Funct. Mater.* **2018**, *28*, 1802473. [CrossRef]
149. Yin, Y.; Han, J.; Zhang, Y.; Zhang, X.; Xu, P.; Yuan, Q.; Samad, L.; Wang, X.; Wang, Y.; Zhang, Z.; et al. Contributions of Phase, Sulfur Vacancies, and Edges to the Hydrogen Evolution Reaction Catalytic Activity of Porous Molybdenum Disulfide Nanosheets. *J. Am. Chem. Soc.* **2016**, *138*, 7965–7972. [CrossRef]
150. Tan, C.; Luo, Z.; Chaturvedi, A.; Cai, Y.; Du, Y.; Gong, Y.; Huang, Y.; Lai, Z.; Zhang, X.; Zheng, L.; et al. Preparation of High-Percentage 1T-Phase Transition Metal Dichalcogenide Nanodots for Electrochemical Hydrogen Evolution. *Adv. Mater.* **2018**, *30*, 1705509. [CrossRef] [PubMed]
151. Liu, J.; Chen, T.; Juan, P.; Peng, W.; Li, Y.; Zhang, F.; Fan, X. Hierarchical Cobalt Borate/MXenes Hybrid with Extraordinary Electrocatalytic Performance in Oxygen Evolution Reaction. *ChemSusChem* **2018**, *11*, 3758–3765. [CrossRef] [PubMed]
152. Liu, J.; Liu, Y.; Xu, D.; Zhu, Y.; Peng, W.; Li, Y.; Zhang, F.; Fan, X. Hierarchical "nanoroll" like MoS_2/Ti_3C_2Tx hybrid with high electrocatalytic hydrogen evolution activity. *Appl. Catal. B Environ.* **2019**, *241*, 89–94. [CrossRef]
153. Chang, K.; Hai, X.; Pang, H.; Zhang, H.; Shi, L.; Liu, G.; Liu, H.; Zhao, G.; Li, M.; Ye, J. Targeted Synthesis of 2H- and 1T-Phase MoS_2 Monolayers for Catalytic Hydrogen Evolution. *Adv. Mater.* **2016**, *28*, 10033–10041. [CrossRef] [PubMed]

 *nanomaterials*

Article

Graphene Nanosheet-Wrapped Mesoporous La$_{0.8}$Ce$_{0.2}$Fe$_{0.5}$Mn$_{0.5}$O$_3$ Perovskite Oxide Composite for Improved Oxygen Reaction Electro-Kinetics and Li-O$_2$ Battery Application

Chelladurai Karuppiah [1,*], Chao-Nan Wei [2], Natarajan Karikalan [3], Zong-Han Wu [1], Balamurugan Thirumalraj [4], Li-Fan Hsu [1], Srinivasan Alagar [5], Shakkthivel Piraman [5], Tai-Feng Hung [1], Ying-Jeng Jame Li [2] and Chun-Chen Yang [1,2,6,*]

1   Battery Research Center of Green Energy, Ming Chi University of Technology, New Taipei City 24301, Taiwan; smartindigoboy@gmail.com (Z.-H.W.); issyokenme@gmail.com (L.-F.H.); taifeng@mail.mcut.edu.tw (T.-F.H.)
2   Department of Chemical Engineering, Ming Chi University of Technology, New Taipei City 24301, Taiwan; xu04m4520@gmail.com (C.-N.W.); yjli@mail.mcut.edu.tw (Y.-J.J.L.)
3   Center of Precision Analysis and Material Research, National Taipei University of Technology, No. 1, Section 3, Chung-Hsiao East Road, Taipei 106, Taiwan; 12karikalan.n@gmail.com
4   Department of Energy & Mineral Resources Engineering, Sejong University, Seoul 05006, Korea; balachem01@gmail.com
5   Sustainable Energy and Smart Materials Research Lab, Department of Nanoscience and Technology, Alagappa University, Karaikudi 630002, India; sbn.chem@gmail.com (S.A.); apsakthivel@yahoo.com (S.P.)
6   Department of Chemical and Materials Engineering, Chang Gung University, Kwei-Shan, Taoyuan 333, Taiwan
*   Correspondence: kcdurai.rmd@gmail.com (C.K.); ccyang@mail.mcut.edu.tw (C.-C.Y.)

Citation: Karuppiah, C.; Wei, C.-N.; Karikalan, N.; Wu, Z.-H.; Thirumalraj, B.; Hsu, L.-F.; Alagar, S.; Piraman, S.; Hung, T.-F.; Li, Y.-J.J.; et al. Graphene Nanosheet-Wrapped Mesoporous La$_{0.8}$Ce$_{0.2}$Fe$_{0.5}$Mn$_{0.5}$O$_3$ Perovskite Oxide Composite for Improved Oxygen Reaction Electro-Kinetics and Li-O$_2$ Battery Application. *Nanomaterials* **2021**, *11*, 1025. https://doi.org/10.3390/nano11041025

Academic Editor: Jaehyun Hur

Received: 9 March 2021
Accepted: 12 April 2021
Published: 16 April 2021

Publisher's Note: MDPI stays neutral with regard to jurisdictional claims in published maps and institutional affiliations.

Abstract: A novel design and synthesis methodology is the most important consideration in the development of a superior electrocatalyst for improving the kinetics of oxygen electrode reactions, such as the oxygen reduction reaction (ORR) and the oxygen evolution reaction (OER) in Li-O$_2$ battery application. Herein, we demonstrate a glycine-assisted hydrothermal and probe sonication method for the synthesis of a mesoporous spherical La$_{0.8}$Ce$_{0.2}$Fe$_{0.5}$Mn$_{0.5}$O$_3$ perovskite particle and embedded graphene nanosheet (LCFM(8255)-gly/GNS) composite and evaluate its bifunctional ORR/OER kinetics in Li-O$_2$ battery application. The physicochemical characterization confirms that the as-formed LCFM(8255)-gly perovskite catalyst has a highly crystalline structure and mesoporous morphology with a large specific surface area. The LCFM(8255)-gly/GNS composite hybrid structure exhibits an improved onset potential and high current density toward ORR/OER in both aqueous and non-aqueous electrolytes. The LCFM(8255)-gly/GNS composite cathode (ca. 8475 mAh g^{-1}) delivers a higher discharge capacity than the La$_{0.5}$Ce$_{0.5}$Fe$_{0.5}$Mn$_{0.5}$O$_3$-gly/GNS cathode (ca. 5796 mAh g^{-1}) in a Li-O$_2$ battery at a current density of 100 mA g^{-1}. Our results revealed that the composite's high electrochemical activity comes from the synergism of highly abundant oxygen vacancies and redox-active sites due to the Ce and Fe dopant in LaMnO$_3$ and the excellent charge transfer characteristics of the graphene materials. The as-developed cathode catalyst performed appreciable cycle stability up to 55 cycles at a limited capacity of 1000 mAh g^{-1} based on conventional glass fiber separators.

Keywords: Ce-doped LaMnO$_3$ perovskite; XPS of LaMnO$_3$; bifunctional activity; probe sonication; carbon-based composite

1. Introduction

In recent decades, green or renewable energy system development has garnered great interest worldwide due to high energy demand and the desire to save the atmosphere from air pollution. Secondary batteries are often considered the best renewable energy resources and alternative energy systems for gasoline fuels to fulfill the energy demand with high energy density [1–4]. Among them, Li-O$_2$ batteries have attracted much attention as the next-generation energy storage system with a high specific energy density that is integrated with low-cost and environmentally friendly air cathode catalysts. Their practical energy density is comparable to gasoline fuel and 10 times higher than commercially available Li-ion batteries [1,5,6]. Molecular oxygen (O$_2$) spread throughout the atmosphere is mainly used as a renewable oxidant source for Li-O$_2$ batteries. However, the practical viability of Li-O$_2$ batteries is still challenging due to the poor rate of oxygen reaction kinetics (O$_2$ reduction/evolution; ORR/OER), electrolyte instability and poor long-term stability [7–10]. On the other hand, noble metal catalysts such as Pt, Ru and Ir have been used as potential air electrodes in Li-O$_2$ batteries. This is owing to their high catalytic activity toward ORR/OER; however, they are very expensive, low-abundance materials, and they deliver poor cyclability due to their poor bifunctional activity, which is also a further reason to limit the commercialization of Li-O$_2$ batteries [6,11]. Therefore, non-precious metals/oxides and carbon-based material catalysts or their composites should be considered in the development of a new air electrode for Li-O$_2$ battery application [12–15].

Perovskite oxides (ABO$_3$) are the most frequently considered air catalysts in recently studied metal–air battery systems [7,16,17]. Because of their high electronic and ionic conducting properties, they have demonstrated excellent bifunctional catalytic activity towards ORR/OER reactions when the A sites and B sites are substituted with other elements [18,19]. Interestingly, lanthanum manganite (LaMnO$_3$) perovskite is a well-known catalyst for ORR in the perovskite family due to its defective cation-deficient lattice and the presence of multiple oxidation states, such as Mn^{3+} and Mn^{4+} [20–22]. Conversely, it has a poorer OER activity than LaCoO$_3$ and LaNiO$_3$ due to the Mn-O bond's lower binding energy on the LaMnO$_3$ surface. However, the bifunctional ORR and OER properties of LaMnO$_3$ can be tuned by the partial substitution of La and Mn sites with alkaline earth or rare-earth and transition-metal cations, respectively. Therefore, metal-ion-doped LaMnO$_3$ perovskites have received enormous attention in Li-air or Li-O$_2$ battery applications as cathode materials to replace the conventional Pt or Pt/C catalysts [23–27]. The discharge and recharge overpotentials of these LaMnO$_3$-based cathodes are significantly reduced as compared to commercial carbon and Pt/C catalysts [28]. However, most of the LaMnO$_3$-based perovskites are prepared by a sol–gel method, which produces non-porous microparticles with a low surface area, resulting in low storage capacity and poor cycle stability. Hence, the development of LaMnO$_3$ perovskites with more porosity and a high surface area is still required to increase the oxygen and Li$^+$ ion diffusion pathways for the high storage capacity and cyclability of Li-O$_2$ batteries [25].

In this study, a catalyst based on cerium- (Ce) and iron (Fe)-co-doped LaMnO$_3$ perovskite was prepared by the hydrothermal synthesis route. Glycine was used as a reducing and pore-generating agent in this process, resulting in mesoporous and large-surface-area perovskite material. Interestingly, the morphology of the particles was changed due to the addition of glycine. Moreover, the catalyst's activity was improved by changing the mole ratio of Ce dopant in the La$_{1-x}$Ce$_x$Fe$_{0.5}$Mn$_{0.5}$O$_3$ (x = 0.2 & 0.5 mol%) perovskite structure. To boost the electronic conductivity and electrocatalytic activity, the La$_{1-x}$Ce$_x$Fe$_{0.5}$Mn$_{0.5}$O$_3$ perovskites were infused with graphene nanosheets (GNS) via a probe sonication method. This is an efficient way to combine the transition-metal oxides with GNS, known to be highly conductive mechanical support materials [5,6], enhancing the composite catalyst's energy conversion and storage properties in scalable applications.

2. Experimental Section

2.1. Chemicals

Analytical-grade La(NO$_3$)$_3$·6H$_2$O and Ce(NO$_3$)$_3$·6H$_2$O (Alfa Aesar, Ward Hill, MA, USA), Fe(NO$_3$)$_3$·9H$_2$O (J.T. Baker, Center Valley, PA, USA), Mn(NO$_3$)$_2$·4H$_2$O (Alfa Aesar, Heysham, England), KOH (Acros Organics, Fair Lawn, NJ, USA), glycine and 5 wt.% Nafion solution (Sigma-Aldrich, Louis, MO, USA) were used as received. GNS powder was purchased from Scientech Corporation, Taiwan.

2.2. Preparation of Porous $La_{1-x}Ce_xFe_{0.5}Mn_{0.5}O_3$ Perovskite Oxides

Porous $La_{1-x}Ce_xFe_{0.3}Mn_{0.7}O_3$ perovskite oxides were prepared via a hydrothermal synthesis method using a glycine–nitrate complex mixture as the precursor material. In brief, stoichiometric amounts of nitrates of La, Ce, Fe and Mn precursors were separately dissolved in de-ionized water (each in 50 mL) and mixed together with constant stirring by a magnetic stirrer for 15 min. Then, a 0.3 M aqueous glycine solution was added dropwise into the mixed precursor solution. The color of the solution changed from pale orange to red-orange, indicating the formation of a glycine–metal nitrate complex. Then, the final solution pH was adjusted to around 8.0–9.0 by 1 M NH_4OH aqueous solution and stirred for 1 h at 25 °C. Finally, the mixed solution was poured into a 350 mL Teflon-lined autoclave for hydrothermal treatment at 180 °C for 6 h. After the temperature cooled down to 25 °C, the product was centrifuged and washed with deionized (DI) water and ethanol several times, and the collected residue was dried at 80 °C overnight. The dried powder was ground well, transferred into an alumina boat and kept inside a muffle furnace for calcination. The final powder was collected after calcining at 800 °C for 3 h in an air atmosphere, as depicted in Scheme 1i.

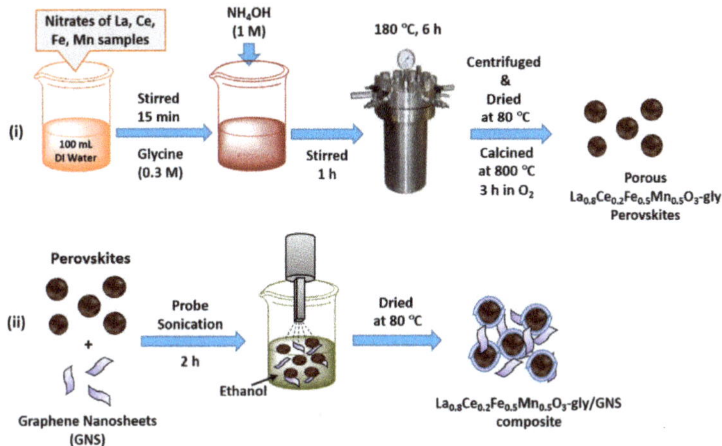

Scheme 1. (**i**) Schematic for the synthesis of LCFM(8255)-gly perovskites. (**ii**) Schematic for the synthesis of LCFM(8255)-gly/GNS composite.

$La_{1-x}Ce_xFe_{0.5}Mn_{0.5}O_3$ perovskite oxide catalysts with different mole concentrations of Ce (x = 0.2 and 0.5) were synthesized in the presence of 0.3 M glycine (gly) to reduce the formation of bulk crystals of CeO_2 and improve electrochemical performance. For comparison, $La_{0.5}Ce_{0.5}Fe_{0.5}Mn_{0.5}O_3$ perovskite was also prepared by the same experimental procedure without the addition of glycine. Finally, the as-synthesized oxide catalyst samples, such as $La_{0.5}Ce_{0.5}Fe_{0.5}Mn_{0.5}O_3$ and $La_{0.8}Ce_{0.2}Fe_{0.5}Mn_{0.5}O_3$, were designated as LCFM(5555)-no gly, LCFM(5555)-gly and LCFM(8255)-gly, respectively.

2.3. Preparation of Porous LCFM(8255)-gly/GNS Composite

The probe sonication method was employed to prepare the GNS-wrapped LCFM(8255)-gly composite materials (designated as LCFM(8255)-gly/GNS). Briefly, 0.7 g of GNS powder was first dispersed in 100 mL ethanol and sonicated using a probe sonicator for 1 h. The probe sonicator was operated at 5 mV amplitude with a 20 min pulse-on and 5 min pulse-off procedure for 1 h. During the process, the probe sonicator's output power and frequency were maintained at 2–4 W and 20 kHz, respectively. Then, about 0.3 g of the as-prepared LCFM(8255)-gly perovskite catalyst was added into the ethanolic GNS solution, followed by probe sonication for 1 h under the same conditions. Finally, the mixture was dried at

60 °C in an air oven; after this, the dried LCFM(8255)-gly/GNS composite catalyst could be used for further analysis, as depicted in Scheme 1ii. To compare with the electrochemical performance of LCFM(8255)-gly/GNS composite, the LCFM(5555)-gly/GNS composite catalyst was also prepared by the above-mentioned synthesis conditions.

2.4. Materials Characterization

Morphology and crystallinity of the synthesized materials were characterized by SEM (Hitachi-S2600, Hitachi Ltd., Tokyo, Japan) and XRD (Bruker D2 PHASER, Karlsruhe, Germany) analysis. TEM (JEM-2100, JEOL Ltd., Tokyo, Japan) equipment was used to confirm the morphology as well as the GNS layer on the perovskite microsphere particles. N_2 adsorption–desorption isotherm analysis (Micromeritics, Gemini VII, Monchegladbach, Germany) was carried out to examine the materials' surface area and porous nature. XPS (VG Scientific ESCALAB 250, Thermo Fisher, CA, USA) analysis was used to characterize the valence state of elements present in the composite samples. A three-electrode configuration for oxygen electrocatalysis study was employed using a CHI405 potentiostat (CHI Instruments, Austin, TX, USA) with a catalyst-modified rotating disk electrode (RDE) as working electrode, Ag/AgCl (sat. KCl) as reference electrode, and Pt wire as the counter electrode. Cyclic voltammetry (potential range = 2.0 to 4.5 V vs. Li/Li$^+$; scan rate = 1 mV s^{-1}) and AC impedance (frequency range = 1 MHz to 0.01 Hz; amplitude = 5 mV) analysis for two-electrode systems were performed using an Autolab PGSTAT30 (Metrohm Autolab B.V., Houten, The Netherlands).

2.5. Preparation of Electrode and Electrochemical Measurements

In order to prepare the oxygen electrode to evaluate the bifunctional activity of the oxide catalyst, the as-prepared composite catalyst (10 mg mL^{-1}) was dispersed in a 1:3 v/v ratio DI water to isopropanol solvent mixture containing 80 µL of Nafion (5 wt.%) binder solution. This mixture was ultrasonicated (ultrasonic cleaner: DC300H; frequency −40 kHz; max. output power −300 W) for 1 h to develop a homogeneous catalyst ink. Then, about 20 µL of the catalyst ink was drop-coated onto the RDE (Geometric area = 0.196 cm^2) surface and dried at 50 °C for 30 min. The as-modified RDE was used to further evaluate the bifunctional (ORR/OER) activity of the catalyst in a 0.1 M KOH electrolyte.

To prepare the oxygen cathode for Li-O$_2$ battery application, a slurry was prepared by mixing the as-prepared catalyst samples, Ketjen black and PVDF in N-methyl pyrrolidone at a weight ratio of 80:10:10 and continuously stirred for 12 h at 25 °C to form a homogeneous mixture. Then, the catalyst mixture slurry was spray-coated (Sono-Tek Corporation, Milton, NY, USA) on both sides of the Ni-foam (2 × 2 cm^2) matrix with a flow rate of 2 mL min^{-1} and dried at 120 °C for 12 h. Finally, the catalyst-coated Ni-foam electrodes were prepared as disks (diameter = 1.3 cm^2) for cell assembly. The catalyst loading of each electrode was approximately 1 mg cm^{-2}. EL-CELL was assembled in an Ar-filled glovebox environment (H$_2$O < 0.5 ppm; O$_2$ < 0.5 ppm) to evaluate the oxygen electrocatalyst's performance toward Li-O$_2$ battery application. Here, the as-prepared composite electrodes were used as a cathode, Li foil as an anode and commercial glass fiber (Whatman; thickness ~260 µm) as a separator for cell preparation. A 1 M Lithium bis(trifluoromethanesulfonyl) imide-Tetraethylene glycol dimethyl ether (LiTFSI-TEGDME)-based electrolyte was used with 0.5 M LiI as an additive. All the cells were tested at the potential cut-off range of 2.0–4.5 V vs. Li/Li$^+$ at 100 mA g^{-1} using the Arbin BT2000 battery test system (Arbin Instruments, College Station, TX, USA). For all these cells, the oxygen supply was passed (10 mL min^{-1}) through a thin microtube (diameter <1 mm) using pure O$_2$ (99.999%) at 1 atm.

3. Results and Discussion

3.1. Surface and Structural Characterization of Perovskite Samples

Glycine-assisted hydrothermal synthesis was performed to obtain the Ce- and Fe-co-doped LaMnO$_3$ perovskites, such as LCFM(5555) and LCFM(8255), for energy conversion and storage application. The surface morphology of the as-prepared perovskite samples was investigated by SEM analysis, as shown in Figure 1A–C. Figure 1A exhibits the SEM image of LCFM(5555)-no gly, which reveals non-uniform particle growth during the hydrothermal reaction. Figure 1B shows the multi-shape morphology with a microscale that consists of LCFM(5555)-gly microspheres and CeO_2 microplate-like structure. The microspheres of perovskite may form due to the efficient chelation chemistry between metal ions and the glycine metal–organic complex. These metal–organic frameworks can favor mesopore formation on the perovskite catalyst surface under the high-temperature calcination process. On the other hand, the high mol% of Ce doping can initiate the growth of aggregated CeO_2 microstructures, resulting in lower electrochemical activity than the nanoparticles of CeO_2. Therefore, we reduced the mol% of Ce to 0.2 to obtain the LCFM(8255) perovskite catalyst. Figure 1C clearly reveals the uniform LCFM(8255) microspheres with the co-existence of CeO_2 nanoparticles. The crystallinity of these synthesized samples was also evaluated by XRD analysis. Figure 1D displays the XRD patterns of (a) LCFM(5555)-no gly, (b) LCFM(5555)-gly and (c) LCFM(8255)-gly. The planes are clearly indexed to partially doped $LaMnO_3$ perovskite with rhombohedral crystal planes (JCPDS no. 89-8775) and a space group of R-3C. All three patterns showed the co-existence of cubic CeO_2 phase (JCPDS no. 04-0593) with a significantly decreasing intensity of (111), (220) and (311) planes related to reducing the mol% of Ce doping. However, the addition of glycine can stabilize particle growth and improve the crystallinity of (b) LCFM(5555)-gly and (c) LCFM(8255)-gly samples, as compared to LCFM(5555)-no gly. In addition, the crystallite sizes of these samples were calculated using the Scherrer equation on the (110) plane; the derived values were 7.5, 16.7 and 10.1 nm for (a) LCFM(5555)-no gly, (b) LCFM(5555)-gly and (c) LCFM(8255)-gly, respectively.

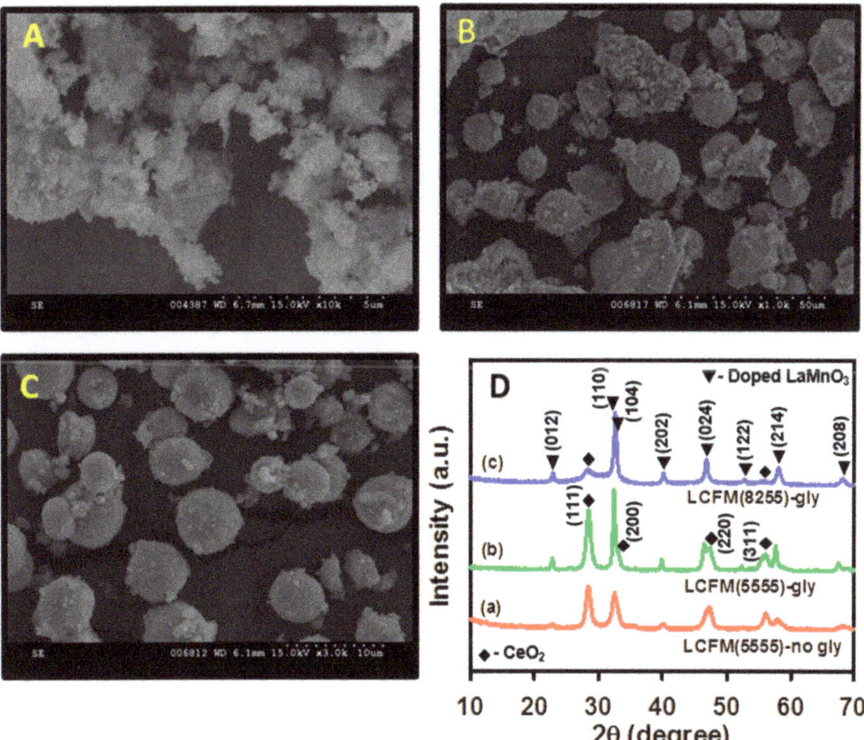

Figure 1. SEM images of (**A**) LCFM(5555)-no gly, (**B**) LCFM(5555)-gly and (**C**) LCFM(8255)-gly. (**D**) XRD patterns of (**a**) LCFM(5555)-no gly, (**b**) LCFM(5555)-gly and (**c**) LCFM(8255)-gly.

The surface area and pore-size distribution of the as-prepared perovskites were studied by using Brunauer–Emmett–Teller (BET) analysis. Figure 2A–C shows the N_2 adsorption–desorption isotherm curves of LCFM(5555)-no gly, LCFM(5555)-gly and LCFM(8255)-gly samples. All the samples are followed the Type IV adsorption isotherm with an H_2 hysteresis loop. The hysteresis loop at the relative pressure range from 0.4 to 1.0 P/P_o is attributed to the mesoporous nature of the sample. Hence, the LCFM(5555)-gly and LCFM(8255)-gly samples are exposed high specific surface areas of 23.81 and 38.38 m^2 g^{-1}, respectively, as compared to the LCFM(5555)-no gly (1.69 m^2 g^{-1}).

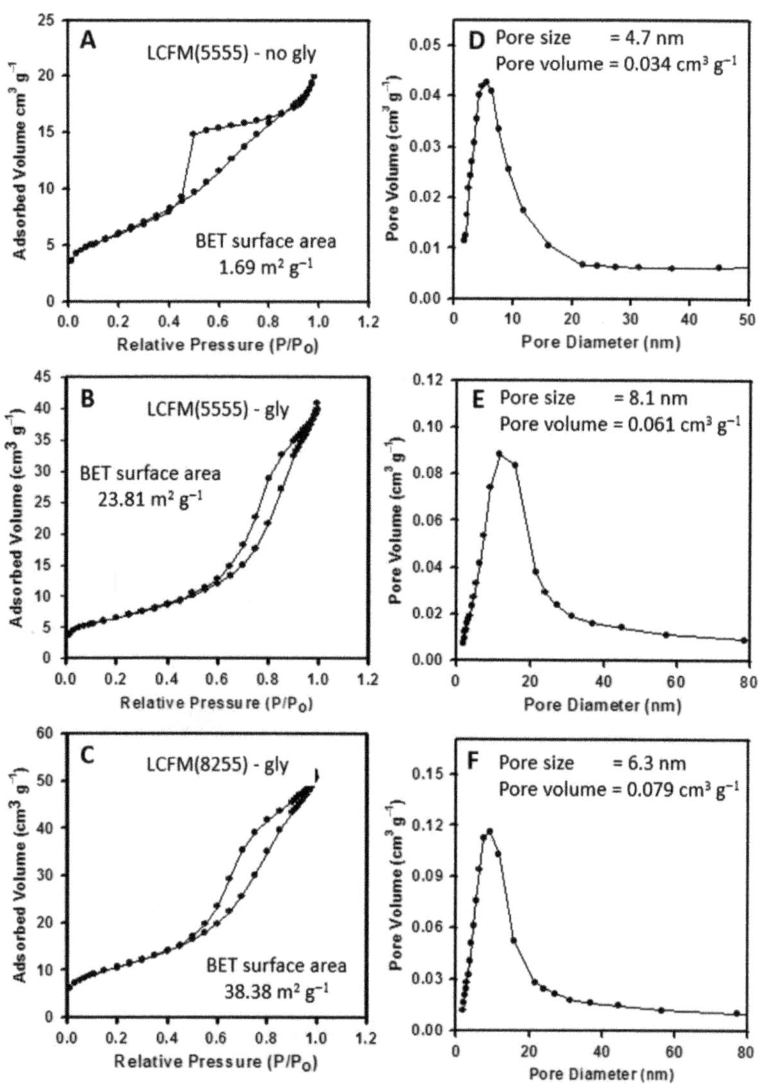

Figure 2. (**A–C**) N_2 adsorption–desorption and (**D–F**) pore-size distribution curves of LCFM(5555)-no gly, LCFM(5555)-gly and LCFM(8255)-gly catalyst samples.

As seen in Figure 2A, a short-range interaction of the hysteresis loop was observed, and the desorption curve was suddenly dropped at 0.46 P/P_0, which indicates that the lowest number of pores may occur in LCFM(5555)-no gly. It is also evidenced from the pore-size distribution curve of LCFM(5555)-no gly (Figure 2D), which shows the pore size and pore volume of about 4.7 nm and 0.034 $m^3 \, g^{-1}$, respectively. The pore size and pore volume are significantly enhanced to approximately 8.1 nm and 0.061 $m^3 \, g^{-1}$ for the LCFM(5555)-gly sample (Figure 2E), which is due to the multi-structure morphology of microspheres and microplates of LCFM(5555)-gly and CeO_2, respectively. On the other hand, the pore size of LCFM(8255)-gly is reduced to 6.3 nm, and the pore volume is increased to approximately 0.071 $m^3 \, g^{-1}$, which confirms the residue of CeO_2 nanoparticles may be positioned onto the pores of the LCFM(8255)-gly microsphere particles (Figure 2F). The obtained specific surface area (38.38 $m^2 \, g^{-1}$) of LCFM(8255)-gly perovskite is much higher than the previously reported $LaMnO_3$ perovskites [16,23–25,29]. Thus, the existence of a large number of pores in LCFM(8255)-gly microspheres can provide support for the efficient diffusion of electrolyte ions and O_2 gas in both O_2 electrocatalysis and Li-O_2 battery performance.

The XPS analysis further confirmed the presence of elements in the as-synthesized LCFM(8255)-gly perovskite catalyst. Figure 3A shows the wide-scan XPS curve of LCFM(8255)-gly perovskite sample, which confirmed the presence of La, Ce, Fe, Mn and O elements. The narrow-scan XPS curves of these elements were further fitted by the Gaussian–Lorentzian (G-L) method (30% of G-L ratio) by using XPSPEAK 4.1 software. Figure 3B shows the high-resolution La 3d XPS curve wherein the spin–orbit splitting (SOS) of $3d_{5/2}$ and $3d_{3/2}$ are separated with 16.8 eV. The multiplet at $3d_{5/2}$ is also separated by 4.2 eV, indicating the +3 oxidation state of lanthanum elements. In Figure 3C, eight peaks appeared for Ce 3d with an SOS of 18.6 eV. The doublet peaks at 885.6 eV ($3d_{5/2}$) and 904.2 eV ($3d_{3/2}$) belong to the Ce^{3+} state, whereas the other three doublet peaks at 883.7/902.3, 891.9/910.5 and 900.8/919.4 eV are related to the Ce^{4+} oxidation state [30]. The high-intensity peak at 919.4 eV also indicates the presence of CeO_2 nanoparticles on the perovskite sample.

Figure 3D,E displays the deconvoluted spectra of Fe 2p and Mn 2p peaks, wherein the SOS of $2p_{3/2}$ and $2p_{1/2}$ are separated by 13.1 and 11.8 eV for Fe and Mn, respectively. The high-intensity peaks at 713.7 and 644.2 eV in Fe $2p_{3/2}$ and Mn $2p_{3/2}$ are corresponding to the +3 oxidation state of Fe and Mn elements in the LCFM(8255)-gly sample. In Mn $2p_{3/2}$, two doublet peaks were observed for the co-existence of the mixed oxidation state of Mn^{3+} and Mn^{4+}. Several Mn^{3+} can be oxidized into Mn^{4+} while doping with other elements on the B site; however, the high intensity of Mn^{3+} indicates that the main component is present in the form of +3 in composite [27]. The ratio of Mn^{4+}/Mn^{3+} was calculated as 0.68, and it was obtained from the peak area of Mn^{4+} and Mn^{3+}. Furthermore, the core-level spectrum of O 1s peak (Figure 3F) is deconvoluted into four peaks wherein the peaks at 530.4 and 532.2 eV are corresponding to the lattice $O_{lat.}$ (O^{2-}) and surface-level adsorbed oxygen species $O_{sur.}$ (O^-, O_2^- or O_2^{2-}). The ratio of $O_{lat.}/O_{total}$ is approximately 0.27, which confirms the presence of oxygen vacancies. Moreover, the high intensity of the $O_{sur.}$ peak indicates the stronger covalence of the B-O bond, which favors the O_2^-/OH^- exchange. The obtained results suggest that the doping of Ce and Fe in $LaMnO_3$ perovskite structures can increase the Mn^{4+} generation and oxygen vacancies, which will improve the kinetics of ORR and OER performance [25–27].

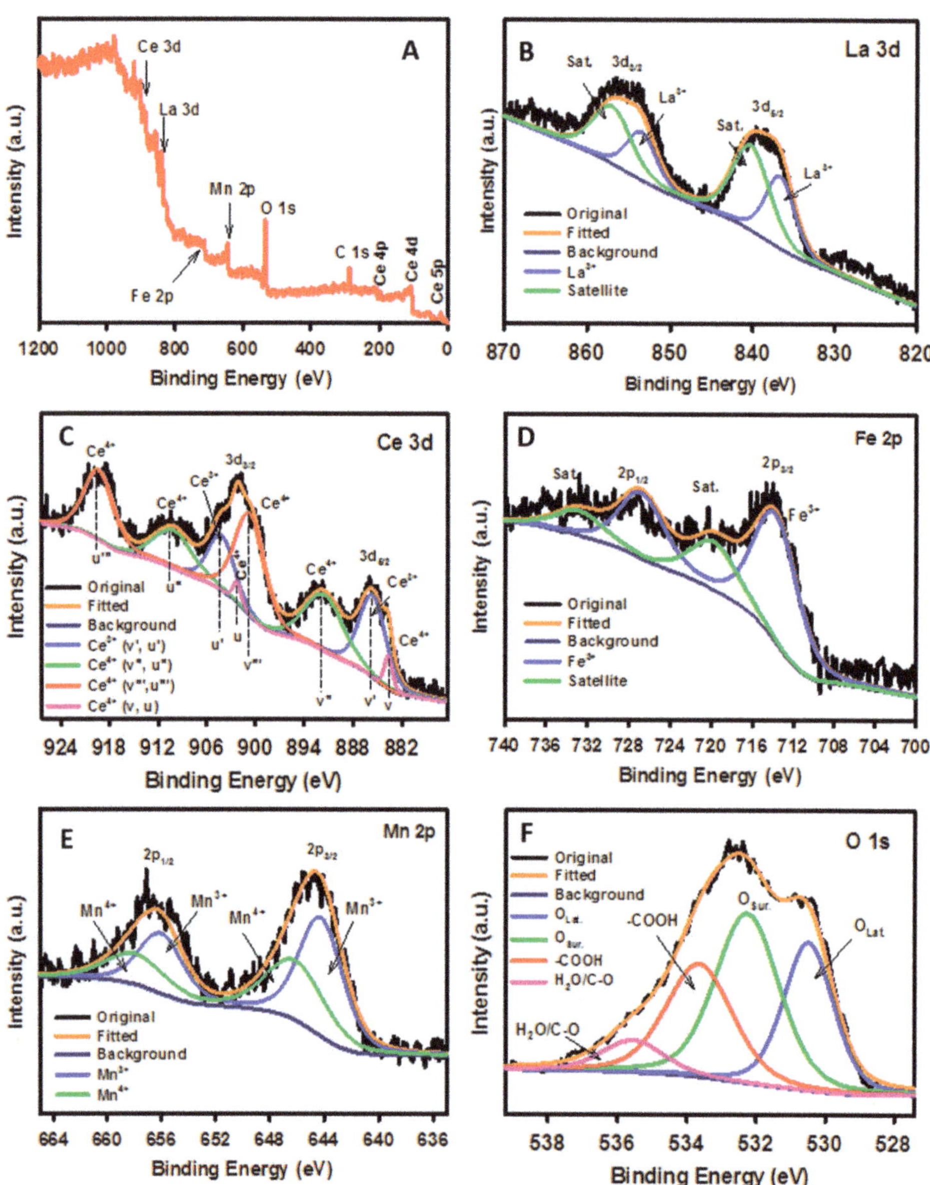

Figure 3. (**A**). XPS survey for elements present in LCFM(8255)-gly, deconvoluted XPS curves of (**B**) La 3d, (**C**) Ce 3d, (**D**) Fe 2p, (**E**) Mn 2p and (**F**) O 1s peak in LCFM(8255)-gly sample.

3.2. Surface and Structural Characterization of LCFM(8255)-gly/GNS Composite Catalyst

In this work, the as-synthesized LCFM(8255)-gly microsphere sample was combined with GNS to improve the electrochemical properties. The composite catalyst was already denoted as LCFM(8255)-gly/GNS. The integration of GNS with LCFM(8255)-gly microspheres is clearly characterized by TEM, XRD and micro-Raman spectroscopy analyses. Figure 4A–D displays the TEM images of LCFM(8255)-gly and LCFM(8255)-gly/GNS composite catalysts. The inset of Figure 4C clearly reveals the decoration of GNS on the LCFM(8255)-gly particles, on which the graphene sheets completely cover the perovskite particles (Figure 4C,D); however, it is not found in the other catalysts presented in Figure 4A,B. The average thickness of the GNS layer on LCFM(8255)-gly particles is ca. 6.5 nm.

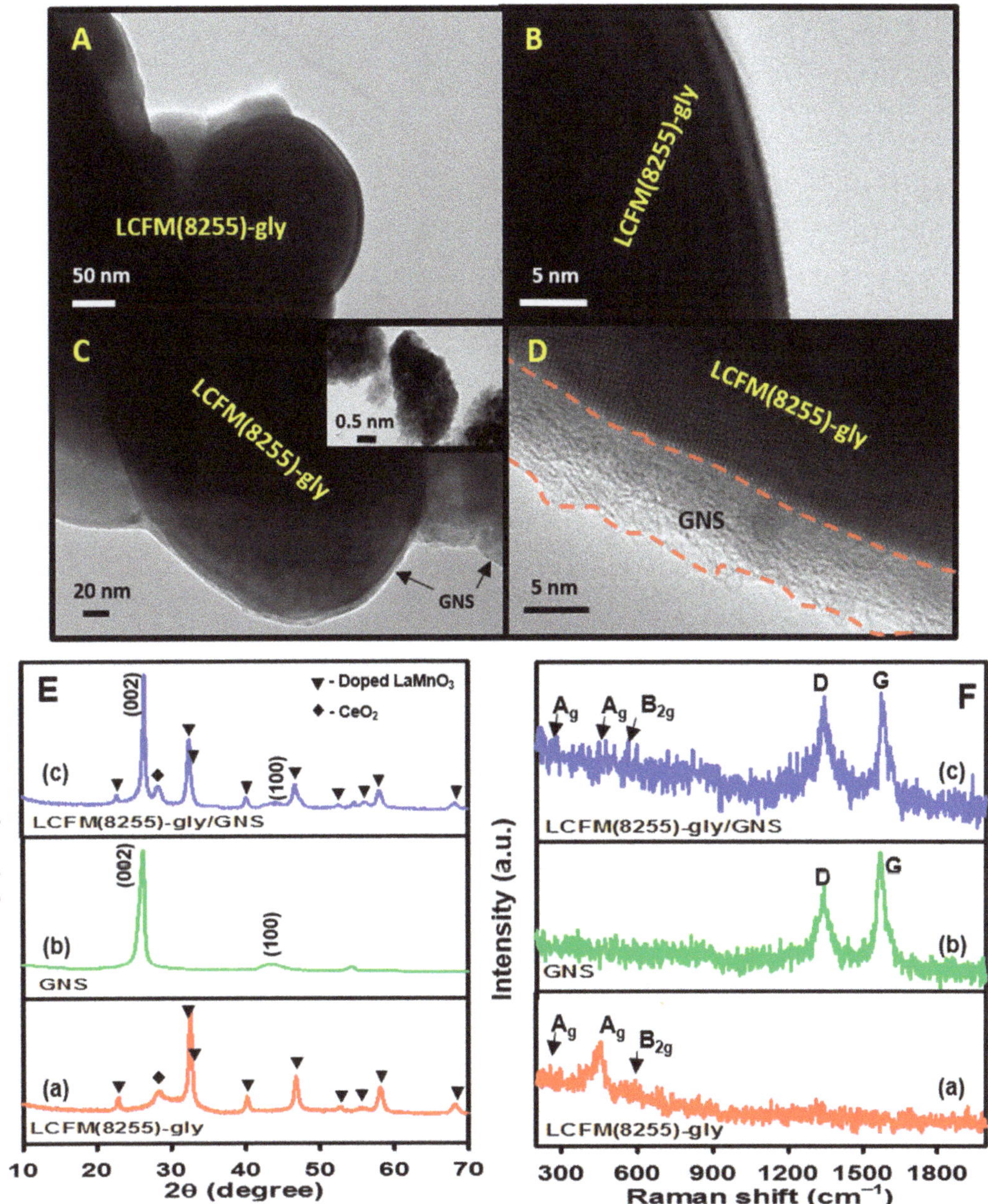

Figure 4. (**A,B**) TEM images of LCFM(8255)-gly and (**C,D**) LCFM(8255)-gly/GNS composite. (**E**) XRD patterns and (**F**) micro-Raman spectra of (**a**) LCFM(8255)-gly, (**b**) GNS and (**c**) LCFM(8255)-gly/GNS composite.

Figure 4E depicts the XRD patterns of (a) LCFM(8255)-gly, (b) GNS and (c) LCFM(8255)-gly/GNS composite catalysts, which confirms the existence of (002) and (100) planes of GNS in the composite sample. The crystallinity of LCFM(8255)-gly sample was clearly discussed in the previous section. However, the diffraction peak of the (002) plane of GNS has shifted to a lower angle in the LCFM(8255)-gly/GNS composite sample, which indicates the disorder of graphene sheets by the interaction with LCFM(8255)-gly particles. Moreover, the disordered properties are further confirmed by micro-Raman spectra (Figure 4F). The typical D band and G band of the GNS appeared around 1347 and 1574 cm^{-1} (Figure 4F(b)), whereas it appeared around 1347 and 1576 cm^{-1} for the LCFM(8255)-gly/GNS composite (Figure 4F(c)). The D/G intensity ratios of GNS and LCFM(8255)-gly/GNS composite catalysts are calculated to ca. 0.98 and 1.0, respectively. These results indicate that the disordered graphene layer is possible in a composite sample and that the basal plane of GNS can be attached strongly to the perovskite microsphere particles that renders more uniform coverage on the surface of the particle. In addition, our micro-Raman spectra also reveal that the presence of LCFM(8255)-gly particles by the appearance of peaks at 264, 476 and 592 cm^{-1} are related to the rotational (A_g) and Jahn–Teller stretching (A_g, B_{2g}) modes of the rhombohedrally distorted LaMnO$_3$ perovskite, respectively [31]. The above results demonstrate clear evidence of the formation of LCFM(8255)-gly/GNS composite catalyst.

3.3. Electrochemical Performances

3.3.1. Oxygen Electrocatalysis

The electrocatalytic ORR studies were performed by cyclic voltammetry (CV) and linear sweep voltammetry (LSV) in 0.1 M KOH electrolyte using different electrodes; the data are displayed in Figure 5A,B. Typical ORR curves were observed from CV scans (20 mV s^{-1}) for all the catalyst-modified electrodes, such as 20 wt.% of Pt/C, GNS, LCFM(8255)-gly, LCFM(5555)-gly/GNS and LCFM(8255)-gly/GNS composite catalysts (Figure 5A). The ORR peak potential (at -0.35 V vs. Ag/AgCl) of LCFM(8255)-gly perovskite is far better than the previously published undoped LaMnO$_3$ perovskite catalyst (at -0.48 V vs. Ag/AgCl) [29]. However, the peak potential of the composite electrodes, i.e., LCFM(5555)-gly/GNS and LCFM(8255)-gly/GNS, was significantly improved, nearly to commercial Pt/C electrode activity, which is due to the high electrical conductivity of GNS. Further, the LSV polarization studies were conducted for the above-mentioned electrodes by the RDE measurements, which operated at a 1600 rpm rotation speed and a 20 mV s^{-1} scan rate in O$_2$-saturated 0.1 M KOH electrolyte. The LCFM(8255)-gly/GNS composite electrode shows good electrocatalytic activity to the reduction of oxygen molecule, as can be observed in Figure 5B, in the presence of N$_2$- and O$_2$-saturated electrolytes. In addition, the LCFM(8255)-gly/GNS composite electrode (at -0.12 V vs. Ag/AgCl; 1.82 mA cm^{-2}) shows higher catalytic activity with respect to the onset potential and limiting current density than that of LCFM(8255)-gly (at -0.26 V vs. Ag/AgCl; 1.11 mA cm^{-2}), GNS (at -0.17 V vs. Ag/AgCl; 1.62 mA cm^{-2}) and LCFM(5555)-gly/GNS composite electrodes (at -0.14 V vs. Ag/AgCl; 1.43 mA cm^{-2}). The obtained results are almost closer to the activity of commercial 20 wt.% Pt/C catalyst (at 0.01 V vs. Ag/AgCl; 1.77 mA cm^{-2}). The half-wave potentials of 20 wt.% Pt/C and LCFM(8255)-gly/GNS composite electrodes are -0.106 and -0.227 V vs. Ag/AgCl, respectively, and the potential difference ($\Delta E_{1/2} = E_{1/2,\text{catalyst}} - E_{1/2,\text{Pt/C}}$) is ca. 121 mV. In addition, the RDE measurements were conducted using the LCFM(8255)-gly/GNS composite electrode at different rotational speeds in the range of 400–2500 rpm (Figure 5C). The electron transfer number (n) can be calculated from these RDE polarization curves, which is one of the major factors for ORR catalyst evaluation [25,27]. To analysis the ORR reaction kinetics, the Koutecky–Levich Equations (1) and (2) can be followed based on the relation of inverted limiting current density J^{-1} versus square roots of rotation speed ($\omega^{-1/2}$), the data can be seen in Figure 5D.

$$\frac{1}{J} = \frac{1}{J_L} + \frac{1}{J_K} = \frac{1}{B\omega^{1/2}} + \frac{1}{J_K} \tag{1}$$

$$B = 0.62nFC_0 D_0^{1/2}v^{-1/6} \tag{2}$$

where J, J_L and J_K are the measured currents, diffusion-limited current and kinetic current densities, respectively, ω is the electrode angular rotation speed, n is the electron transfer number, F is the Faraday constant (96485 A·s mol^{-1}), v is the kinematic viscosity (0.01 cm^2 s^{-1}), C_0 is the bulk concentration (1.9 $\times$ 10^{-5} mol cm^{-3}) and D_0 is the diffusion coefficient (1.2 $\times$ 10^{-6} mol cm^{-3}) of dissolved O$_2$ in 0.1 M KOH electrolyte. From the slope of the linear fit in Figure 5D, the average electron transfer number of the LCFM(8255)-gly/GNS composite catalyst is calculated as ca. 3.85, which indicates the one-pot, four-electron transfer pathway. The reaction mechanism for the four-electron transfer of the ORR process can be expressed as follows [25]:

$$O_2 + H_2O + 4e^- \rightarrow 4OH^- \tag{3}$$

The above-obtained results indicate that the LCFM(8255)-gly/GNS composite sample is the best ORR active catalyst and a better replacement to the high-cost conventional Pt/C in fuel cells and metal–air batteries.

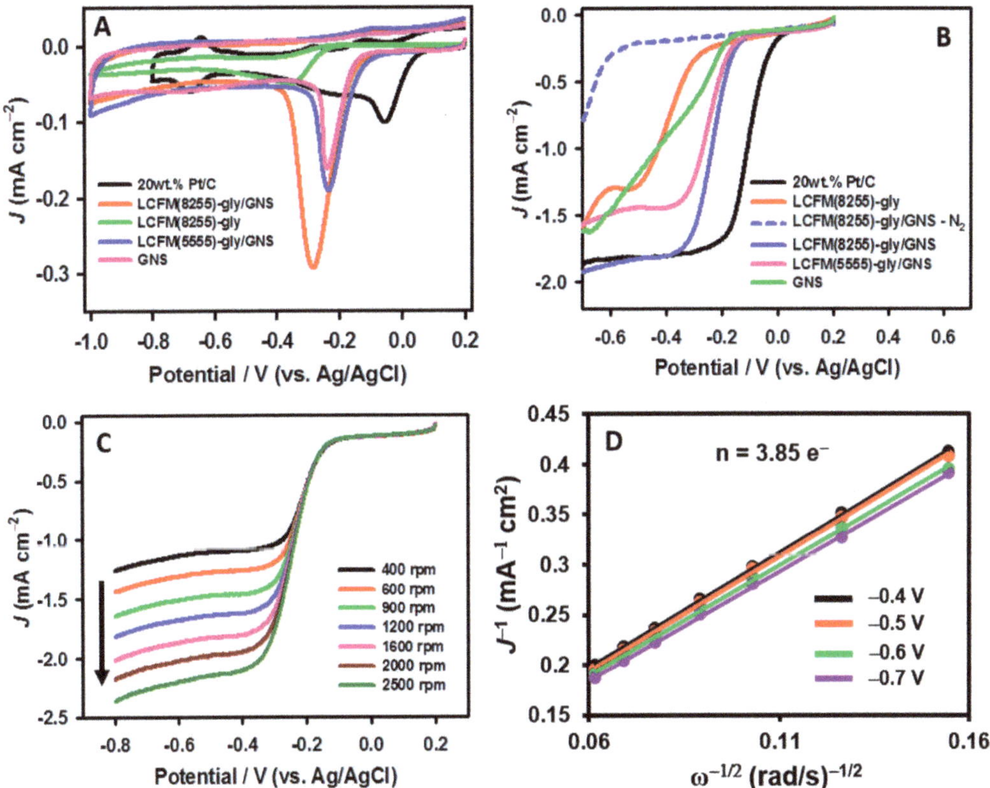

Figure 5. (**A**). CVs and (**B**). LSVs for ORR kinetics at various electrodes recorded in O$_2$-saturated 0.1 M KOH; Scan rate: 20 mV s^{-1}; Rotation speed: 1600 rpm for LSV measurements. (**C**). LSVs of RDE measurements using LCFM(8255)-gly/GNS composite electrode at various rotation speeds from 400 to 2500 rpm in O$_2$-saturated 0.1 M KOH; Scan rate: 20 mV s^{-1}, (**D**). Koutecky–Levich plot of j^{-1} vs. $\omega^{-1/2}$.

3.3.2. Bifunctional ORR/OER Activity

To further evaluate the bifunctional characteristics, LSV polarization curves were obtained using (a) 20 wt.% Pt/C, (b) LCFM(8255)-gly/GNS, (c) LCFM(5555)-gly/GNS and (d). GNS electrodes in O_2-saturated 0.1 M KOH electrolyte (Figure 6A). The total overpotential difference ($\Delta E = E_{OER} - E_{ORR}$) between ORR and OER curves of those electrodes was measured at 0.5 and 1 mA cm^{-2}, respectively. The ΔE value of LCFM(8255)-gly/GNS composite electrode is ca. 0.94 V vs. Ag/AgCl, which is very close to the 20 wt.% Pt/C (at 0.92 V vs. Ag/AgCl) and lower than the LCFM(5555)-gly/GNS (at 1.02 V vs. Ag/AgCl) and bare GNS (at 1.06 V vs. Ag/AgCl) electrodes. Hence, the results revealed that the LCFM(8255)-gly/GNS composite electrode is a superior bifunctional catalyst, which can be used as a cathode in Li-O_2 battery application instead of a conventional Pt/C catalyst and other carbon-based cathodes.

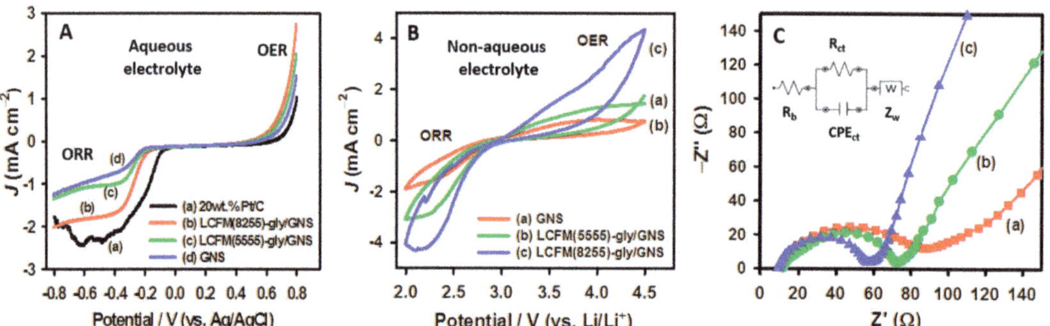

Figure 6. (**A**). LSVs for ORR and OER kinetics at various electrodes recorded in O_2-saturated 0.1 M KOH; Scan rate: 20 mV s^{-1}; Rotation speed: 1600 rpm, (**B**). CVs for ORR and OER kinetics using (a) GNS, (b) LCFM(5555)-gly/GNS and (c) LCFM(8255)-gly/GNS-based Li-O_2 cells in 1 M LiTFSI-TEGDME + 0.5 M LiI at a scan rate of 1 mV s^{-1}, (**C**). EIS data for electron transfer properties of (a) GNS, (b) LCFM(5555)-gly/GNS and (c) LCFM(8255)-gly/GNS-based Li-O_2 cells.

Prior to use these catalysts as Li-O_2 battery cathode, their bifunctionality is to be further examined in a non-aqueous electrolyte (1 M LiTFSI-TEGDME + 0.5 M LiI) system. Figure 6B shows the CVs of ORR and OER activity of (a) GNS, (b) LCFM(5555)-gly/GNS, and (c) LCFM(8255)-gly/GNS-based cathodes in O_2-saturated non-aqueous electrolyte at a scan rate of 1 mV s^{-1}. It is clearly indicated that the LCFM(8255)-gly/GNS composite cathode shows much lower onset overpotential and high peak current density than the GNS and LCFM(5555)-gly/GNS electrodes. This superior ORR and OER activity can be achieved due to the intrinsic electronic conductivity of LCFM(8255)-gly/GNS composite catalyst materials. The Nyquist plot (Figure 6C) of EIS data were taken to evaluate the charge transfer properties of the (a) GNS, (b) LCFM(5555)-gly/GNS and (c) LCFM(8255)-gly/GNS cathodes. The inset of Figure 6C shows the equivalent-circuit model for the fitting of the Nyquist plots. It was found that the lower charge transfer impedance can be observed at LCFM(8255)-gly/GNS cathodes, as compared to GNS and LCFM(5555)-gly/GNS. As noted, the charge transfer characteristics of perovskite and graphene composite are significantly improved due to the interaction between the redox-active perovskite structure and conductive graphene support.

3.3.3. Li-O$_2$ Battery Performance

Figure 7A shows the discharge (charge storage) curves of (a) GNS, (b) LCFM(5555)-gly/GNS and (c) LCFM(8255)-gly/GNS-composite-based Li-O$_2$ battery cathodes, operated in the voltage range of 2000–4500 mV vs. Li/Li$^+$ at a current density of 100 mA g^{-1}, which shows capacities around 4860, 5796 and 8475 mAh g^{-1}, respectively. The discharge–charge overpotential difference of the cathodes was also studied at the limiting discharge–charge capacity of 1000 mAh g^{-1} at a current density of 100 mA g^{-1}. As clearly seen in Figure 7B, the LCFM(8255)-gly/GNS composite cathode shows a very low overpotential difference of about 271 mV vs. Li/Li$^+$, as compared to LCFM(5555)-gly/GNS (around 365 mV vs. Li/Li$^+$) and GNS cathodes (around 420 mV vs. Li/Li$^+$). The obtained results are in strong accordance with the CV and RDE measurements. To further evaluate the performance of Li-O$_2$ batteries, the cycle stability curves of the as-developed cathodes were obtained at 100 mA g^{-1} with limiting capacities of 1000 mAh g^{-1}; the data can be seen in Figure 7 ((C) GNS, (D) LCFM(5555)-gly/GNS and (E) LCFM(8255)-gly/GNS samples). The comparison of cycle stability with respect to discharge-limiting capacity was also shown in Figure 7D. As noted in cycle stability curves, the discharge–charge polarization overpotential rapidly increases to the high cut-off voltage for GNS (up to only 20 cycles) rather than composite cathodes, even in the presence of LiI redox mediator in the electrolyte system, confirms the poor, sluggish ORR/OER kinetics. This is also due to the parasitic reaction between the electrolyte ions and carbon cathodes. Conversely, the LCFM(5555)-gly/GNS composite cathode exhibits stable polarization for up to 35 cycles, and the charge potential is below 3500 mV vs. Li/Li$^+$; however, the discharge overpotential reached cut-off voltage. Therefore, a sudden drop in discharge capacity after 35 cycles was observed (Figure 7C), due to the less porous nature of LCFM(5555)-gly particles. On the other hand, the battery with LCFM(8255)-gly/GNS cathode also shows stable polarization up to 55 cycles, as the charge potential does not increase beyond 3100 mV vs. Li/Li$^+$, indicating the excellent catalytic activity of the LCFM(8255)-gly/GNS composite. The charge potential on both composite catalysts is still lower upon increasing the cycle life of the battery; this is mainly due to the LiI-soluble catalyst, which can also help to decompose the Li$_2$O$_2$ discharge product during the charging process. On the other hand, the increasing discharge potential range towards the cut-off discharge voltage is due to the blocking of oxygen and Li$^+$ ion diffusion sites by the continuous deposition of Li$_2$O$_2$ during the discharge process, resulting in poor ORR activity of the catalyst in prolonged cycles. Thus, the results suggest that the air cathode should have a large number of pores and high surface area catalysts, which may enhance the oxygen and Li$^+$ ion diffusion property that facilitates the high-performance battery lifespan. Therefore, as compared to previous reports [23–27] demonstrating the perovskite oxide catalyst prepared by citric acid and urea as fuel or a chelating agent, the present synthesis method with glycine can increase the surface area due to the presence of a large volume of mesopores and the integration of conductive GNS as a support, which can further improve the electrocatalytic properties of the perovskite materials.

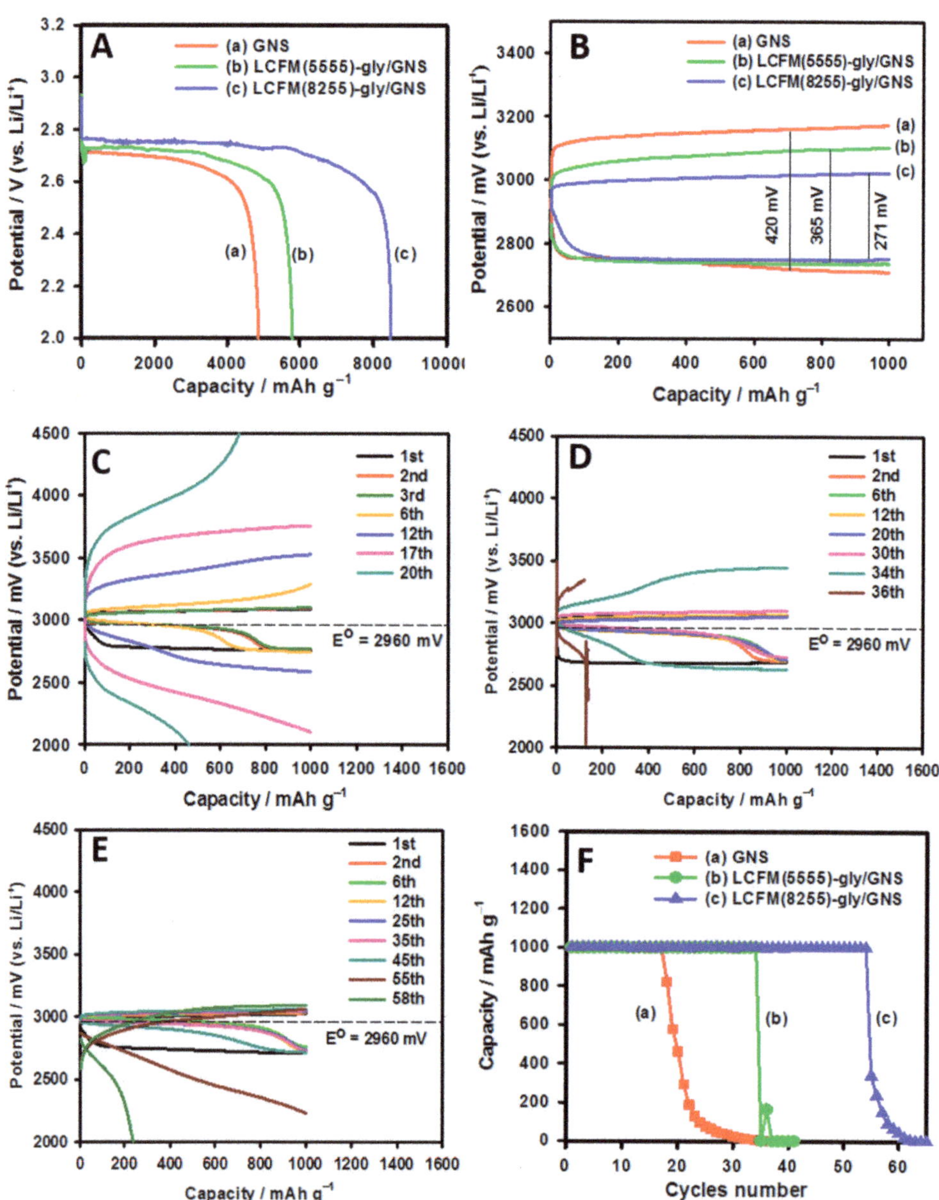

Figure 7. (**A**) Initial discharge curves of (**a**) GNS, (**b**) LCFM(5555)-gly/GNS and (**c**) LCFM(8255)-gly/GNS-based Li-O$_2$ cells at a current density of 100 mA g^{-1}. Discharge–charge curves of (**a**) GNS, (**b**) LCFM(5555)-gly/GNS and (**c**) LCFM(8255)-gly/GNS-based Li-O$_2$ cells for (**B**) overpotential difference and (**C–E**) cycling stability analysis. (**F**) Cycle life of (**a**) GNS, (**b**) LCFM(5555)-gly/GNS and (**c**) LCFM(8255)-gly/GNS-based Li-O$_2$ cells. Current density: 100 mA g^{-1}; Limited discharge capacity: 1000 mAh g^{-1}; Electrolyte: 1 M LiTFSI-TEGDME + 0.5 M LiI.

4. Conclusions

A simple synthesis methodology has been developed for synthesizing a highly porous, high-surface-area A-site- and B-site-doped LaMnO$_3$ perovskite. XRD analysis confirmed the distorted rhombohedral crystal structure for this perovskite is due to Ce^{3+} and Fe^{3+} cation doping, respectively. The CV results confirmed that the LCFM(8255)-gly/GNS composite electrode shows better ORR onset potential and peak current density than the bare LCFM(8255)-gly and LCFM(5555)-gly/GNS composite electrodes. The electrochemical ORR/OER kinetics of the proposed catalyst can be significantly improved by incorporating conductive GNS as support and reducing the mol% of Ce ion doping on the A site. Furthermore, the effect of dual cation doping on the $LaMnO_3$ structure increases the ratio of Mn^{4+}/Mn^{3+} species and generates more oxygen vacancies on the perovskite crystal structure. Finally, the as-developed LCFM(8255)-gly/GNS composite catalyst achieved efficient bifunctional activity comparable to commercial Pt/C catalysts. Therefore, the proposed synthesis method can derive low-cost and highly active perovskite materials, including a large-surface-area and highly porous structure, as the air cathode for energy storage and conversion applications.

Author Contributions: C.K.—designed the work, conducted material synthesis and characterization, wrote final draft; C.-N.W.—experimental section and characterization; N.K.—characterization and data analysis; Z.-H.W.—experimental section and characterization; B.T.—participated in the writing process; L.-F.H.—experimental section and characterization; S.A.—materials synthesis and characterization; S.P.—review and editing; T.-F.H.—data analysis, review and editing; Y.-J.J.L.—review and editing, and C.-C.Y. contributed in supervision, funding acquisition, review and editing the draft. All authors have read and agreed to the published version of the manuscript.

Funding: This research was funded by Ministry of Science and Technology (MOST), Taiwan, project number is MOST 109-3116-F-131-001-CC1 and the APC was funded by Ming Chi University of Technology (MCUT), New Taipei City, Taiwan.

Institutional Review Board Statement: Not applicable.

Informed Consent Statement: Not applicable.

Data Availability Statement: No new data were created or analyzed in this study.

Acknowledgments: Authors grateful to acknowledge the Battery Research Center of Green Energy, MCUT, New Taipei City, Taiwan for the valuable support to conduct this work.

Conflicts of Interest: The authors declare no conflict of interest.

References

1. Liu, X.; Gong, H.; Wang, T.; Guo, H.; Song, L.; Xia, W.; Gao, B.; Jiang, Z.; Feng, L.; He, J. Cobalt-Doped Perovskite-Type Oxide $LaMnO_3$ as Bifunctional Oxygen Catalysts for Hybrid Lithium–Oxygen Batteries. *Chem. Asian J.* **2018**, *13*, 528–535. [CrossRef] [PubMed]
2. Zhang, X.; Gong, Y.; Li, S.; Sun, C. Porous Perovskite $La_{0.6}Sr_{0.4}Co_{0.8}Mn_{0.2}O_3$ Nanofibers Loaded with RuO_2 Nanosheets as an Efficient and Durable Bifunctional Catalyst for Rechargeable Li-O_2 Batteries. *ACS Catal.* **2017**, *7*, 7737–7747. [CrossRef]
3. Yoo, H.D.; Markevich, E.; Salitra, G.; Sharon, D.; Aurbach, D. On the challenge of developing advanced technologies for electrochemical energy storage and conversion. *Mater. Today* **2014**, *17*, 110–121. [CrossRef]
4. Karuppiah, C.; Hsieh, Y.C.; Beshahwured, S.L.; Wu, X.W.; Wu, S.H.; Jose, R.; Lue, S.J.; Yang, C.C. Poly(vinyl alcohol)/Melamine Composite Containing LATP Nanocrystals as a High-Performing Nanofibrous Membrane Separator for High-Power, High-Voltage Lithium-Ion Batteries. *ACS Appl. Energy Mater.* **2020**, *3*, 8487–8499. [CrossRef]
5. Lai, Y.; Chen, W.; Zhang, Z.; Qu, Y.; Gan, Y.; Li, J. Fe/Fe_3C decorated 3-D porous nitrogen-doped graphene as a cathode material for rechargeable Li-O_2 batteries. *Electrochim. Acta* **2016**, *191*, 733–742. [CrossRef]
6. Chen, X.; Chen, S.; Nan, B.; Jia, F.; Lu, Z.; Deng, H. In situ, facile synthesis of $La_{0.8}Sr_{0.2}MnO_3$/nitrogen-doped graphene: A high-performance catalyst for rechargeable Li-O_2 batteries. *Ionics* **2017**, *23*, 2241–2250. [CrossRef]
7. Wei, C.; Karuppiah, C.; Yang, C.; Shih, J.; Jessie, S. Journal of Physics and Chemistry of Solids Bifunctional perovskite electrocatalyst and PVDF/PET/PVDF separator integrated split test cell for high performance Li-O_2 battery. *J. Phys. Chem. Solids* **2019**, *133*, 67–78. [CrossRef]
8. Lin, S.; Li, Y.; Luo, S.; Ren, X.; Deng, L.; Mi, H.; Zhang, P.; Sun, L.; Gao, Y. 3D-ordered porous nitrogen and sulfur Co-Doped carbon supported PdCuW nanoparticles as efficient catalytic cathode materials for Li-O_2 batteries. *Electrochim. Acta* **2018**, *272*, 33–43. [CrossRef]
9. Wang, L.; Ara, M.; Wadumesthrige, K.; Salley, S.; Ng, K.Y.S. Graphene nanosheet supported bifunctional catalyst for high cycle life Li-air batteries. *J. Power Sources* **2013**, *234*, 8–15. [CrossRef]

10. Cao, C.; Xie, J.; Zhang, S.; Pan, B.; Cao, G.; Zhao, X. Graphene-like δ-MnO_2 decorated with ultrafine CeO_2 as a highly efficient catalyst for long-life lithium-oxygen batteries. *J. Mater. Chem. A* **2017**, *5*, 6747–6755. [CrossRef]

11. Kumar, S.; Munichandraiah, N. Nanoparticles of a Pt_3Ni alloy on reduced graphene oxide (RGO) as an oxygen electrode catalyst in a rechargeable Li-O_2 battery. *Mater. Chem. Front.* **2017**, *1*, 873–878. [CrossRef]

12. Liu, S.; Zhu, Y.; Xie, J.; Huo, Y.; Yang, H.Y.; Zhu, T.; Cao, G.; Zhao, X.; Zhang, S. Direct growth of flower-like δ-MnO_2 on three-dimensional graphene for high-performance rechargeable Li-O_2 batteries. *Adv. Energy Mater.* **2014**, *4*, 1–9. [CrossRef]

13. Zhu, Y.; Liu, S.; Jin, C.; Bie, S.; Yang, R.; Wu, J. MnO_x decorated CeO_2 nanorods as cathode catalyst for rechargeable lithium-air batteries. *J. Mater. Chem. A* **2015**, *3*, 13563–13567. [CrossRef]

14. Li, J.; Shu, C.; Hu, A.; Ran, Z.; Li, M.; Zheng, R.; Long, J. Tuning oxygen non-stoichiometric surface via defect engineering to promote the catalysis activity of Co_3O_4 in Li-O_2 batteries. *Chem. Eng. J.* **2020**, *381*. [CrossRef]

15. Shui, J.; Lin, Y.; Connell, J.W.; Xu, J.; Fan, X.; Dai, L. Nitrogen-Doped Holey Graphene for High-Performance Rechargeable Li-O_2 Batteries. *ACS Energy Lett.* **2016**, *1*, 260–265. [CrossRef]

16. Karuppiah, C.; Rani, K.K.; Wang, S.F.; Devasenathipathy, R.; Yang, C.C. Dry particle coating preparation of highly conductive $LaMnO_3$@C composite for the oxygen reduction reaction and hydrogen peroxide sensing. *J. Taiwan Inst. Chem. Eng.* **2018**, *93*, 94–102. [CrossRef]

17. Ashok, A.; Kumar, A.; Bhosale, R.R.; Almomani, F.; Malik, S.S.; Suslov, S.; Tarlochan, F. Combustion synthesis of bifunctional $LaMO_3$ (M = Cr, Mn, Fe, Co, Ni) perovskites for oxygen reduction and oxygen evolution reaction in alkaline media. *J. Electroanal. Chem.* **2018**, *809*, 22–30. [CrossRef]

18. Islam, M.; Jeong, M.; Oh, I.; Nam, K.; Jung, H. Role of strontium as doping agent in $LaMn_{0.5}Ni_{0.5}O_3$ for oxygen electro-catalysis. *J. Ind. Eng. Chem.* **2020**, *85*, 94–101. [CrossRef]

19. Wang, C.C.; Cheng, Y.; Ianni, E.; Jiang, S.P.; Lin, B. A highly active and stable $La_{0.5}Sr_{0.5}Ni_{0.4}Fe_{0.6}O_{3-\delta}$ perovskite electrocatalyst for oxygen evolution reaction in alkaline media. *Electrochim. Acta* **2017**, *246*, 997–1003. [CrossRef]

20. Khellaf, N.; Kahoul, A.; Naamoune, F.; Alonso-Vante, N. Electrochemistry of Nanocrystalline $La_{0.5}Sr_{0.5}MnO_3$ Perovskite for the Oxygen Reduction Reaction in Alkaline Medium. *Electrocatalysis* **2017**, *8*, 450–458. [CrossRef]

21. Yan, L.; Lin, Y.; Yu, X.; Xu, W.; Salas, T.; Smallidge, H.; Zhou, M.; Luo, H. $La_{0.8}Sr_{0.2}MnO_3$-Based Perovskite Nanoparticles with the A-Site Deficiency as High Performance Bifunctional Oxygen Catalyst in Alkaline Solution. *ACS Appl. Mater. Interfaces* **2017**, *9*, 23820–23827. [CrossRef]

22. Sunarso, J.; Torriero, A.A.J.; Zhou, W.; Howlett, P.C.; Forsyth, M. Oxygen reduction reaction activity of La-based perovskite oxides in alkaline medium: A thin-film rotating ring-disk electrode study. *J. Phys. Chem. C* **2012**, *116*, 26108. [CrossRef]

23. Fu, Z.; Lin, X.; Huang, T.; Yu, A. Nano-sized $La_{0.8}Sr_{0.2}MnO_3$ as oxygen reduction catalyst in nonaqueous Li/O_2 batteries. *J. Solid State Electrochem.* **2012**, *16*, 1447–1452. [CrossRef]

24. Lu, F.; Wang, Y.; Jin, C.; Li, F.; Yang, R.; Chen, F. Microporous $La_{0.8}Sr_{0.2}MnO_3$ perovskite nanorods as efficient electrocatalysts for lithium-air battery. *J. Power Sources* **2015**, *293*, 726–733. [CrossRef]

25. Zhao, Y.; Hang, Y.; Zhang, Y.; Wang, Z.; Yao, Y.; He, X.; Zhang, C.; Zhang, D. Strontium-doped perovskite oxide $La_{1-x}Sr_xMnO_3$ (x = 0, 0.2, 0.6) as a highly efficient electrocatalyst for nonaqueous Li-O_2 batteries. *Electrochim. Acta* **2017**, *232*, 296–302. [CrossRef]

26. Zhao, Y.; Liu, T.; Shi, Q.; Yang, Q.; Li, C.; Zhang, D.; Zhang, C. Perovskite oxides $La_{0.4}Sr_{0.6}Co_xMn_{1-x}O_3$ (x = 0, 0.2, 0.4) as an effective electrocatalyst for lithium-air batteries. *Green Energy Environ.* **2018**, *3*, 78–85. [CrossRef]

27. Lv, Y.; Li, Z.; Yu, Y.; Yin, J.; Song, K.; Yang, B.; Yuan, L.; Hu, X. Copper/cobalt-doped $LaMnO_3$ perovskite oxide as a bifunctional catalyst for rechargeable Li-O_2 batteries. *J. Alloys Compd.* **2019**, *801*, 19–26. [CrossRef]

28. Han, X.; Hu, Y.; Yang, J.; Cheng, F.; Chen, J. Porous perovskite $CaMnO_3$ as an electrocatalyst for rechargeable Li-O_2 batteries. *Chem. Commun.* **2014**, *50*, 1497–1499. [CrossRef] [PubMed]

29. Karuppiah, C.; Thirumalraj, B.; Alagar, S.; Piraman, S.; Li, Y.J.J.; Yang, C.C. Solid-state ball-milling of Co_3O_4 nano/microspheres and carbon black endorsed $LaMnO_3$ perovskite catalyst for bifunctional oxygen electrocatalysis. *Catalysts* **2021**, *11*, 76. [CrossRef]

30. Hu, Q.; Yue, B.; Shao, H.; Yang, F.; Wang, J.; Wang, Y.; Liu, J. Facile syntheses of cerium-based $CeMO_3$ (M = Co, Ni, Cu) perovskite nanomaterials for high-performance supercapacitor electrodes. *J. Mater. Sci.* **2020**, *55*, 8421–8434. [CrossRef]

31. Merten, S.; Shapoval, O.; Damaschke, B.; Samwer, K.; Moshnyaga, V. Magnetic-Field-Induced Suppression of Jahn-Teller Phonon Bands in $(La_{0.6}Pr_{0.4})_{0.7}Ca_{0.3}MnO_3$: The Mechanism of Colossal Magnetoresistance shown by Raman Spectroscopy. *Sci. Rep.* **2019**, *9*, 1–7. [CrossRef] [PubMed]

 *nanomaterials*

Article

A Facile Chemical Method Enabling Uniform Zn Deposition for Improved Aqueous Zn-Ion Batteries

Congcong Liu [†], Qiongqiong Lu *,[†] , Ahmad Omar and Daria Mikhailova *

Leibniz Institute for Solid State and Materials Research (IFW) Dresden e.V., 01069 Dresden, Germany; congcong.liu@ifw-dresden.de (C.L.); a.omar@ifw-dresden.de (A.O.)
* Correspondence: q.lu@ifw-dresden.de (Q.L.); d.mikhailova@ifw-dresden.de (D.M.)
† These authors contributed equally to this work.

Abstract: Rechargeable aqueous Zn-ion batteries (ZIBs) have gained great attention due to their high safety and the natural abundance of Zn. Unfortunately, the Zn metal anode suffers from dendrite growth due to nonuniform deposition during the plating/stripping process, leading to a sudden failure of the batteries. Herein, Cu coated Zn (Cu–Zn) was prepared by a facile pretreatment method using $CuSO_4$ aqueous solution. The Cu coating transformed into an alloy interfacial layer with a high affinity for Zn, which acted as a nucleation site to guide the uniform Zn nucleation and plating. As a result, Cu–Zn demonstrated a cycling life of up to 1600 h in the symmetric cells and endowed a stable cycling performance with a capacity of 207 mAh g^{-1} even after 1000 cycles in the full cells coupled with a V_2O_5-based cathode. This work provides a simple and effective strategy to enable uniform Zn deposition for improved ZIBs.

Keywords: zinc metal anode; copper coating; alloy interfacial layer; uniform Zn deposition; aqueous zinc-ion battery

Citation: Liu, C.; Lu, Q.; Omar, A.; Mikhailova, D. A Facile Chemical Method Enabling Uniform Zn Deposition for Improved Aqueous Zn-Ion Batteries. *Nanomaterials* **2021**, *11*, 764. https://doi.org/10.3390/nano11030764

Academic Editor: Jaehyun Hur

Received: 26 February 2021
Accepted: 16 March 2021
Published: 18 March 2021

Publisher's Note: MDPI stays neutral with regard to jurisdictional claims in published maps and institutional affiliations.

1. Introduction

The significantly growing consumption of fossil fuels worldwide has led to severe global warming. To reduce the usage of fossil fuels and sustainably develop renewable energy utilization, e.g., solar energy and wind energy, advanced energy storage systems (such as batteries and supercapacitors) are in high demand [1–4]. For example, lithium-ion batteries (LIBs) have seen tremendous success as one of the most common types of power source in the portable electronics market, due to their high energy density [5–7]. However, the fire risk of the flammable organic electrolyte, high cost, as well as limited reserves of lithium, severely restrict the largescale implementation of LIBs in automotive and stationary storage applications [8,9]. In this regard, developing alternative battery technology is important and essential. Recently, rechargeable aqueous Zn-ion batteries (ZIBs) have been increasingly investigated, due to their low cost, the high theoretical capacity of zinc (819 mAh g^{-1}), and the low redox potential (-0.762 V vs. SHE). More importantly, the aqueous electrolyte possesses the merits of nonflammable, nontoxic, and environmental benignity [10–12]. However, ZIBs suffer from a challenging issue related to the zinc anode, which is afflicted with uncontrollable dendrite formation during the Zn stripping/plating process. The problem is generally attributed to what is known as the "tip effect", where, due to charge aggregation near the protuberances on the inhomogeneous zinc surface, zinc ions are easily absorbed onto the protruded tips and nucleate preferentially on these spots, thereby triggering continuous growth and finally form Zn dendrites [13–16].

To address this issue, various methods have been developed, such as optimizing the electrolytes [17], stabilizing the structure by employing host materials [18], and modifying the surface [19–24]. Among surface modification methods, an efficient strategy to stabilize the Zn anode during the stripping/plating process is a surface coating with a metal such as In [19], Au [20], Ag [21], and Cu [21] coating, which acts as a nucleation site to guide uniform Zn deposition. For instance, Zhang et al. developed Ag and Cu coatings on Zn metal through the thermal evaporation method [21]. They found out that Ag–Zn and Cu–Zn alloys were formed after cycling, which improved the affinity to Zn and further contributed to the uniform nucleation and deposition of Zn. Owing to its easy alloy formation with Zn and low cost, Cu is a promising material for Zn stabilization as compared with noble metals such as Ag and Au. However, the high temperature required in the thermal evaporation method would not enable cost efficiency and would limit largescale applicability. Consequently, developing direct, economically viable methods to fabricate an effective Cu coating on Zn metal is required for the practical application of ZIBs. From the point of view of efficacy and cost, chemical methods may be a good alternative. Therefore, Cu coating on Zn metal prepared by a chemical route to guide uniform Zn nucleation and deposition is a meaningful route to explore.

In this work, Cu coated Zn (Cu–Zn) was prepared by a facial chemical method. Cu coating was transformed to Cu–Zn alloy after cycling, which acted as the nucleation sites to guide uniform Zn nucleation and deposition. As a result, Cu–Zn not only showed improved cycling life in the symmetric cells, but also enabled a stable cycling performance in the full cells coupled with V_2O_5-based cathode. The results demonstrated that our strategy was facile and efficient to guide uniform Zn deposition for improved ZIBs.

2. Materials and Methods

2.1. Preparation of the Cu Coated Zn

A pristine zinc foil (125 μm in thickness, Goodfellow GmbH, Bad Nauheim, Germany) was polished with sandpaper and then cut into discs with a diameter of 12 mm. A portion of 100 μL of 0.1 M copper sulfate ($CuSO_4$, 99.99%, Sigma-Aldrich, St Louis, MO, USA) aqueous solution was dropped on the zinc disc surface and kept for 3 min to get a one-side-coated Zn foil. The treated zinc metal was washed with deionized water a few times and stored in the air to dry naturally.

2.2. Preparation of Poly(3,4-ethylenediophene)-Coated V_2O_5 (V_2O_5-PEDOT) Cathode

For the synthesis of V_2O_5-PEDOT, 7 g of commercial V_2O_5 powder (Sigma-Aldrich, St Louis, MO, USA) was dispersed in 70 mL of deionized water, then 1 mL of 3,4-ethylenediophene (EDOT, Aladdin, Shanghai, China) was added dropwise. The mixture was continuously stirred for 6 days and filtered. The obtained powder was dried in a vacuum oven at 70 °C for overnight [25]. The cathode was prepared by coating the slurry of V_2O_5-PEDOT, Super C65 (TIMCAL, Bodio, Switzerland), and polyvinylidene fluoride (PVDF, Solef 21216, Solvay, Milan, Italy) in N-methyl pyrrolidone (99%, NMP, Sigma-Aldrich, St Louis, MO, USA) at a mass ratio of 8:1:1 on to stainless steel mesh (500 pores per linear inch, wire diameter of 0.2 mm, Gelon lib group, Linyi, China) and dried in a vacuum at 80 °C overnight. The mass loading of V_2O_5-PEDOT was about 1 mg·cm^{-1}.

2.3. Characterization

A scanning electron microscope (FFG-SEM, Zeiss-Leo Gemini 1530, Carl Zeiss NTS GmbH, Oberkochen, Germany) was employed to characterize the surface morphologies of zinc electrodes at different stages of the experiment process. Elemental mappings were performed using energy dispersive X-ray spectroscopy with a Bruker XFlash 6 detector (Bruker, Karlsruhe, Germany). X-ray diffraction (XRD) was carried out on a Panalytical X'pert Pro diffractometer device (Panalytical, Almelo, Netherlands) operating with Co Kα radiation in reflection mode. The X-ray photoelectron spectroscopy (XPS) analysis was performed in a PHI 1600 ESCA (PerkinElmer, Waltham, MA, USA) spectrometer with a monochromatic Al-Kα source. The binding energies were calibrated using the C 1s peak at 284.8 eV.

2.4. Electrochemical Characterization

The ZIBs were assembled using Swagelok cells under ambient conditions in air. 3 M Zn (CF$_3$SO$_3$)$_2$ (98%, Sigma-Aldrich, St Louis, MO, USA) aqueous solution and glass fiber (Whatman GF/D, Whatman, Clifton, NJ, USA) were used as electrolyte and separator, respectively. The galvanostatic cycling of symmetric cells was performed with different current densities using LAND CT2001A (Wuhan Land Electronic Co., Ltd., Wuhan, China) potentiostat. The galvanostatic cycling performance of full cells was tested in the voltage range of 0.3 V $\leq$ U $\leq$ 1.6 V vs. Zn/Zn$^+$ using LAND potentiostat.

3. Results

Cu-coated Zn (Cu–Zn) was prepared via an in situ chemical method by dropping an aqueous CuSO$_4$ solution on bare zinc foils. The treated Zn foils were subsequently washed and dried. Due to the potential difference between Zn^{2+}/Zn and Cu^{2+}/Cu, a spontaneous replacement reaction occurred and the surface of Zn changed to a black color after treatment (insets of Figure 1a,b). Scanning electron microscopy (SEM) was conducted to characterize the morphology of bare Zn and Cu–Zn. The bare zinc had a rough and scratched surface with a unique texture due to the polish process (Figure 1a). Cu–Zn clearly showed a coated surface (Figure 1b). The elemental mapping of Cu–Zn showed Cu uniformly distributed on the surface of Zn (Figure 2). The thickness of the Cu coating layer was about 20 μm by the cross-section SEM image (Figure 1c). The high resolution Cu 2p XP spectrum of Cu–Zn showed two peaks located at 932.5 eV and 952.3 eV, assigned to metallic Cu (Figure 1d) [26]. It should be noted that it was challenging to distinguish between the metallic Cu and Cu(I) by binding energy of Cu 2p, as the Cu 2p signals overlapped. Although the Cu LMM Auger peak is recommended to be used for the identification and analysis of Cu(I), unfortunately, the intensity of the signal in the Cu LMM Auger region of Cu–Zn was too weak for further analysis [27]. However, in our case since the reaction represented a chemical reduction of Cu(II), the formation of Cu(I) was highly unlikely.

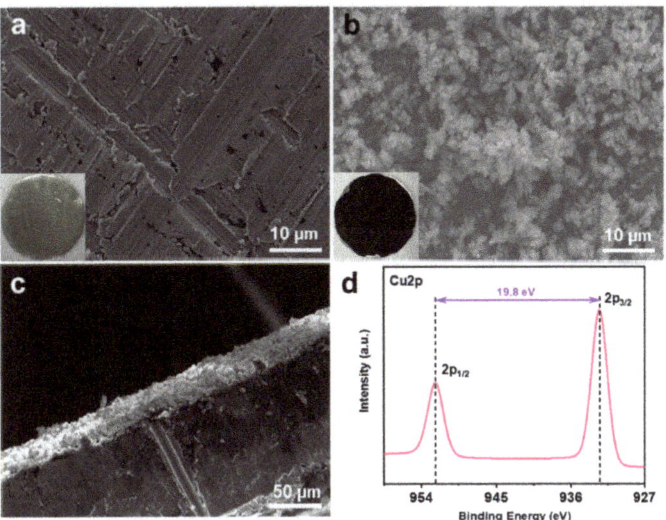

Figure 1. SEM images and optical images (insets) of (**a**) bare Zn foil and (**b**) Cu–Zn foil, (**c**) cross-section SEM images of Cu–Zn foil, (**d**) high resolution Cu 2p XP spectrum measured on Cu–Zn foil.

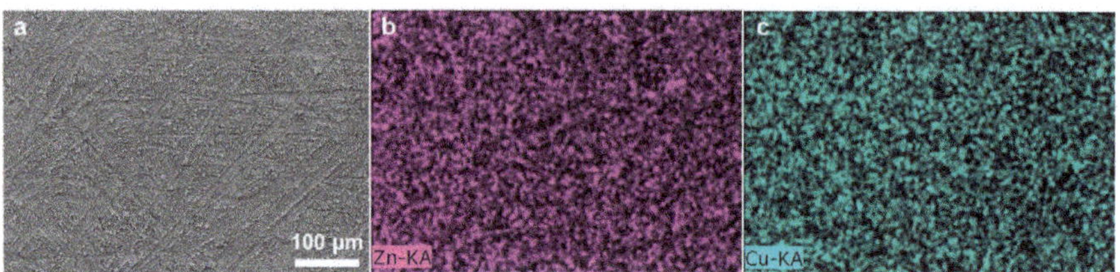

Figure 2. (**a**) SEM image and (**b,c**) the corresponding elemental mappings of Cu–Zn.

In order to evaluate the effect of the Cu coating on the Zn plating/stripping behavior, symmetric cells comprised of bare Zn and Cu–Zn (henceforth referred to Zn//Zn and Cu-Zn//Cu–Zn respectively) were assembled and tested. Galvanostatic charge/discharge was performed at different current densities. At 0.2 mA cm^{-2}, the Zn//Zn cell exhibited an increasing overpotential after 70 h cycling, subsequently showing an abnormal voltage drop at 110 h, indicating a short circuit of the battery due to the zinc dendrite. In contrast, at 0.2 mA cm^{-2}, Cu-Zn//Cu–Zn showed a markedly prolonged cycling life of over 1600 h with a lower overpotential (Figure 3a). Even when the current density was increased to 0.5 mA cm^{-2} and 1 mA cm^{-2}, Cu-Zn//Cu–Zn cells showed a prolonged cycle life of at least 600 h and 290 h, respectively, whereas Zn//Zn cells exhibited a cycling life of only 300 h and 145 h, respectively (Figure 3b,c). Moreover, the cycling life of Cu–Zn symmetric cells was much better than that for most of the reported works (Table 1). The results highlight the effectiveness of the Cu-coating strategy via the chemical route towards improving the cycling stability and prolonging the cycle life, while at the same time involving a rather facile process.

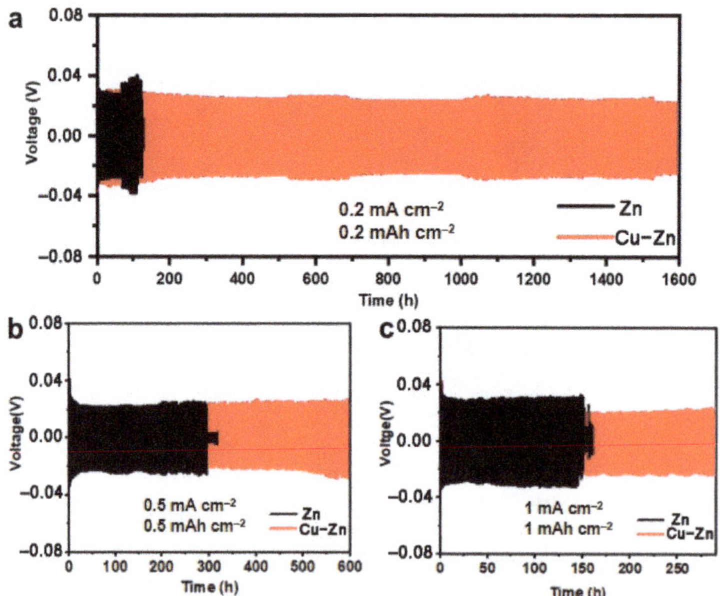

Figure 3. Long-term galvanostatic discharge/charge profiles of symmetric cells with bare Zn and Cu–Zn at current density of (**a**) 0.2 mA cm^{-2}, (**b**) 0.5 mA cm^{-2}, and (**c**) 1 mA cm^{-2}.

Table 1. Comparison of the performance of Cu–Zn symmetric cells with recent literature on various Zn surface modification strategies.

Protective Layers	Current Density (mA cm^{-2})	Capacity (mAh cm^{-2})	Life (h)	Reference
Cu coating	0.2	0.2	1600	This work
	0.5	0.5	600	
	1	1	290	
In coating	0.2	0.2	1500	[19]
Au coating	0.25	0.05	2000	[17]
MXene	0.2	0.2	800	[22]
CaCO$_3$ coating	0.25	0.05	840	[24]
TiO$_2$ coating	1	1	150	[23]

The nucleation overpotential is related to the kinetics of the Zn nucleation and deposition process, and the Zn nucleation consequently determines the quality of Zn deposition [28]. Thus, the Zn nucleation overpotential was measured and compared. As shown in Figure 4a, Cu–Zn showed a lower nucleation potential of 12 mV as compared to 44 mV for Zn, indicating a lower nucleation barrier for Cu–Zn which contributed to a uniform Zn nucleation. To further study the effect of Cu coating on the Zn deposition, the morphology of Cu and Cu–Zn after Zn deposition was characterized (Figure 4b,c). Bare Zn shows a huge amount of Zn microclusters with a porous structure, demonstrating uneven deposition. In contrast, Cu–Zn exhibited a dense and uniform Zn deposit, confirming the beneficial role of the Cu coating.

Figure 4. (a) The voltage-time curves during Zn nucleation and deposition on Zn and Cu–Zn at 1 mA cm^{-2}, SEM images of (b) Zn and (c) Cu–Zn after Zn-depositing with a capacity of 2 mAh cm^{-2} at current density of 1 mA cm^{-2}.

To further investigate the Zn electrodeposition behavior, the morphology of Zn and Cu–Zn electrodes after 30 cycles at a high current density of 5 mA cm^{-2} was characterized (Figure 5a,b). Bare Zn showed a bulk Zn deposition morphology while Cu–Zn exhibited a uniform and compact morphology. Elemental mapping of Cu–Zn after 50 cycles at a high current density of 5 mA cm^{-2}, confirmed that the uniform distribution of the Cu coating was maintained (Figure 6). It was expected that Cu–Zn alloy was formed during the stripping/plating, due to the negative Gibbs free energy of the reaction [29,30]. In order to check this, a Cu–Zn electrode after 1 cycle was characterized by XRD. As is shown in Figure 7a, two weak peaks at 2θ = 44.1° and 49.5° additionally appeared in the XRD pattern of Cu–Zn after cycling, corresponding to the CuZn$_5$ phase (PDF Number 00-035-1152). The XRD pattern of Cu–Zn after 100 cycles showed that the Cu–Zn alloy was retained, confirming the durability of Cu–Zn alloy during cycling (Figure 7b). The binding energy of Zn-CuZn$_5$ (−1.94 eV) was higher than that of Zn-Cu (−1.58 eV), demonstrating a high Zn affinity of the formed CuZn$_5$ [30]. Thus, the formed alloy could effectively reduce the activation energy of zinc nucleation and the plating resistance of zinc. Consequently, the zinc grew in a smaller size and achieved uniform nucleation without the formation of long and disordered dendrites.

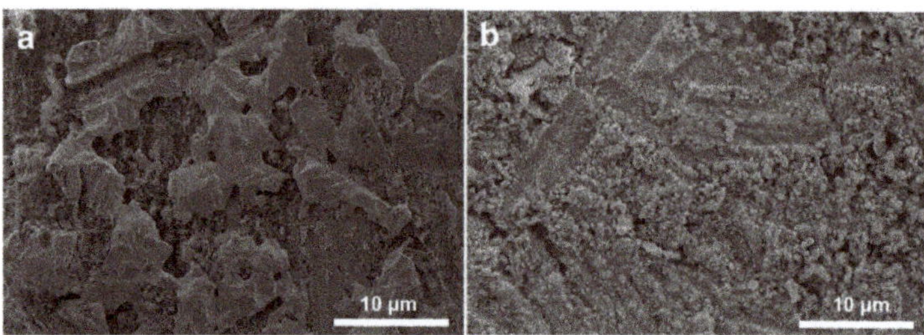

Figure 5. SEM images of (**a**) Zn and (**b**) Cu–Zn after 30 cycles at a current density of 5 mA cm^{-2} with a capacity of 1 mAh cm^{-2}.

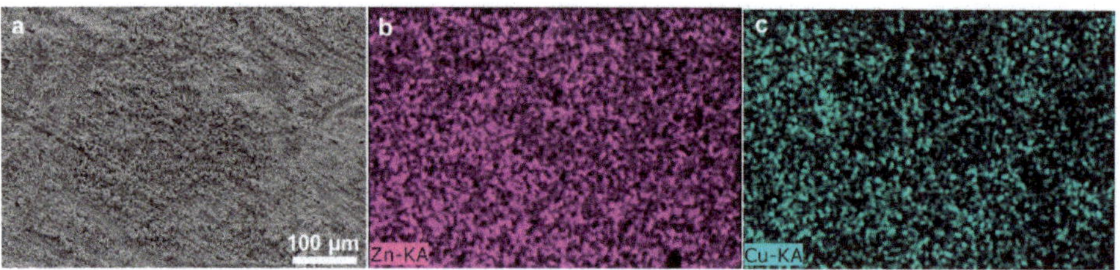

Figure 6. (**a**) SEM image and (**b,c**) corresponding elemental mappings of Cu–Zn after 50 cycles at current density of 5 mA cm^{-2} with a capacity of 1 mAh cm^{-2}.

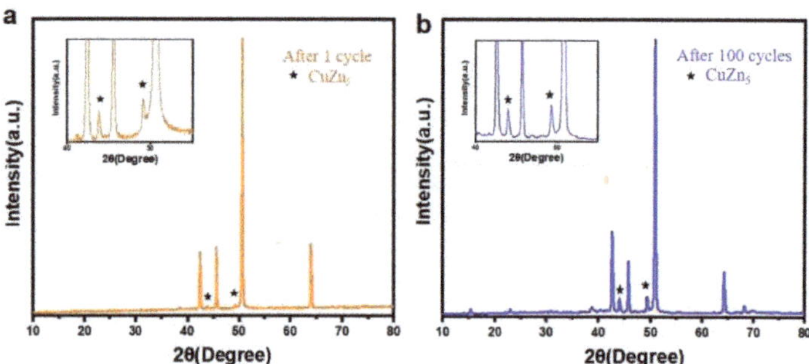

Figure 7. XRD pattern of Cu–Zn (**a**) after 1 cycle and (**b**) after 100 cycles at current density of 5 mA cm^{-2} with a capacity of 1 mAh cm^{-2}. ★ represents the XRD peak of CuZn$_5$.

In order to validate the practical application of Cu–Zn, it is necessary to evaluate full cells with cathodes. V_2O_5-based materials have been used as the cathode for ZIBs due to their high capacity, and V_2O_5-PEDOT offers improved and stable performance [25,31–34]. Therefore, V_2O_5-PEDOT was synthesized based on previously reported work, and full batteries where Zn anodes coupled with V_2O_5-PEDOT cathodes were assembled and tested (henceforth referred to as V_2O_5-PEDOT//Zn and V_2O_5-PEDOT//Cu–Zn, accordingly). The rate performance of the full cells was investigated, as shown in Figure 8a,b. At 0.1, 0.2, 0.5, 1, 2, 3, 4, 5 A·g^{-1}, V_2O_5-PEDOT//Cu–Zn cell revealed a high capacity of 323, 311, 266, 246, 227, 215, 205, 195 mAh g^{-1} respectively, while the V_2O_5-PEDOT//Zn battery showed a capacity of 284, 255, 231, 223, 211, 201, 194, 187 mAh g^{-1}, respectively. When the current density was set back to 0.5 A·g^{-1}, the capacity of the V_2O_5-PEDOT//Zn battery only reached a value of 208 mAh g^{-1}. In contrast, the capacity of the V_2O_5-PEDOT//Cu–Zn battery recovered significantly to a capacity of 250 mAh g^{-1}, indicating high reversibility. The long-term cycling was also tested at the current density of 5 A·g^{-1} (Figure 8c). During the first 60 cycles, both cells showed an enhancement of capacity due to activation. V_2O_5-PEDOT//Zn underwent a capacity fade after 120 cycles, which was ascribed to the formation of zinc dendrites resulting in "dead Zn" with a subsequent cycling, thereby increasing the internal resistance of the batteries. In sharp contrast, V_2O_5-PEDOT//Cu–Zn showed a relatively high and stable capacity over 1000 cycles. The stable cycling performance of the full cells with Cu–Zn anodes highlighted the efficiency of Cu in suppressing zinc dendrite and "dead Zn" formation by guiding uniform Zn deposition, and demonstrated high potential for practical applications.

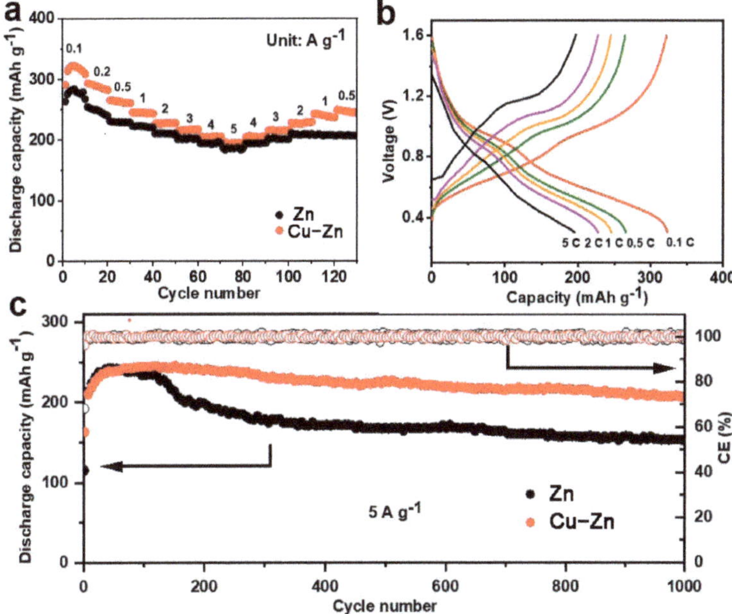

Figure 8. (a) Rate performance of V_2O_5-PEDOT//Zn and V_2O_5-PEDOT//Cu-Zn, (b) charge/discharge curves of V_2O_5-PEDOT//Cu–Zn at different current densities, (c) cycling performance of V_2O_5-PEDOT//Zn and V_2O_5-PEDOT//Cu–Zn at a current density of 5 A·g⁻¹.

4. Conclusions

In conclusion, we developed a Cu-coated Zn by a straightforward $CuSO_4$ aqueous solution treatment strategy. A $CuZn_5$ alloy was formed after cycling in the battery, which guided the uniform Zn nucleation suppressing the formation of large size Zn dendrites thus improving cycling stability. This strategy not only enabled a remarkable improvement in the cycling life of symmetric cells, but also endowed a high capacity and a stable cycling performance of the full cells coupled with the V_2O_5-PEDOT cathode. Therefore, coupled with the easy and scalable Zn treatment route, the approach is highly viable for practical implementation in ZIBs. Moreover, this work should open up a promising direction for modifying and protecting other metallic electrodes of rechargeable aqueous battery systems.

Author Contributions: Conceptualization, C.L. and Q.L.; methodology, Q.L.; validation, C.L. and Q.L.; formal analysis, C.L. and Q.L.; investigation, C.L., and Q.L.; data curation, C.L. and Q.L.; visualization, C.L. and Q.L.; writing—original draft preparation, C.L. and Q.L.; writing—review and editing, A.O. and D.M.; supervision, A.O. and D.M.; All authors have read and agreed to the published version of the manuscript.

Funding: Q.L. thanks the financial support from China Scholarship Council (No. 201808080137). The APC was funded by Leibniz Open Access Publishing Fund.

Data Availability Statement: Data is contained within the article.

Acknowledgments: Tianbing He and Xiaoliang Han are acknowledged for help with XRD testing.

Conflicts of Interest: The authors declare no conflict of interest.

References

1. Zhang, N.; Chen, X.; Yu, M.; Niu, Z.; Cheng, F.; Chen, J. Materials chemistry for rechargeable zinc-ion batteries. *Chem. Soc. Rev.* **2020**, *49*, 4203–4219. [CrossRef] [PubMed]
2. Lu, Q.; Wang, X.; Cao, J.; Chen, C.; Chen, K.; Zhao, Z.; Niu, Z.; Chen, J. Freestanding carbon fiber cloth/sulfur composites for flexible room-temperature sodium-sulfur batteries. *Energy Storage Mater.* **2017**, *8*, 77–84. [CrossRef]
3. Wu, T.; Zhang, W.; Yang, J.; Lu, Q.; Peng, J.; Zheng, M.; Xu, F.; Liu, Y.; Liang, Y. Architecture engineering of carbonaceous anodes for high-rate potassium-ion batteries. *Carbon Energy* **2021**, 1–28. [CrossRef]
4. Xu, F.; Zhai, Y.; Zhang, E.; Liu, Q.; Jiang, G.; Xu, X.; Qiu, Y.; Liu, X.; Wang, H.; Kaskel, S. Ultrastable Surface-Dominated Pseudocapacitive Potassium Storage Enabled by Edge-Enriched N-Doped Porous Carbon Nanosheets. *Angew. Chem. Int. Ed.* **2020**, *59*, 19460–19467. [CrossRef] [PubMed]
5. Jiang, G.; Qiu, Y.; Lu, Q.; Zhuang, W.; Xu, X.; Kaskel, S.; Xu, F.; Wang, H. Mesoporous Thin-Wall Molybdenum Nitride for Fast and Stable Na/Li Storage. *ACS Appl. Mater. Interfaces* **2019**, *11*, 41188–41195. [CrossRef]
6. Xu, F.; Ding, B.; Qiu, Y.; Dong, R.; Zhuang, W.; Xu, X.; Han, H.; Yang, J.; Wei, B.; Wang, H.; et al. Generalized Domino-Driven Synthesis of Hollow Hybrid Carbon Spheres with Ultrafine Metal Nitrides/Oxides. *Matter* **2020**, *3*, 246–260. [CrossRef]
7. Lu, Q.; Wang, X.; Omar, A.; Mikhailova, D. 3D Ni/Na metal anode for improved sodium metal batteries. *Mater. Lett.* **2020**, *275*, 128206. [CrossRef]
8. Chao, D.; Zhou, W.; Xie, F.; Ye, C.; Li, H.; Jaroniec, M.; Qiao, S.-Z. Roadmap for advanced aqueous batteries: From design of materials to applications. *Sci. Adv.* **2020**, *6*, eaba4098. [CrossRef] [PubMed]
9. Zhu, M.; Hu, J.; Lu, Q.; Dong, H.; Karnaushenko, D.D.; Becker, C.; Karnaushenko, D.; Li, Y.; Tang, H.; Qu, Z. A Patternable and In Situ Formed Polymeric Zinc Blanket for a Reversible Zinc Anode in a Skin-Mountable Microbattery. *Adv. Mater.* **2021**, *33*, 2007497. [CrossRef] [PubMed]
10. Yao, M.; Yuan, Z.; Li, S.; He, T.; Wang, R.; Yuan, M.; Niu, Z. Scalable Assembly of Flexible Ultrathin All-in-One Zinc-Ion Batteries with Highly Stretchable, Editable, and Customizable Functions. *Adv. Mater.* **2021**, 2008140. [CrossRef] [PubMed]
11. Wang, X.; Qin, X.; Lu, Q.; Han, M.; Omar, A.; Mikhailova, D. Mixed phase sodium manganese oxide as cathode for enhanced aqueous zinc-ion storage. *Chin. J. Chem. Eng.* **2020**, *28*, 2214–2220. [CrossRef]
12. Zhu, M.; Schmidt, O.G. Tiny robots and sensors need tiny batteries—Here's how to do it. *Nature* **2021**, *589*, 195–197. [CrossRef] [PubMed]
13. Wan, F.; Zhou, X.; Lu, Y.; Niu, Z.; Chen, J. Energy Storage Chemistry in Aqueous Zinc Metal Batteries. *ACS Energy Lett.* **2020**, *5*, 3569–3590. [CrossRef]
14. Blanc, L.E.; Kundu, D.; Nazar, L.F. Scientific Challenges for the Implementation of Zn-Ion Batteries. *Joule* **2020**, *4*, 771–799. [CrossRef]
15. Tang, B.; Shan, L.; Liang, S.; Zhou, J. Issues and opportunities facing aqueous zinc-ion batteries. *Energy Environ. Sci.* **2019**, *12*, 3288–3304. [CrossRef]
16. Ma, L.; Li, Q.; Ying, Y.; Ma, F.; Chen, S.; Li, Y.; Huang, H.; Zhi, C. Toward Practical High-Areal-Capacity Aqueous Zinc-Metal Batteries: Quantifying Hydrogen Evolution and a Solid-Ion Conductor for Stable Zinc Anodes. *Adv. Mater.* **2021**, 2007406. [CrossRef]
17. Wang, F.; Borodin, O.; Gao, T.; Fan, X.; Sun, W.; Han, F.; Faraone, A.; Dura, J.A.; Xu, K.; Wang, C. Highly reversible zinc metal anode for aqueous batteries. *Nat. Mater.* **2018**, *17*, 543–549. [CrossRef]
18. Kang, Z.; Wu, C.; Dong, L.; Liu, W.; Mou, J.; Zhang, J.; Chang, Z.; Jiang, B.; Wang, G.; Kang, F. 3D porous copper skeleton supported zinc anode toward high capacity and long cycle life zinc ion batteries. *ACS Sustain. Chem. Eng.* **2019**, *7*, 3364–3371. [CrossRef]
19. Han, D.; Wu, S.; Zhang, S.; Deng, Y.; Cui, C.; Zhang, L.; Long, Y.; Li, H.; Tao, Y.; Weng, Z.; et al. A Corrosion-Resistant and Dendrite-Free Zinc Metal Anode in Aqueous Systems. *Small* **2020**, *16*, e2001736. [CrossRef]
20. Cui, M.; Xiao, Y.; Kang, L.; Du, W.; Gao, Y.; Sun, X.; Zhou, Y.; Li, X.; Li, H.; Jiang, F.; et al. Quasi-Isolated Au Particles as Heterogeneous Seeds to Guide Uniform Zn Deposition for Aqueous Zinc-Ion Batteries. *ACS Appl. Energy Mater.* **2019**, *2*, 6490–6496. [CrossRef]
21. Zhang, Y.; Wang, G.; Yu, F.; Xu, G.; Li, Z.; Zhu, M.; Yue, Z.; Wu, M.; Liu, H.-K.; Dou, S.-X. Highly reversible and dendrite-free Zn electrodeposition enabled by a thin metallic interfacial layer in aqueous batteries. *Chem. Eng. J.* **2020**, 128062. [CrossRef]
22. Zhang, N.; Huang, S.; Yuan, Z.; Zhu, J.; Zhao, Z.; Niu, Z. Direct Self-Assembly of MXene on Zn Anodes for Dendrite-Free Aqueous Zinc-Ion Batteries. *Angew. Chem. Int. Ed.* **2020**, *60*, 2861–2865. [CrossRef]
23. Zhao, K.; Wang, C.; Yu, Y.; Yan, M.; Wei, Q.; He, P.; Dong, Y.; Zhang, Z.; Wang, X.; Mai, L. Ultrathin surface coating enables stabilized zinc metal anode. *Adv. Mater. Interfaces* **2018**, *5*, 1800848. [CrossRef]
24. Kang, L.; Cui, M.; Jiang, F.; Gao, Y.; Luo, H.; Liu, J.; Liang, W.; Zhi, C. Nanoporous $CaCO_3$ coatings enabled uniform Zn stripping/plating for long-life zinc rechargeable aqueous batteries. *Adv. Energy Mater.* **2018**, *8*, 1801090. [CrossRef]

25. Du, Y.; Wang, X.; Sun, J. Tunable oxygen vacancy concentration in vanadium oxide as mass-produced cathode for aqueous zinc-ion batteries. *Nano Res.* **2021**, *14*, 754–761. [CrossRef]
26. Marcus, P.; Hinnen, C. XPS study of the early stages of deposition of Ni, Cu and Pt on HOPG. *Surf. Sci.* **1997**, *392*, 134–142. [CrossRef]
27. Vogel, Y.B.; Zhang, J.; Darwish, N.; Ciampi, S. Switching of Current Rectification Ratios within a Single Nanocrystal by Facet-Resolved Electrical Wiring. *ACS Nano* **2018**, *12*, 8071–8080. [CrossRef] [PubMed]
28. Pei, A.; Zheng, G.; Shi, F.; Li, Y.; Cui, Y. Nanoscale nucleation and growth of electrodeposited lithium metal. *Nano Lett.* **2017**, *17*, 1132–1139. [CrossRef]
29. Cai, Z.; Ou, Y.; Wang, J.; Xiao, R.; Fu, L.; Yuan, Z.; Zhan, R.; Sun, Y. Chemically resistant Cu–Zn/Zn composite anode for long cycling aqueous batteries. *Energy Storage Mater.* **2020**, *27*, 205–211. [CrossRef]
30. Zhang, Q.; Luan, J.; Fu, L.; Wu, S.; Tang, Y.; Ji, X.; Wang, H. The three-dimensional dendrite-free zinc anode on a copper mesh with a zinc-oriented polyacrylamide electrolyte additive. *Angew. Chem. Int. Ed.* **2019**, *58*, 15841–15847. [CrossRef]
31. Dong, Y.; Jia, M.; Wang, Y.; Xu, J.; Liu, Y.; Jiao, L.; Zhang, N. Long-Life Zinc/Vanadium Pentoxide Battery Enabled by a Concentrated Aqueous ZnSO$_4$ Electrolyte with Proton and Zinc Ion Co-Intercalation. *ACS Appl. Energy Mater.* **2020**, *3*, 11183–11192. [CrossRef]
32. Zhang, N.; Jia, M.; Dong, Y.; Wang, Y.; Xu, J.; Liu, Y.; Jiao, L.; Cheng, F. Hydrated layered vanadium oxide as a highly reversible cathode for rechargeable aqueous zinc batteries. *Adv. Funct. Mater.* **2019**, *29*, 1807331. [CrossRef]
33. Zhang, N.; Dong, Y.; Jia, M.; Bian, X.; Wang, Y.; Qiu, M.; Xu, J.; Liu, Y.; Jiao, L.; Cheng, F. Rechargeable aqueous Zn–V$_2$O$_5$ battery with high energy density and long cycle life. *ACS Energy Lett.* **2018**, *3*, 1366–1372. [CrossRef]
34. Qin, X.; Wang, X.; Sun, J.; Lu, Q.; Omar, A.; Mikhailova, D. Polypyrrole Wrapped V$_2$O$_5$ Nanowires Composite for Advanced Aqueous Zinc-Ion Batteries. *Front. Energy Res.* **2020**, *8*, 199. [CrossRef]

nanomaterials

Article

Ag Nanoparticle-Decorated MoS$_2$ Nanosheets for Enhancing Electrochemical Performance in Lithium Storage

Thang Phan Nguyen and Il Tae Kim *

Department of Chemical and Biological Engineering, Gachon University, Seongnam-si 13120, Gyeonggi-do, Korea; phanthang87@gmail.com
* Correspondence: itkim@gachon.ac.kr

Abstract: Metallic phase 1T MoS$_2$ is a well-known potential anode for enhancing the electrochemical performance of lithium-ion batteries owing to its mechanical/chemical stability and high conductivity. However, during the lithiation/delithiation process, MoS$_2$ nanosheets (NSs) tend to restack to form bulky structures that deteriorate the cycling performance of bare MoS$_2$ anodes. In this study, we prepared Ag nanoparticle (NP)-decorated 1T MoS$_2$ NSs via a liquid exfoliation method with lithium intercalation and simple reduction of AgNO$_3$ in NaBH$_4$. Ag NPs were uniformly distributed on the MoS$_2$ surface with the assistance of 3-mercapto propionic acid. Ag NPs with the size of a few nanometers enhanced the conductivity of the MoS$_2$ NS and improved the electrochemical performance of the MoS$_2$ anode. Specifically, the anode designated as Ag3@MoS$_2$ (prepared with AgNO$_3$ and MoS$_2$ in a weight ratio of 1:10) exhibited the best cycling performance and delivered a reversible specific capacity of 510 mAh·g^{-1} (approximately 73% of the initial capacity) after 100 cycles. Moreover, the rate performance of this sample had a remarkable recovery capacity of ~100% when the current decreased from 1 to 0.1 A·g^{-1}. The results indicate that the Ag nanoparticle-decorated 1T MoS$_2$ can be employed as a high-rate capacity anode in lithium-ion storage applications.

Keywords: MoS$_2$; Ag; nanosheets; nanoparticle; lithium-ion battery; high rate

Citation: Nguyen, T.P.; Kim, I.T. Ag Nanoparticle-Decorated MoS$_2$ Nanosheets for Enhancing Electrochemical Performance in Lithium Storage. *Nanomaterials* **2021**, *11*, 626. https://doi.org/10.3390/nano11030626

Academic Editors: Diego Cazorla-Amorós and Christophe Detavernier

Received: 5 February 2021
Accepted: 27 February 2021
Published: 3 March 2021

Publisher's Note: MDPI stays neutral with regard to jurisdictional claims in published maps and institutional affiliations.

1. Introduction

Recently, there has been increasing interest among researchers in transition metal chalcogenides (TMCs), which are graphene-like two-dimensional (2D) materials consisting of a transition metal atom layer sandwiched between two chalcogenide atom layers. Each monolayer of TMC is formed as a 2D structured layer, and these layers are bonded to each other by van der Waals forces in the bulk structure. Therefore, the TMCs can be easily exfoliated into a single layer or a few layers. Moreover, various properties of these 2D materials have been investigated, and they have been found to be superior to bulk materials in strength, physical and chemical stability, and conductivity [1–4]. Therefore, TMCs have been employed in several electronic, optical, and energy conversion/storage applications, for example, in energy applications such as solar cells, light emitting diodes, hydrogen evolution reactions, and metal-ion batteries [5–12]. Among TMC materials, MoS$_2$ shows great potential for easy processing and high stability. Therefore, numerous studies have reported on its characteristics and applications. In particular, with a direct band-gap structure, the 1T phase of MoS$_2$ nanosheets (NSs) is attractive owing to its high mechanical/chemical stability and high conductivity [13–16]. Recently, MoS$_2$ NSs have been used as potential candidates for anodes in lithium-ion batteries (LIBs) [17–22]. With a nanosheet structure, MoS$_2$ has a large surface area and flexibility for the lithiation and delithiation processes and enhances the electrochemical performance of LIBs. Furthermore, the 1T phase of MoS$_2$, with metallic properties, can afford high conductivity, which facilitates the processes of lithiation and delithiation. Various methods have been used to enhance the electrochemical performance of MoS$_2$ in LIBs by using metals doping, metal particles or metal oxides. Zhu et al. used TiO$_2$ nanoparticles (NPs) decorating on 2H- MoS$_2$ NS via hydrothermal method to achieve the reversible capacity of 604 mAh·g^{-1} after 100 cycles [23]. Pan et al. developed Ag methanesulfonic-acid capped NPs with 2H-MoS$_2$ NSs by sonication method to get high reversible capacity of ~920 mAh·g^{-1} after 50 cycles [24]. In addition, Li et al. synthesized lithium molten salt of MoS$_2$ as a precursor at 1050 °C for liquid exfoliation of 1T MoS$_2$ [25]. This report showed that the superior properties of 1T MoS$_2$ to 2H MoS$_2$ due to the existence of abundant monolayer structures, providing diffusion path for lithium ion insertion/desertion. Wang et al. reported vertically aligned MoS$_2$ NSs patterned on graphene for LIBs, which exhibited high-rate energy storage [26]. This structure also enables sodium-ion storage capability. Tang et al. developed hollow 1T MoS$_2$ grown on carbon cloth and demonstrated high rate performance, high capacity, and good stability in sodium-ion batteries [27]. Li et al. combined 1T MoS$_2$ with MnO in lithium molten salts assisted with a ball milling method to develop a high-stability LIB anode. This MoS$_2$/MnO composite anode retained a high capacity of ~589 mAh·g^{-1} after 2000 cycles [28]. Bai et al. fabricated a 1T MoS$_2$/C hybrid anode material through a hydrothermal method [20]. These carbon-covered MoS$_2$ NS materials also exhibited a high rate performance in LIBs. Therefore, 1T phase MoS$_2$ could be a potential anode for high capacity and high rate performance in LIBs. However, the commercialization of MoS$_2$ anode materials requires an easy fabrication process and further improvement in stability and rate performance.

In this study, we successfully investigated Ag NP-decorated 1T MoS$_2$ nanosheets as a potential anode for high-rate performance and stable LIBs. MoS$_2$ was prepared by a liquid chemical exfoliation method with lithium intercalation. By adding 3-mercapto propionic acid (MPA) as a functional group, Ag NPs were uniformly decorated on the MoS$_2$ surface. The presence of Ag NPs not only improves the specific capacity but also significantly enhances the rate performance and stability of the anode material in lithium storage. Notably, the Ag3@MoS$_2$ anode can restore ~100% capacity after high-rate cycling.

2. Materials and Methods

2.1. Chemical Materials

Molybdenum (IV) sulfide (MoS$_2$, powder), silver nitrate (AgNO$_3$, 99%), MPA (99%), polyvinylidene fluoride (PVDF, MW 534,000), *N*-methyl-2-pyrrolidinone (NVP, anhydrous, 99.5%), and a 2.5 M solution of n-butyllithium ion hexane and sodium borohydride (NaBH$_4$, 99%) were purchased from Sigma-Aldrich Inc. (St. Louis, MO, USA). Super P amorphous carbon black (C, approximately 40 nm, 99.99%) was purchased from Alpha Aesar Inc. (Haverhill, MA, USA).

2.2. Synthesis MoS$_2$ NSs

MoS$_2$ NSs were prepared as described in our previous report [17]. The loading of butyllithium was conducted in Ar-filled glovebox to prevent the reactions between butyllithium and oxygen/moister. First, 1.0 g of MoS$_2$ powder was added to 3 mL of 2.5 M butyllithium in hexane. Then, the solution was sealed with parafilm, taken out to be sonicated for 1 h, and kept for two days to obtain lithium-intercalated MoS$_2$ (Li$_x$MoS$_2$) in glove box. The excess lithium was removed by washing with hexane. The obtained Li$_x$MoS$_2$ was placed in 100 mL of deionized (DI) water. The interlayer lithium reacted with DI water to break the layer structure of the bulk MoS$_2$ and form MoS$_2$ NSs. The solution was further sonicated for 1 h and stirred for 1 h to obtain a complete dispersion of the MoS$_2$ NS. Finally, the dispersion was centrifuged and washed four times to remove excess lithium.

2.3. Ag-Decorated MoS$_2$ NS

The surface of the NSs was modified by MPA in order to obtain Ag-decorated MoS$_2$ NSs. The prepared LixMoS$_2$, after being washed with hexane, was added to 100 mL of 0.045 M MPA solution. This process was similar to the synthesis of MoS$_2$ NSs. After sonication and washing with DI water, MPA-modified MoS$_2$ NSs were redispersed in DI water via sonication. Then, amounts of $AgNO_3$ with different weight ratios (1:50, 1: 20, and 1:10) to MoS_2 were added to the solution during stirring, and the samples were denoted as Ag1@MoS_2, Ag2@MoS_2, and Ag3@MoS_2, respectively. Naturally, MPA contains both thiol and carboxyl groups. The thiol group is able to exchange with the S atom on MoS_2 with the appearance of Li ions [29,30]. Meanwhile, the carboxyl group induces a partial negative charge, which attracts Ag^+ ions in the solution. Then, a solution of 0.5 M $NaBH_4$ was added to the aforementioned solution to reduce Ag^+ to Ag nanoparticles. The solution was further washed via centrifugation with DI water three times to remove $NaBO_2$. The final product was obtained after drying at 70 °C for 12 h.

2.4. Characterization

Scanning electron microscopy (SEM; Hitachi S4700, Tokyo, Japan) and transmission electron microscopy (TEM; TECNAI G2F30, FEI Corp., Hillsboro, OR, USA) were used to analyze the morphologies and sizes of the as-prepared materials. Samples were coated a few-nanometers Pt layer via magnetron sputtering system for high quality SEM images. A high-resolution X-ray diffractometer (XRD; SmartLab, Rigaku, Tokyo, Japan) was used to investigate material structures. XRD patterns were recorded over the 2θ range 10–70°.

2.5. Electrochemical Measurements

Anode materials were assembled in a half-cell LIB using coin-type cells (CR 2032, Rotech Inc., Gwangju, Korea). Typically, the anode electrode was prepared using a doctor blade on a Cu foil using a slurry of 70% active material, 15% PVDF, and 15% Super P in NVP. Then, the electrodes were dried in a vacuum oven at 70 °C for 24 h before use. The anodes were punched into 12 mm diameter circular disks. The loading of active materials was ~0.7–1.0 mg cm^{-2}. Then, battery half-cell structures were assembled in an Ar-filled glovebox with positive pressure (>1.0 atm). Lithium foil and polyethylene were used as the reference electrode and separator, respectively. A solution of 1 M $LiPF_6$ in ethylene carbonate-diethylene carbonate (1:1 by volume) was employed as the electrolyte. Galvanostatic electrochemical discharge/charge analysis of the different cells was performed using a battery cycle tester (WBCS3000, WonAtech Co., Ltd., Seocho-gu, Seoul, Korea) over the voltage range 0.01–3.00 V versus Li/Li+. Electrochemical impedance spectroscopy (EIS) and cyclic voltammetry (CV) tests were performed using a ZIVE MP1 apparatus (WonAtech Co., Ltd., Seocho-gu, Seoul, Korea). EIS measurements were recorded at 3.0 V over the frequency range between 100 kHz and 0.1 Hz. CV tests were performed over the voltage range 0.01–3.0 V at a scan rate of 0.1 mV·s^{-1}.

3. Results and Discussion

The MoS_2 NSs were fabricated using a liquid exfoliation method [17]. MoS_2 bulk powder was mixed with butyllithium in hexane to form intercalated lithium ions in MoS_2 as interlayer structures. The lithium ions easily filled the MoS_2, forming inter-layers between two MoS_2 layers by the following process:

$$MoS_2 + xLi \rightarrow Li_xMoS_2 \tag{1}$$

According to Dines [31], the x value is in the range 1.1–1.5. This implies that the lithium ions easily fill the bulk MoS_2. Then, reactions between the intercalated lithium and water create hydroxide ions and hydrogen gas. This reaction and the generated gas exfoliate the MoS_2 layers. Finally, the bulk MoS_2 is cleaved into MoS_2 NSs.

Figure 1a,b are SEM images of exfoliated MoS_2 NSs at different scale bars of 5 μm and 500 nm, respectively. The lateral size of the MoS_2 NSs was between 100 nm and 3 μm. This wide range of MoS_2 sizes is because of the strong reaction with Li(1) and the random shape of the bulk MoS_2 powder. According to previous reports, the butyllithium intercalation process converts the 2H MoS_2 to 1T due to the effect of lithium ion [10,32]. In this phase, the material could have high conductivity (10–100 S cm^{-1}), which is about 10^5 times that of 2H-MoS_2 [25,33]. Therefore, it is thought that highly conductive 1T MoS_2 phase would generate different electrochemical performance (discussed later). Besides, $AgNO_3$ was selected as the Ag source in order to decorate Ag nanoparticles on the MoS_2. MoS_2 NSs were functionalized by MPA to prevent random decoration and achieve uniform decoration. Using this functional group, metal ions are easily attracted to the partial negative charge of $-COO^-$ to uniformly distribute the Ag^+ ions. Then, H_2 generated by the hydrolysis of $NaBH_4$ reduces Ag^+ ions into Ag NPs, as illustrated in Figure 1c. Three samples with different $AgNO_3$:MoS_2 weight ratios—1:50, 1:20, and 1:10—were prepared, and, as aforementioned, they are designated herein as Ag1@MoS_2, Ag2@MoS_2, and Ag3@MoS_2, respectively. Figure 1d, showing the Ag1@MoS_2 sample and an inset image of its magnified surface, indicates that the surface of MoS_2 has only a few Ag NPs owing to the small amount of Ag^+ used. When the amount of $AgNO_3$ was increased in Ag2@MoS_2, the Ag NPs were more uniformly decorated with higher density. Figure 1e and its inset image show uniform Ag NPs on the MoS_2 surface; however, they do not fully cover the surface. The 1:10 ratio of the Ag3@MoS_2 sample reveals the highest coverage of Ag nanoparticles with a size of <10 nm on the MoS_2 nanosheets, as shown in Figure 1f.

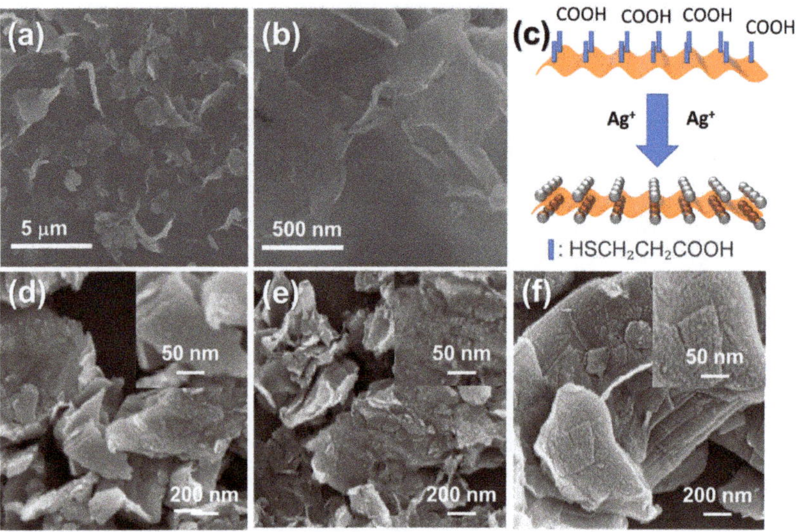

Figure 1. (**a**,**b**) Scanning electron microscopy (SEM) images of MoS_2 nanosheets (NSs); (**c**) illustration of Ag-decorated MoS_2 NS; SEM images of (**d**) Ag1@MoS_2, (**e**) Ag2@MoS_2, and (**f**) Ag3@MoS_2 NSs.

The structure of MoS_2 and Ag-decorated MoS_2 were analyzed by XRD. Figure 2 shows the XRD patterns of bulk MoS_2, Li_xMoS_2, and Ag-decorated MoS_2 NSs. It is noted that the bulk MoS_2 powder contains many peaks contributed by various planes in the lattice. However, after lithium ions are intercalated in the MoS_2 layers, the XRD pattern of Li_xMoS_2 has a main peak for the (002) plane of MoS_2 at ~14.36° and a peak at 15.1°, which is related to the interlayer plane of Li between the MoS_2 layers. The other peaks were much reduced and negligible due to the high intensity of (002) plane. After exfoliation and Ag decoration, the sample shows some main peaks for MoS_2 at ~14.4°, ~29.0°, and 32.7°, which correspond to the (002), (004), and (100) planes according to the #JCPDS card number 00-037-1492. The XRD peaks of MoS_2 are broadened in comparison to those of the bulk material. According to the Scherrer equation, the average size of the crystals can be calculated by $D = 0.9\lambda/\beta\cos\theta$, where D is the average size, λ is the X-ray wavelength, β is the full width at half maximum of the peak, and θ is the diffraction angle. Thus, the broadening of these peaks indicates small crystal sizes in those planes, which implies that the bulk MoS_2 material was exfoliated into nanosheets. These results are similar to other synthesized MoS_2 nanosheets by bottom-up methods such as hydrothermal method and hot injection method [34–37]. The appearance of Ag peaks for the (111) and (200) planes confirms the successful reduction of $AgNO_3$ into Ag nanoparticles.

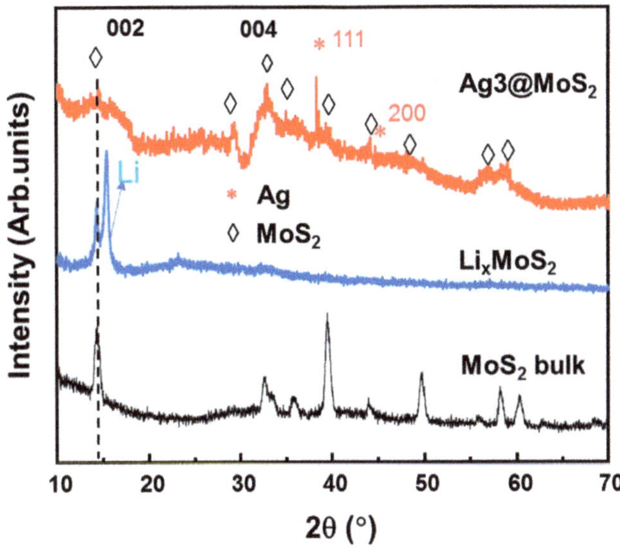

Figure 2. X-ray diffraction patterns of bulk MoS_2, Li_xMoS_2, and Ag3@MoS$_2$ materials.

The structure of Ag-decorated MoS_2 was further confirmed by TEM measurements. Figure 3a,b shows TEM images with individual and overlaid elemental mapping images of Mo (K), S (K), and Ag (L) elements. The Mo and S atoms are clearly shown with a high density of purple and orange colors, indicating the formation of MoS_2 NSs. Ag atoms are uniformly distributed in the MoS_2 area, confirming the successful decoration of Ag nanoparticles on MoS_2. Figure 3c,f show TEM images and high-resolution TEM (HRTEM) images with an inset of the selected area electron diffraction (SAED) pattern of the MoS_2 NSs. The MoS_2 NS surface shows a clear lattice spacing distance of 0.32 nm, which corresponds to the (004) plane. Figure 3d,e shows TEM and HRTEM images of Ag3@MoS$_2$ with an inset figure. The TEM image has many dark points representing Ag particles decorated on MoS_2. The HRTEM image was also analyzed to reveal the lattice spacing of 0.24 nm resulting from the Ag lattice structure. The size of the Ag NPs is a few nanometers. The SAED pattern of the Ag-decorated sample is changed in comparison to that of the pure MoS_2 sample. The pattern shows that the reflective planes reveal wide, blurred points, which could be because of the overlap of Ag NPs on the MoS_2 NSs. The high uniformity of Ag decoration on MoS_2 NSs is, thus, confirmed.

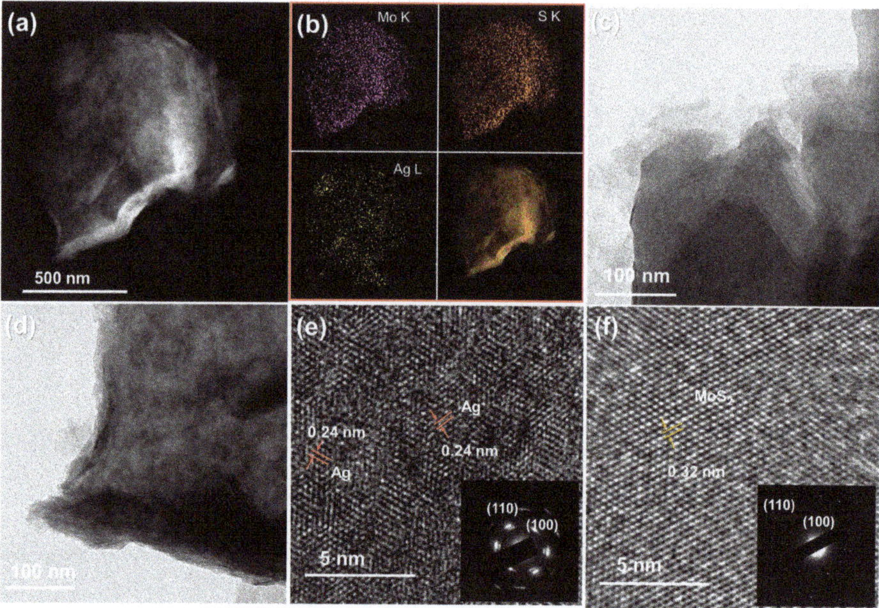

Figure 3. (**a**,**b**) Transmission electron microscopy (TEM) image and element mapping images of Ag3@MoS$_2$ materials; (**c**,**f**) TEM, high-resolution TEM (HRTEM) with inset selected area electron diffraction (SAED) pattern of MoS$_2$ NS; (**d**,**e**) TEM, HRTEM with inset SAED pattern of Ag3@MoS$_2$ materials.

To investigate the effect of Ag-decorated MoS$_2$ NSs on the electrochemical properties in lithium batteries, CV tests were performed at a scan rate of 0.1 mV·s^{-1} over the range 0.01–3.00 V (vs. Li/Li$^+$). Figure 4a–d shows the initial three CV curves for MoS$_2$ NSs and Ag1/2/3@MoS$_2$ anodes. The electrochemical processes in the anode can be expressed by the following reactions:

$$MoS_2 + xLi^+ + xe^- \rightarrow Li_xMoS_2 \tag{2}$$

$$Li_xMoS_2 \rightarrow Li_2S + Mo \tag{3}$$

$$Li^+ + e^- + electrolyte \rightarrow SEI\ layer \tag{4}$$

$$Ag + yLi^+ + ye^- \rightarrow Li_yAg \tag{5}$$

In the cathodic process, the MoS$_2$ NSs anode shows a peak at 1.3 V (vs. Li/Li$^+$), which is the intercalation process of lithium ions into MoS$_2$ NSs to form Li$_x$MoS$_2$, corresponding to reaction (2) [26]. The strong peak at 0.5–1.0 V (vs. Li/Li$^+$) is related to the strong formation of a solid electrolyte interface (SEI) layer and the transformation of Li$_x$MoS$_2$ to metallic Mo nanoparticles and the Li$_2$S conversion reaction, as shown in Equations (3) and (4) [17,19]. In the 2nd and 3rd cycles, the SEI layer is stable; therefore, the peak at 0.5–1.0 V is significantly reduced. The peak at 1.3 V is strong and stable, indicating stable lithium intercalation process. In the anodic process, the peaks at 1.8 and 2.3 V are associated with the oxidation of Mo to Mo^{+4} and the delithiation of Li$_2$S to sulfur, respectively. In the case of the Ag1@MoS$_2$ anode, the CV curves indicate some different electrochemical processes. The intercalation peak in the first cycle is shifted to ~1.0 V (vs. Li/Li$^+$). The CV profiles mainly have two pairs of redox peaks at 1.1/1.7 V and 1.75/2.3 V, which are related to the multiple steps of lithiation and delithiation. When the amount of Ag is increased, the reduction peaks shift to higher potentials. In the Ag2@MoS$_2$ anode, there are two pairs of redox peaks at 1.3/1.75 and 1.75/2.4 V. The peak at ~0.3 V is the formation of Ag–Li alloys [38]. The dealloying peak of Ag–Li is not shown as a clear peak; however, the hump at 0.2–0.5 V may indicate the multiple Ag–Li phases of the de-alloying process [38]. Notably, when the Ag content is increased, the oxidation peak position shifts to a higher potential, and the shape is broadened. This might be because of the enhanced multiple steps of the oxidation process for Li_2S [39].

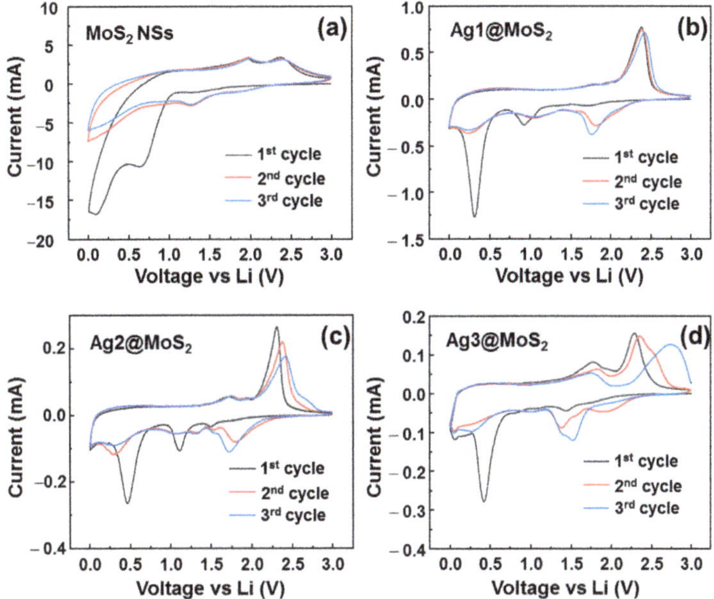

Figure 4. Cyclic voltammetry (CV) profiles of (**a**) MoS_2 NSs, (**b**) $Ag1@MoS_2$, (**c**) $Ag2@MoS_2$ and (**d**) $Ag3@MoS_2$ anodes, over the initial three cycles.

The initial voltage profiles of the as-prepared anodes are shown in Figure 5. It can be observed that with Ag decoration, the discharge/charge capacity of the MoS_2 anodes is improved from ~500 to ~900 $mAh \cdot g^{-1}$, while the MoS_2 NS reveals a low initial discharge/charge capacity of only ~490/466 $mAh \cdot g^{-1}$. Notably, the capacity does not decrease significantly during the initial cycles. The $Ag1@MoS_2$ demonstrates a high initial discharge/charge capacity of 866/780 $mAh \cdot g^{-1}$ and ~800/773 $mAh \cdot g^{-1}$ at the 2nd and 3rd cycles, which is >92% of that of the first cycle. The $Ag2@MoS_2$ anode exhibits a high discharge/charge capacity of 937/720 $mAh \cdot g^{-1}$ at the 1st cycle and ~773/739 $mAh \cdot g^{-1}$ at the 2nd and 3rd cycles. The enhancement of the lithium storage capacity could be owing to the Ag decoration improving the conductivity of the materials, thus, facilitating the lithiation/delithiation process. For the $Ag3@MoS_2$ anode, the discharge/charge capacity is slightly reduced to ~840/660 at the 1st cycle and 694/679 $mAh \cdot g^{-1}$ at the 2nd and 3rd cycles, which can be attributed to the higher amount of Ag NPs, leading to stable cyclic performance. Meanwhile, the initial voltage profile of the MoS_2 anode shows a sloping plateau at ~1.8 V, which is ascribed to the insertion of Li ions into MoS_2, according to Equation (2) [26]. A sloping plateau at ~1.2 V corresponds to the reaction of lithium with sulfur in Equation (4). At below 0.5 V, the sloping plateau is related to the deep conversion reaction of lithium with MoS_2 and formation of the SEI layer. The MoS_2 NSs, $Ag1@MoS_2$, and $Ag2@MoS_2$ have similar reaction potentials. In contrast, the $Ag3@MoS_2$ electrode has a higher plateau voltage at 2 V. The charging process of the $Ag3@MoS_2$ anode also shows a higher plateau at ~2.4 V, higher than the plateau at ~2.2 V for the other electrodes. This process is because of a shift in the redox potential, which was indicated by the CV profiles of these anodes.

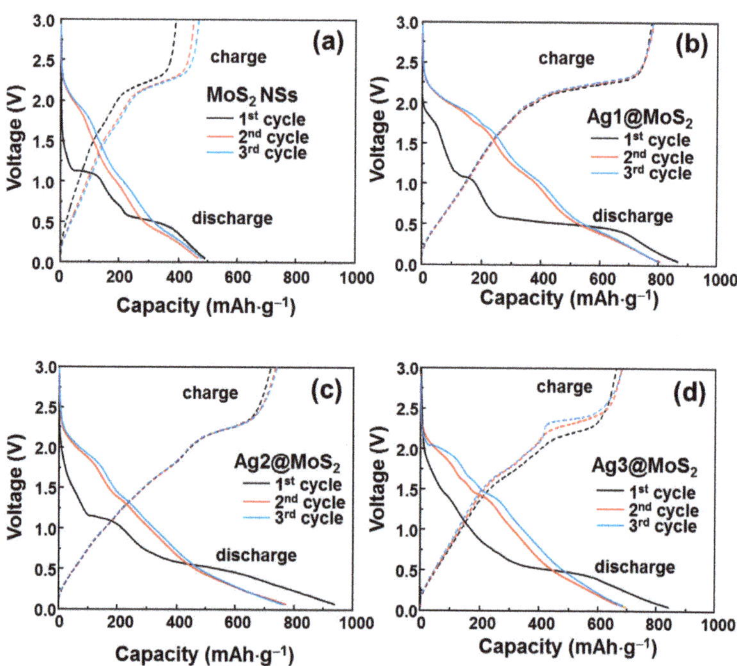

Figure 5. Initial voltage profiles of (**a**) MoS$_2$ NSs, (**b**) Ag1@MoS$_2$, (**c**) Ag2@MoS$_2$, and (**d**) Ag3@MoS$_2$ anodes.

To further investigate the long-term cyclability, the performance of these anodes was analyzed in half-cells for 100 charge-discharge cycles at a current rate of 100 mA·g^{-1}, as shown in Figure 6a–d. The MoS$_2$ NS anode exhibits a stable cycling performance during the initial 15 cycles; however, the capacity gradually decays thereafter and remarkably fades from the 30th cycle to the 100th cycle to a capacity of ~100 mAh·g^{-1}. This can be attributed to the formation of a broken MoS$_2$ structure after the cycling process and a restacking of the MoS$_2$ NS layers. By addition of a small amount of Ag nanoparticles, the Ag1@MoS$_2$ anode improves the electrochemical performance for the initial 30 cycles; however, a dramatic capacity decay still occurs after that. A small amount of Ag enhances the conductivity of MoS$_2$; however, the applied amount in the Ag1@MoS$_2$ anode does not appear to be enough to protect the entire MoS$_2$ structure. Thus, after 40 cycles, the MoS$_2$ NS structure collapses, and the lithium storage capability worsens. A further increase in the amount of Ag diminishes the collapse of the MoS$_2$ structure, leading to an enhancement in the cycling performance of the anode. The Ag2@MoS$_2$ anode retains a capacity of 330 mAh·g^{-1} up to 100 cycles. The Ag3@MoS$_2$ anode exhibits the best cycling performance; it retains a specific capacity of ~510 mAh·g^{-1} after 100 cycles, corresponding to a capacity retention of ~73%.

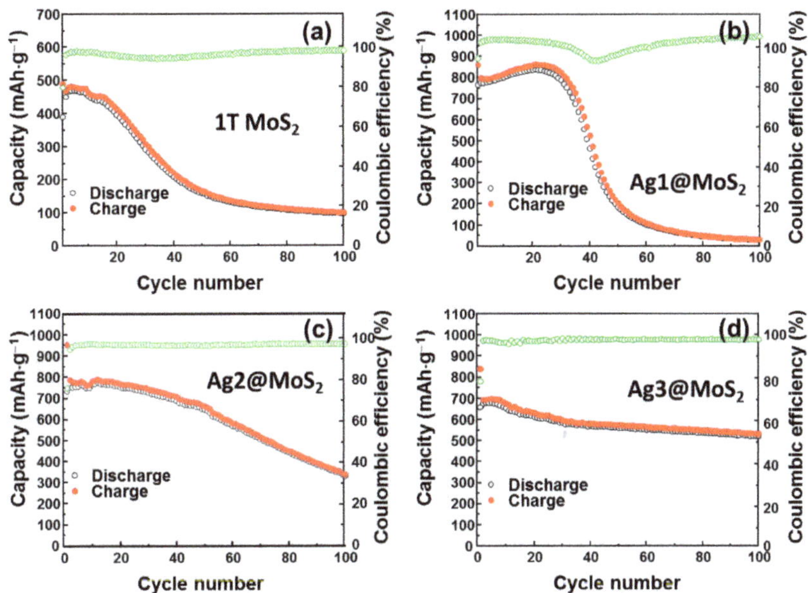

Figure 6. Cyclic performance of (**a**) MoS$_2$ NSs, (**b**) Ag1@MoS$_2$, (**c**) Ag2@MoS$_2$, and (**d**) Ag3@MoS$_2$ anodes.

The discharge/charge rate performance of MoS$_2$ NS without and with Ag nanoparticle decoration is shown in Figure 7a–c. These cycles were recorded at 0.1, 0.2, 0.5, and 1.0 A·g^{-1}. As observed, bare MoS$_2$ NSs show an inferior rate performance with a dramatic decrease in capacity from ~600 to 300, 180, and ~100 mAh·g^{-1}, which corresponds to capacity retentions of 50, 30, and 17%, respectively. The restored capacity when going back from 1 to 0.1 A·g^{-1} is ~430 mAh·g^{-1}, which is approximately 72%. When Ag nanoparticles are introduced, however, the rate performance of the anodes significantly improves. The Ag1@MoS$_2$ anode shows capacity reductions from ~730 to 690, 650, and 550 mAh·g^{-1}, respectively, which correspond to capacity retentions of 94, 89, and 75%, respectively. The restored capacity reaches ~676 mAh·g^{-1} (~92% of the initial capacity). Moreover, the Ag3@MoS$_2$ electrode demonstrates the best rate performance, exhibiting capacity retentions of 98%, 96%, and 92%, which correspond to capacity values from 700 to 690, 677, and 646 mAh·g^{-1}, respectively. The capacity after the high-rate test at 1 A·g^{-1} was 100% recovered when returning to 0.1 A·g^{-1}. To further investigate the effect of Ag in MoS$_2$, EIS measurements were performed to evaluate the change in the charge-transfer resistance, as illustrated in Figure 7d. The equivalent circuit using the modified Randles model contains a series resistance, SEI resistance, charge-transfer resistance, and a Warburg impedance element, and this was used to simulate the Nyquist plot [40]. The extracted charge-transfer resistances of the MoS$_2$ NS and Ag1/2/3@MoS$_2$ anodes are 210.5, 152.3, 99.1, and 95.8 Ω, respectively. The presence of Ag NPs clearly leads to an improvement in the anode conductivity [41,42]. Between the Ag2@MoS$_2$ and the Ag3@MoS$_2$ anode, the charge-transfer resistance is not significantly reduced, which indicates that the amount of Ag NPs is sufficient to decorate and enhance the electronic properties of the MoS$_2$ NSs. Thus, the 1:10 weight ratio of AgNO$_3$:MoS$_2$ can contribute to the best performance in lithium-ion storage of MoS$_2$ anode materials.

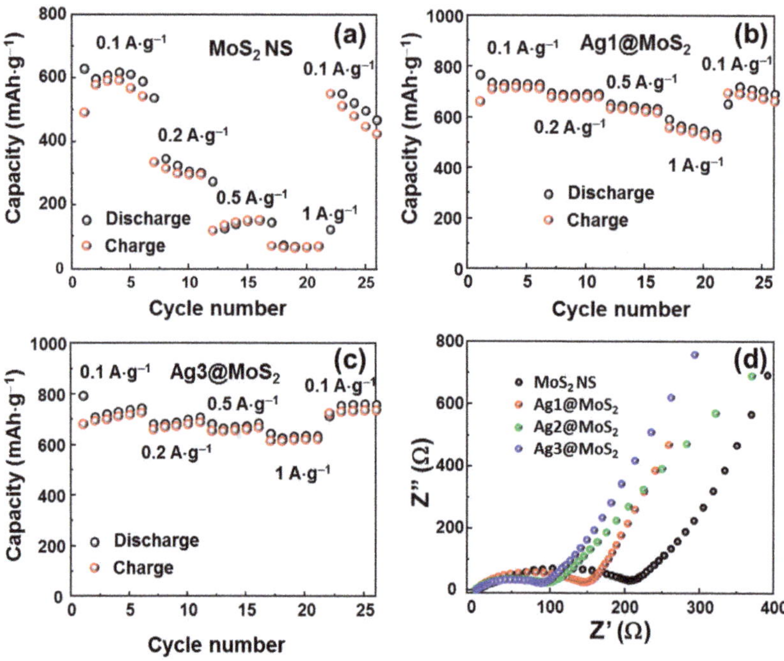

Figure 7. Rate performance of (**a**) MoS$_2$ NSs, (**b**) Ag1@MoS$_2$, (**c**) Ag3@MoS$_2$, and (**d**) Nyquist plots of MoS$_2$ without/with Ag-decorated anodes.

Recent works are summarized in Table 1. The reversible capacity of modified MoS$_2$ NSs can deliver up to ~1000 mAh·g^{-1}. From our method, the MoS$_2$ bulk was exfoliated into few layer MoS$_2$, with decoration of Ag on MoS$_2$ NSs. The Ag decorated MoS$_2$ NSs exhibited stable cyclability and high-rate performance. Moreover, the butyllithium assisted technique and uniform decorating technique for metal-particles can be easily scaled up to industrial purpose. This work can be further improved by optimizing and modifying the synthesis of MoS$_2$ to develop uniform MoS$_2$ single layer with the insertion of butyllithium by applying the pressure or temperature.

Table 1. Recent research of MoS$_2$ nanosheets (NSs) for Li-ion storage.

Materials	Method	Phase of MoS$_2$	Reversible Capacity after 100 Cycles at 0.1 A·g^{-1} (mAh·g^{-1})	Reference
Ag/MoS$_2$ nanohybrids	Sonication	2H	~920 (after 50 cycles)	[24]
Sn/MoS$_2$ composite	Hydrothermal	-	~1087	[43]
MoS$_2$/reduced graphene oxide	Hydrothermal	-	~667	[44]
TiO$_2$ decorated MoS$_2$	Hydrothermal	2H	~604	[23]
Fe$_2$O$_3$@Carbon nanofiber/MoS$_2$	Electrospinning and hydrothermal	2H	~900 (at 0.2 Ah·g^{-1})	[45]
1T MoS$_2$	Liquid exfoliation assisted lithium molten salt at 1050 °C	1T	~855	[25]
Ag nanoparticles-decorated MoS$_2$ NSs	Liquid exfoliation method	1T	~510	[This work]

4. Conclusions

In this study, we successfully prepared MoS_2 NS with Ag NP decoration, using the assistance of MPA functionalization. The structure and morphology of the Ag NPs on the MoS_2 NSs were confirmed by SEM, XRD, and TEM measurements. The size of the MoS_2 NSs was from 100 nm to ~3 μm. Ag NPs with a size of a few nm were decorated on the surface of the MoS_2 NSs. The MoS_2 NS shows inferior cycling performance of lithium storage capacity (~500 mAh·g^{-1}) and rate performance. By incorporating Ag NPs, the storage capacity and rate performance of anodes were significantly improved. Among the three anodes prepared, the Ag3@MoS_2 anode demonstrated the best cycling performance retention capacity of 73% compared to that in the first cycle after 100 cycles. Moreover, this anode could restore ~100% of the capacity after high rate performance. These results suggest that Ag-decorated MoS_2 can be a potential anode for a high-rate and high-stability anode in lithium storage applications in the future.

Author Contributions: T.P.N.: Conceptualization, Methodology, Validation, Visualization, Writing–review & editing. I.T.K.: Project administration, Funding acquisition, Review & editing. Both authors have read and agreed to the published version of the manuscript. All authors have read and agreed to the published version of the manuscript.

Funding: This work was supported by the National Research Foundation of Korea (NRF) grant funded by the Korean government (MSIT) (NRF-2020R1F1A1048335). This research was also supported by the Basic Science Research Capacity Enhancement Project through a Korea Basic Science Institute (National Research Facilities and Equipment Center) grant funded by the Ministry of Education (2019R1A6C1010016).

Institutional Review Board Statement: Not applicable.

Informed Consent Statement: Not applicable.

Data Availability Statement: Data is contained within the article.

Conflicts of Interest: The authors declare no conflict of interest.

References

1. Song, X.F.; Hu, J.L.; Zeng, H.B. Two-dimensional semiconductors: Recent progress and future perspectives. *J. Mater. Chem. C* **2013**, *1*, 2952–2969. [CrossRef]
2. Nguyen, T.P.; Sohn, W.; Oh, J.H.; Jang, H.W.; Kim, S.Y. Size-Dependent Properties of Two-Dimensional MoS_2 and WS_2. *J. Phys. Chem. C* **2016**, *120*, 10078–10085. [CrossRef]
3. Li, H.; Lu, G.; Yin, Z.; He, Q.; Li, H.; Zhang, Q.; Zhang, H. Optical Identification of Single- and Few-Layer MoS_2 Sheets. *Small* **2012**, *8*, 682–686. [CrossRef] [PubMed]
4. Joensen, P.; Frindt, R.F.; Morrison, S.R. Single-layer MoS_2. *Mater. Res. Bull.* **1986**, *21*, 457–461. [CrossRef]
5. Radisavljevic, B.; Radenovic, A.; Brivio, J.; Giacometti, V.; Kis, A. Single-layer MoS_2 transistors. *Nat. Nanotechnol.* **2011**, *6*, 147–150. [CrossRef]
6. Li, S.; Chen, Z.; Zhang, W. Dye-sensitized solar cells based on WS_2 counter electrodes. *Mater. Lett.* **2012**, *72*, 22–24. [CrossRef]
7. Kim, C.; Nguyen, T.P.; Le, Q.V.; Jeon, J.M.; Jang, H.W.; Kim, S.Y. Performances of Liquid-Exfoliated Transition Metal Dichalcogenides as Hole Injection Layers in Organic Light-Emitting Diodes. *Adv. Funct. Mater.* **2015**, *25*, 4512–4519. [CrossRef]
8. Nguyen, T.P.; Le, Q.V.; Choi, S.; Lee, T.H.; Hong, S.-P.; Choi, K.S.; Jang, H.W.; Lee, M.H.; Park, T.J.; Kim, S.Y. Surface extension of MeS_2 (Me=Mo or W) nanosheets by embedding MeS_x for hydrogen evolution reaction. *Electrochim. Acta* **2018**, *292*, 136–141. [CrossRef]
9. He, Q.; Wang, L.; Yin, K.; Luo, S. Vertically Aligned Ultrathin 1T-WS_2 Nanosheets Enhanced the Electrocatalytic Hydrogen Evolution. *Nanoscale Res. Lett.* **2018**, *13*, 167. [CrossRef]
10. Voiry, D.; Salehi, M.; Silva, R.; Fujita, T.; Chen, M.; Asefa, T.; Shenoy, V.B.; Eda, G.; Chhowalla, M. Conducting MoS_2 nanosheets as catalysts for hydrogen evolution reaction. *Nano Lett.* **2013**, *13*, 6222–6227. [CrossRef]
11. Nguyen, Q.H.; Hur, J. MoS_2-C-TiC Nanocomposites as New Anode Materials for High-Performance Lithium-Ion Batteries. *J. Nanosci. Nanotechnol.* **2019**, *19*, 996–1000. [CrossRef] [PubMed]
12. Nguyen, T.P.; Kim, I.T. W_2C/WS_2 Alloy Nanoflowers as Anode Materials for Lithium-Ion Storage. *Nanomaterials (Basel)* **2020**, *10*, 1336. [CrossRef] [PubMed]
13. El Beqqali, O.; Zorkani, I.; Rogemond, F.; Chermette, H.; Chaabane, R.B.; Gamoudi, M.; Guillaud, G. Electrical properties of molybdenum disulfide MoS_2. Experimental study and density functional calculation results. *Synth. Met.* **1997**, *90*, 165–172. [CrossRef]

14. Molina-Sanchez, A.; Wirtz, L. Phonons in single-layer and few-layer MoS_2 and WS_2. *Phys. Rev. B* **2011**, *84*, 155413. [CrossRef]
15. Castellanos-Gomez, A.; Poot, M.; Steele, G.A.; Van Der Zant, H.S.J.; Agrait, N.; Rubio-Bollinger, G. Elastic properties of freely suspended MoS_2 nanosheets. *Adv. Mater.* **2012**, *24*, 24. [CrossRef] [PubMed]
16. Gao, D.; Si, M.; Li, J.; Zhang, J.; Zhang, Z.; Yang, Z.; Xue, D. Ferromagnetism in freestanding MoS_2 nanosheets. *Nanoscale. Res. Lett.* **2013**, *8*, 129–136. [CrossRef]
17. Nguyen, T.P.; Kim, I.T. Self-Assembled Few-Layered MoS_2 on SnO_2 Anode for Enhancing Lithium-Ion Storage. *Nanomaterials* **2020**, *10*, 2558. [CrossRef]
18. Stephenson, T.; Li, Z.; Olsen, B.; Mitlin, D. Lithium ion battery applications of molybdenum disulfide (MoS_2) nanocomposites. *Energy Environ. Sci.* **2014**, *7*, 209–231. [CrossRef]
19. Cha, E.; Patel, M.D.; Park, J.; Hwang, J.; Prasad, V.; Cho, K.; Choi, W. 2D MoS_2 as an efficient protective layer for lithium metal anodes in high-performance Li–S batteries. *Nat. Nanotechnol.* **2018**, *13*, 337–344. [CrossRef]
20. Bai, J.; Zhao, B.; Zhou, J.; Si, J.; Fang, Z.; Li, K.; Ma, H.; Dai, J.; Zhu, X.; Sun, Y. Glucose-Induced Synthesis of 1T-MoS_2/C Hybrid for High-Rate Lithium-Ion Batteries. *Small* **2019**, *15*, 1805420. [CrossRef]
21. Chen, Y.; Lu, J.; Wen, S.; Lu, L.; Xue, J. Synthesis of SnO_2/MoS_2 composites with different component ratios and their applications as lithium ion battery anodes. *J. Mater. Chem. A* **2014**, *2*, 17857–17866. [CrossRef]
22. Tripathi, K.M.; Ahn, H.T.; Chung, M.; Le, X.A.; Saini, D.; Bhati, A.; Sonkar, S.K.; Kim, M.I.; Kim, T. N, S, and P-Co-doped Carbon Quantum Dots: Intrinsic Peroxidase Activity in a Wide pH Range and Its Antibacterial Applications. *ACS Biomater. Sci. Eng.* **2020**, *6*, 5527–5537. [CrossRef]
23. Zhu, X.; Yang, C.; Xiao, F.; Wang, J.; Su, X. Synthesis of nano-TiO_2-decorated MoS_2 nanosheets for lithium ion batteries. *New J. Chem.* **2015**, *39*, 683–688. [CrossRef]
24. Pan, L.; Liu, Y.T.; Xie, X.M.; Zhu, X.D. Coordination-driven hierarchical assembly of silver nanoparticles on MoS_2 nanosheets for improved lithium storage. *Chem. Asian J.* **2014**, *9*, 1519–1524. [CrossRef] [PubMed]
25. Li, Y.; Chang, K.; Sun, Z.; Shangguan, E.; Tang, H.; Li, B.; Sun, J.; Chang, Z. Selective Preparation of 1T- and 2H-Phase MoS_2 Nanosheets with Abundant Monolayer Structure and Their Applications in Energy Storage Devices. *ACS Appl. Energy Mater.* **2020**, *3*, 998–1009. [CrossRef]
26. Wang, G.; Zhang, J.; Yang, S.; Wang, F.; Zhuang, X.; Müllen, K.; Feng, X. Vertically Aligned MoS^2 Nanosheets Patterned on Electrochemically Exfoliated Graphene for High-Performance Lithium and Sodium Storage. *Adv. Energy Mater.* **2018**, *8*, 1702254. [CrossRef]
27. Tang, W.-j.; Wang, X.-l.; Xie, D.; Xia, X.-h.; Gu, C.-d.; Tu, J.-p. Hollow metallic 1T MoS_2 arrays grown on carbon cloth: A freestanding electrode for sodium ion batteries. *J. Mater. Chem. A* **2018**, *6*, 18318–18324. [CrossRef]
28. Li, Z.; Sun, P.; Zhan, X.; Zheng, Q.; Feng, T.; Braun, P.V.; Qi, S. Metallic 1T phase MoS_2/MnO composites with improved cyclability for lithium-ion battery anodes. *J. Alloys Compd.* **2019**, *796*, 25–32. [CrossRef]
29. Presolski, S.; Pumera, M. Covalent functionalization of MoS_2. *Mater. Today* **2016**, *19*, 140–145. [CrossRef]
30. Zhou, L.; He, B.; Yang, Y.; He, Y. Facile approach to surface functionalized MoS_2 nanosheets. *RSC Adv.* **2014**, *4*, 32570–32578. [CrossRef]
31. Dines, M.B. Lithium intercalation via -Butyllithium of the layered transition metal dichalcogenides. *Mater. Res. Bull.* **1975**, *10*, 287–291. [CrossRef]
32. Eda, G.; Yamaguchi, H.; Voiry, D.; Fujita, T.; Chen, M.; Chhowalla, M. Photoluminescence from Chemically Exfoliated MoS_2. *Nano Lett.* **2011**, *11*, 5111–5116. [CrossRef] [PubMed]
33. Lei, Z.; Zhan, J.; Tang, L.; Zhang, Y.; Wang, Y. Recent Development of Metallic (1T) Phase of Molybdenum Disulfide for Energy Conversion and Storage. *Adv. Energy Mater* **2018**, *8*, 1703482. [CrossRef]
34. Takahashi, Y.; Nakayasu, Y.; Iwase, K.; Kobayashi, H.; Honma, I. Supercritical hydrothermal synthesis of MoS_2 nanosheets with controllable layer number and phase structure. *Dalton Trans.* **2020**, *49*, 9377–9384. [CrossRef]
35. Li, F.; Li, J.; Cao, Z.; Lin, X.; Li, X.; Fang, Y.; An, X.; Fu, Y.; Jin, J.; Li, R. MoS_2 quantum dot decorated RGO: A designed electrocatalyst with high active site density for the hydrogen evolution reaction. *J. Mater. Chem. A* **2015**, *3*, 21772–21778. [CrossRef]
36. Altavilla, C.; Sarno, M.; Ciambelli, P. A Novel Wet Chemistry Approach for the Synthesis of Hybrid 2D Free-Floating Single or Multilayer Nanosheets of MS2@oleylamine (M=Mo, W). *Chem. Mater.* **2011**, *23*, 3879–3885. [CrossRef]
37. Zhang, X.; Ding, P.; Sun, Y.; Wang, Y.; Wu, Y.; Guo, J. Shell-core MoS_2 nanosheets@Fe_3O_4 sphere heterostructure with exposed active edges for efficient electrocatalytic hydrogen production. *J. Alloys Compd.* **2017**, *715*, 53–59. [CrossRef]
38. Taillades, G.; Sarradin, J. Silver: High performance anode for thin film lithium ion batteries. *J. Power Sources* **2004**, *125*, 199–205. [CrossRef]
39. Ma, Y.; Ma, Y.; Giuli, G.; Diemant, T.; Behm, R.J.; Geiger, D.; Kaiser, U.; Ulissi, U.; Passerini, S.; Bresser, D. Conversion/alloying lithium-ion anodes—Enhancing the energy density by transition metal doping. *Sustain. Energy Fuels* **2018**, *2*, 2601–2608. [CrossRef]
40. Vo, T.N.; Kim, D.S.; Mun, Y.S.; Lee, H.J.; Ahn, S.-k.; Kim, I.T. Fast charging sodium-ion batteries based on Te-P-C composites and insights to low-frequency limits of four common equivalent impedance circuits. *Chem. Eng. J.* **2020**, *398*, 125703. [CrossRef]
41. Hasani, A.; Nguyen, T.P.; Tekalgne, M.; Van Le, Q.; Choi, K.S.; Lee, T.H.; Jung Park, T.; Jang, H.W.; Kim, S.Y. The role of metal dopants in WS_2 nanoflowers in enhancing the hydrogen evolution reaction. *Appl. Catal. A.* **2018**, *567*, 73–79. [CrossRef]
42. Kadam, A.N.; Bhopate, D.P.; Kondalkar, V.V.; Majhi, S.M.; Bathula, C.D.; Tran, A.-V.; Lee, S.-W. Facile synthesis of Ag-ZnO core–shell nanostructures with enhanced photocatalytic activity. *J. Ind. Eng. Chem.* **2018**, *61*, 78–86. [CrossRef]

43. Lu, L.; Min, F.; Luo, Z.; Wang, S.; Teng, F.; Li, G.; Feng, C. Synthesis and electrochemical properties of tin-doped MoS_2 (Sn/MoS_2) composites for lithium ion battery applications. *J. Nanopart. Res.* **2016**, *18*, 357. [CrossRef]
44. Zhong, Y.; Shi, T.; Huang, Y.; Cheng, S.; Chen, C.; Liao, G.; Tang, Z. Three-dimensional MoS_2/Graphene Aerogel as Binder-free Electrode for Li-ion Battery. *Nanoscale. Res. Lett.* **2019**, *14*, 85. [CrossRef] [PubMed]
45. Huang, X.; Cai, X.; Xu, D.; Chen, W.; Wang, S.; Zhou, W.; Meng, Y.; Fang, Y.; Yu, X. Hierarchical Fe_2O_3@CNF fabric decorated with MoS_2 nanosheets as a robust anode for flexible lithium-ion batteries exhibiting ultrahigh areal capacity. *J. Mater. Chem. A* **2018**, *6*, 16890–16899. [CrossRef]

 *nanomaterials*

Review

Inorganic Fillers in Composite Gel Polymer Electrolytes for High-Performance Lithium and Non-Lithium Polymer Batteries

Vo Pham Hoang Huy, Seongjoon So and Jaehyun Hur *

Department of Chemical and Biological Engineering, Gachon University, Seongnam 13120, Korea;
vophamhoanghuy@yahoo.com.vn (V.P.H.H.); tjdwns7594@naver.com (S.S.)
* Correspondence: jhhur@gachon.ac.kr

Abstract: Among the various types of polymer electrolytes, gel polymer electrolytes have been considered as promising electrolytes for high-performance lithium and non-lithium batteries. The introduction of inorganic fillers into the polymer-salt system of gel polymer electrolytes has emerged as an effective strategy to achieve high ionic conductivity and excellent interfacial contact with the electrode. In this review, the detailed roles of inorganic fillers in composite gel polymer electrolytes are presented based on their physical and electrochemical properties in lithium and non-lithium polymer batteries. First, we summarize the historical developments of gel polymer electrolytes. Then, a list of detailed fillers applied in gel polymer electrolytes is presented. Possible mechanisms of conductivity enhancement by the addition of inorganic fillers are discussed for each inorganic filler. Subsequently, inorganic filler/polymer composite electrolytes studied for use in various battery systems, including Li-, Na-, Mg-, and Zn-ion batteries, are discussed. Finally, the future perspectives and requirements of the current composite gel polymer electrolyte technologies are highlighted.

Keywords: inorganic filler; gel polymer electrolytes; TiO_2; Al_2O_3; SiO_2; ZrO_2; CeO_2; $BaTiO_3$; lithium polymer batteries

Citation: Hoang Huy, V.P.; So, S.; Hur, J. Inorganic Fillers in Composite Gel Polymer Electrolytes for High-Performance Lithium and Non-Lithium Polymer Batteries. *Nanomaterials* **2021**, *11*, 614. https://doi.org/10.3390/nano11030614

Academic Editor: Christian M. Julien

Received: 19 January 2021
Accepted: 26 February 2021
Published: 1 March 2021

Publisher's Note: MDPI stays neutral with regard to jurisdictional claims in published maps and institutional affiliations.

1. Introduction

Electrolytes serve as the transportation medium for charge carriers between a pair of electrodes that are ubiquitous in electrolyte cells, fuel cells, and batteries [1]. For several decades, liquid electrolytes (LEs) have been employed extensively in electrochemical devices owing to their high electrolytic conductance (10^1–10^2 mS cm^{-1} for aqueous electrolytes and 10^0–10^1 mS cm^{-1} for organic electrolytes). However, the safety issues associated with LEs in terms of electrolyte leakage from flammable organic solvents have hindered the commercialization of lithium-metal (Li-metal) electrodes in lithium-ion batteries (LIBs) [2–11]. Another limitation associated with LEs is the inevitable dendrite growth of lithium due to uneven current when using porous separators [12–18]. Furthermore, increasing the energy density in LIBs using high-voltage materials often leads to electrode degradation as it requires electrode–electrolyte compatibility [19–25]. Therefore, the development of new electrolytes is essential to overcome the aforementioned issues in LIB applications.

Solid polymer electrolytes (SPEs) without a liquid solvent have the potential to overcome the limitations associated with Les [26,27]. One of the interesting concepts recently demonstrated is "polymer-in-ceramic" configuration. Zhang et al. synthesized SPE with a flexible and self-standing membrane by incorporating 75 wt% ceramic particles into polymer matrix [28]. Poly(ε-caprolactone) (PCL) was proved to be a great polymer matrix that can accommodate high content of ceramic particles compared to other polymers due to its good mechanical strength and ability of structural rearrangement. The high concentration of ceramic in polymer-in-ceramic can be favorable in enhancing the ionic conductivity. For example, in poly(ethylene oxide)-garnet electrolyte SPE, Li$^+$ ion transport is highly enhanced when the content of ceramic garnet reaches the percolation threshold.

While the conductivity of the SPE is dominated by the polymer chain movement when the ceramic concentration is below the percolation threshold, the conductivity is boosted when ceramic concentration reaches the percolation threshold. This is because of the additional Li^+ ion transport pathway formed by the ionic conducting ceramic. Therefore, above the percolation threshold, the presence of ceramic in SPE can dominantly affect the ionic conductivity [29]. However, the utilization of SPEs in electrochemical cells is still limited owing to their low ionic conductivities (10^{-5}–10^{-2} mS cm^{-1}), poor contact at the electrode-electrolyte interface, and narrow electrochemical window, resulting in a degenerated cyclic performance [30,31]. Recently, plastic crystalline electrolytes (PCEs) have been extensively studied for their high electrical conductivity, soft texture, and good thermal stability. Succinonitrile (SN)-based PCE is synthesized as an alternate layer to resolve the interfacial instability between solid electrolyte (SE) and Li metal, allowing for the widespread use of Li metal anode in the solid-state lithium batteries. Besides, chemical compatibility between SE and PCE ensures the prolonged life of solid-state batteries [32,33]. Tong et al. synthesized the SE interface by incorporating SN-based PCE in combination with salt and $Li_7La_3Zr_{1.4}Ta_{0.6}O_{12}$ (LLZTO), which enhances surface stability for Li metal anode and $Li_{1.5}Al_{0.5}Ge_{1.5}(PO_4)_3$ (LAGP)-based ceramic electrolyte [34]. The presence of the PCE interface helps to protect the ceramic electrolyte from reduction and promotes good interface stability with Li metal anode and LAGP due to its soft texture. In the absence of the PCE interface, the cell achieved a discharge capacity of 80 mAh g^{-1} at 0.05 C, whereas the cell with PCE interface showed an increase in discharge capacity of 126 mAh g^{-1} at 0.05 C. Thus, PCE that can act as a combination agent between the LAGP and Li metal anode prevents the penetration of dendrites into LAGP-based ceramic electrolyte. Compared with SPEs, gel polymer electrolytes (GPEs) are promising candidates for LIBs and non-LIBs as they combine the advantages of both LEs and SPEs in terms of ionic conductivity and mechanical properties [1,35–41]. Figure 1 shows the general advantages and disadvantages of LEs, SPEs, and GPEs.

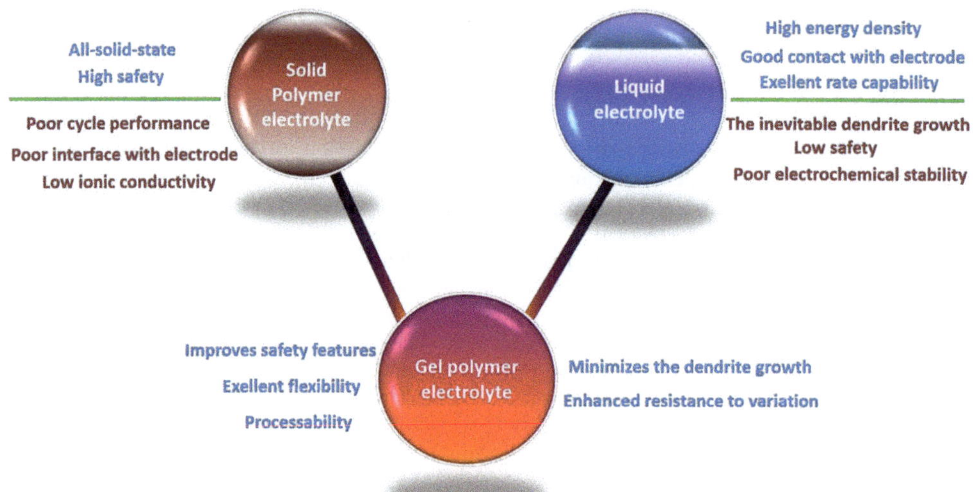

Figure 1. Advantages and disadvantages of solid polymer electrolytes (SPEs), liquid electrolytes (LEs), and gel polymer electrolytes (GPEs).

GPEs generally comprise a polymer matrix-lithium salt (Li-salt) system, small amount of integrated liquid plasticizer, and/or solvent, as shown in Figure 2 [42]. GPEs are characterized as homogeneous (uniform) and heterogeneous (phase-separated) gels. Typically, heterogeneous GPEs consist of a polymer framework of which the interconnected pores are filled with LEs. Thus, lithium ion (Li$^+$ ion) transport mainly proceeds in the swollen gel phase or liquid phase in heterogeneous GPEs, which has a higher electrolytic conductance than SPEs. In addition, owing to their high safety and flexibility, GPEs are increasingly utilized for the manufacturing of advanced energy storage devices [43,44]. Several types of polymer matrices have been investigated as frameworks for GPEs, including polyethylene oxide (PEO), polyacrylonitrile (PAN), poly(vinylidene fluoride-co-hexafluoropropylene) (PVDF-HFP), poly(ethylene oxide-co-ethylene carbonate) (P(EO-EC)), poly(methyl methacrylate) (PMMA), poly(vinyl alcohol) (PVA), poly(vinyl chloride) (PVC), poly(propylene glycol) (PPG) [42–45]. In addition, the elastomeric polymer (polydimethylsiloxane (PDMS)) was used in the GPE nanocomposite due to its mechanical flexibility as well as chemical and thermal stability; thus, the nanocomposite PDMS-based membrane provides good electrochemical performance with high mechanical flexibility [46–49]. Owing to the combination of a polymer-salt system with a plasticizer, the mechanical strength of GPEs is mainly determined by the polymer matrix, while the plasticizer reduces the crystallized phase of the polymer matrix. This promotes segmental motion of the polymer matrix and affects the ionic conductivity of the GPE [50].

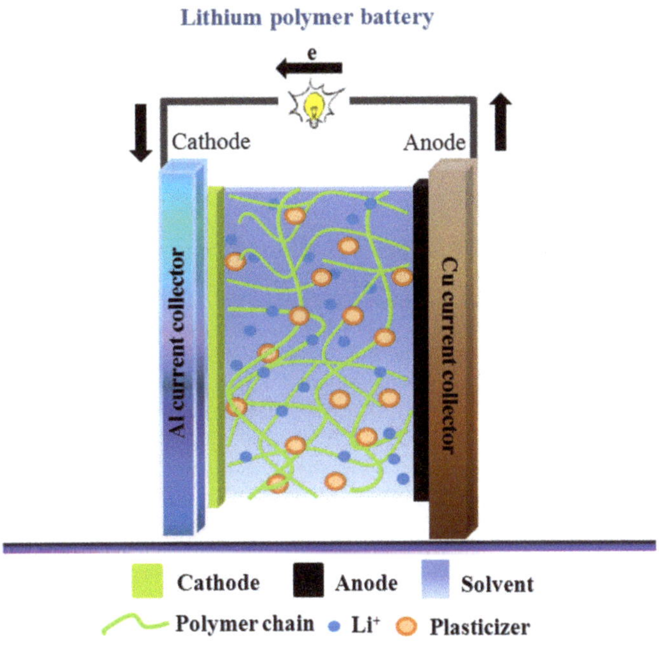

Figure 2. Schematic of a lithium polymer battery based on GPEs.

However, when incorporating an excess of plasticizer, it can deteriorate the mechanical strength of the film and its thermal stability, resulting in safety hazards [13,51]. Generally, blending, copolymerization, and crosslinking are used to improve the properties of polymer matrices and produce GPEs that perform well in LIBs. However, more importantly, the use of appropriate inorganic fillers in GPEs has recently emerged as one of the most promising methods to enhance the strength of the membranes, ionic conductivity, and Li$^+$ ion transfer, which results in the GPEs performing well in LIBs [43].

Recently, a series of new approaches for incorporation of inorganic fillers have been developed to improve the electrochemical properties of GPEs. Besides, the incorporation of inorganic fillers in shape memory polymer (SMP) has been studied as an effective method for enhancing the mechanical properties and allowing multiple functionalities.

SMP is a kind of material that can highly interact with stimulants such as temperature, light, and electromagnetic field to recover its original shape, suggesting this material as an important smart polymer which widely applied in industry [52]. In an attempt to enhance SMP's recovery, Park et al. studied the effect of SiO_2 fillers on polyurethane (PU) properties [53]. A cross-linking formed between the PU chains and hydroxide on the surface of SiO_2 has improved the shape memory effect and enhanced the mechanical properties with 0.2 wt% SiO_2. Besides, with the advantages of mechanical strengths as well as high elastic modulus, CNTs were used as effective fillers in improving the recovery of SMPs. CNTs-based nanocomposite showed excellent shape fixation ability with the recovery up to 90%, much higher than pristine SMP. Thus, SMP composite can be widely utilized in the industry [54,55]. GPEs have become one of the most effective electrolytes for applications such as wearable devices that require multiple functionalities, including flexibility, deformability, stretchability, and compressibility. Nevertheless, there has been a lack of discussion regarding the fundamental aspects of GPEs, including the materials used, preparation processes, their properties (mechanical, optical, and electrochemical), and their mechanism [56–59]. To obtain an overall comprehension of the recent studies on inorganic fillers and highlight their representative achievements, it is necessary to summarize the important findings for future studies on GPEs with an aim at developing high electrochemical energy storage properties.

In this review, the historical development of inorganic fillers in GPEs is first introduced. In the following sections, the details of inorganic fillers applied in GPEs along with various synthetic routes are presented. Subsequently, the mechanism of conductivity enhancement in the presence of inorganic fillers is discussed in detail. Finally, the application of GPEs in various battery systems (Li, Na, Mg, and Zn batteries) is discussed.

2. Historical Overview of GPEs

GPEs can be divided into three different categories based on the constituents of the mixture: plasticizer-added GPEs, inorganic filler-added GPEs, and a combination of plasticizer and inorganic filler-added GPEs. Lithium salt is responsible for the transportation of ions in the polymer framework, whereas the polymer accommodates the electrolyte to provide mechanical strength. The development of inorganic fillers in GPEs goes back to the early 1980s, when an attempt was made to improve the mechanical stability of the polymer matrix using an aprotic solution containing an alkali metal [60]. In the 1990s, the role of inert fillers in GPE composite systems started to be actively recognized. The addition of fillers to polymer segments has been reported to enhance Li^+-polymer interactions and Li^+ ion transport because of the dominant Li^+ ion movement along the filler surface rather than through the polymer segment [61,62]. Since then, research on the incorporation of inorganic fillers into GPEs has rapidly expanded. In the 2000s, the effect of particle size of inorganic fillers in GPEs was extensively studied. Wang et al. prepared GPEs containing PVDF-HFP with a novel hierarchical mesoporous SiO_2 network, which exhibited mechanical stability and higher ionic conductivity compared with GPEs without SiO_2, or with fumed SiO_2 [63]. Yang et al. synthesized SiO_2 (m-SBA15) with enhanced ionic conductivity owing to the liquid electrolyte being trapped by the mesoporous structure and the large specific surface area of m-SBA15 [64]. In the 2010s, another effective method, surface modification, was proposed to improve the dispersion and affinity of inorganic fillers within organic compounds. A significant effort has been made to develop GPEs based on inorganic fillers, such as $BaTiO_3$ [65], Al_2O_3 [66], TiO_2 [67], ZnS [68], and CeO_2 [69]. Figure 3 shows a historical overview of the development of inorganic fillers in GPEs.

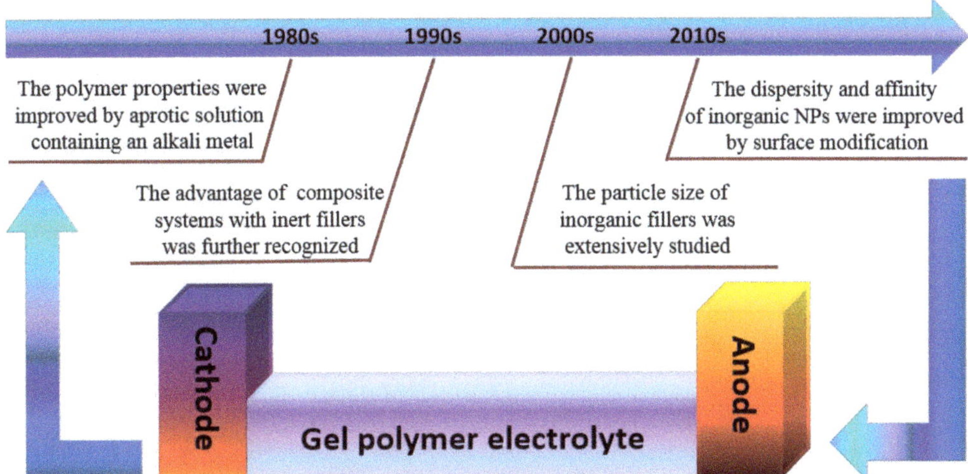

Figure 3. Historical overview of the developments of inorganic fillers in GPEs.

The implementation of GPEs in a variety of battery systems, including LIBs and Li-sulfur batteries, has been increasingly studied. The presence of plasticizers in GPEs increases Li^+ ion transfer; however, it simultaneously deteriorates the mechanical properties of the polymer matrix. Furthermore, electrode degradation may occur because of redox reactions between the plasticizer and electrodes. Consequently, to overcome the limitations of plasticizers, inorganic fillers have been proposed as additive materials to increase the electrolytic conductance and mechanical stability of GPEs. Osinska et al. [70] modified the surface of inorganic fillers to improve electrolyte absorption, thereby providing a more favorable medium for ion transport. A new ion transport pathway was discovered by Kumar et al. [71] by introducing spatially charged layers that tend to overlap with the concentrated filler grains dispersed in GPEs. The polymer chain containing active sites and surface groups of the inorganic fillers are affected by Lewis acid–base interactions. This interaction mostly occurs between the carbonyl groups of poly(acrylamide) (PAM) and surface groups of Al_2O_3, resulting in changes in the morphology of the GPEs [72].

3. Details of the Inorganic Fillers Applied in GPEs

The key functionalities of inorganic fillers in GPEs are to enhance Li^+ ion transfer and mechanical stability, wherein the polymer provides the conduction pathway for the ions, whereas the fillers influence the physical durability of the polymer to support ion transport. In addition, the inorganic filler particles can be used as a "solid plasticizer," which reduces the crystallinity of the host polymer and increases the transport properties. In this section, the addition of different types of inorganic fillers to enhance the mechanical properties and electrolytic conductance of GPEs is described in detail. Various inorganic fillers are listed according to research prevalence from the early 2000s to the present.

3.1. Titanium Dioxide (TiO$_2$)

Chung et al. [73] investigated the effect of TiO_2 nanoparticle (NP) addition in (PEO)-$LiClO_4$ for the improvement of electrochemical performance of GPEs. This study provided a model for the effects of inorganic fillers on the overall Li^+ ion transport in nanocomposite electrolytes. In addition, the specific role of inorganic filler was interpreted in terms of Lewis acid-base interactions. Over a wide temperature range, two structural modifications occurred at the ceramic surface. First, the morphology of the polymer was modified by the structural groups on the surface of the particles, which provided crosslinking opportunities for the PEO segments and X-anion. This resulted in a reduction in the energy barrier of the reorganized PEOs, where appropriate Li^+ ion-conducting pathways were established at the ceramic surface. Second, an "ion-ceramic complex" was formed through the dissociation of salt due to the interaction between polar groups on the surface of the fillers and ions of the electrolyte. These two structural variations can account for the improvement in the electrolytic conductance of inorganic fillers in GPEs. Liu et al. [67] reported that GPEs with a TiO_2 ceramic filler exhibited higher Li^+ ion transfer numbers than GPEs without TiO_2. The interaction between the fillers, anions, and polymer chains enhanced Li^+ ion transfer. Kim et al. [74] investigated the influence of filler content on the morphology of GPE membranes. As the TiO_2 content increased, the surface of the membrane coarsened, and small aggregates appeared; however, the TiO_2 NPs remained well distributed over the entire surface area of the membrane with an increase in the content to 60 wt% (Figure 4). In addition, the ionic conductivity was enhanced owing to the nanopores in the liquid medium, as well as the effective ion transport achieved by the presence of TiO_2. Therefore, the addition of rutile TiO_2 NPs not only enhanced the dispersion of the constituents but also improved the physical and electrochemical properties of the GPE. Karlsson et al. [75] studied polymer kinetics using quasi-elastic neutron scattering experiments to observe the effect of the filler on the crystallinity and structural changes in the polymer. The results showed that the improvement in ionic conductivity was because of the filler rather than the enhanced polymer dynamics based on the presence of a 5-nm-thick immobilized polymer layer around the filler particles. Kwak et al. [76] prepared a viscous $P(EO-EC)/LiCF_3SO_3/TiO_2$ polymer electrolyte mixture with a porous $P(VdF-HFP)/P(EO-EC)/TiO_2$ membrane. The TiO_2 content used was 0.0, 0.5, 1.0, 1.5, 2.0, 5.0, and 10.0 wt% in the polymer electrolyte and 0.0, 10.0, 20.0, 30.0, and 40.0 wt% in the porous membrane with a blend composition of 6:4 $P(VdF-HFP)$ to $P(EO-EC)$. The stress and tensile modulus values showed an increase for the membrane up to 2.0 and 41.0 MPa, respectively, using 30 wt% TiO_2, and a decrease up to 1.2 and 34.5 MPa, respectively, when adding 40 wt% TiO_2. These results indicated that the presence of high concentrations of TiO_2 up to 20 wt% improved the mechanical properties of the membrane owing to the interaction between the NPs and the host polymer. In addition, an increase in ionic conductivity was observed up to 4.7×10^{-2} mS cm^{-1} at 25 °C for the GPEs containing 1.5 wt% TiO_2 owing to the interaction between the oxide groups of the polymer and hydroxyl groups of the TiO_2 filler.

Hwang et al. [77] reported that the electrolytic conductance of GPEs was increased when reducing the particle size of the TiO_2. The ionic conductivity of a GPE containing TiO_2 NPs was higher than that of a GPE without TiO_2 at 30 °C. Agnihotry et al. [78] examined the effects of different concentrations of nanosized TiO_2 in a PMMA-based GPE. This study demonstrated that the ionic conductivity was enhanced when using an optimum TiO_2 loading (2 wt%) in the GPE. Byung et al. [79] synthesized a GPE by blending poly(acrylonitrile)-poly(ethylene glycol diacrylate) (PAN-PEGDA) with Li-salt and TiO_2 NPs. The high surface area of inorganic fillers can improve the interfacial resistance and ionic conductivity of Li metal owing to an increase in chemical stability and possible retention of organic solvents in the micropores; this assists in the regulation of possible side reactions associated with Li metal. In addition, nanosized inorganic fillers were evenly dispersed and increased the mechanical stability of the polymer matrix. Walkowial et al. [80] and Kurc et al. [81,82] modified the surface of a new hybrid TiO_2/SiO_2 filler for GPEs. The original hybrid fillers were modified by grafting functional groups, such as methacryloxy or vinyl groups, on the surface of the fillers. The surface modification chemistry of the filler seemed to be another factor affecting the overall performance of the GPE in terms of solvent absorption, specific conductivity, and intercalation of lithium on graphite. As shown in the SEM images (Figure 5a), the hybrid TiO_2/SiO_2 spherical NPs are homogeneously dispersed. Figure 5b shows that the hybrid TiO_2-SiO_2 predominantly contains rutile TiO_2. Figure 5c indicates that the mean pore diameter of the TiO_2-SiO_2 hybrid is 3.8 nm, which represents the mesoporous components, and the surface area is 12.5 m^2/g, which suggests an intermediate surface activity level. Hybrid TiO_2/SiO_2

powder with a moderate degree of surface functionalization was considered as a potential candidate for GPE applications in LIBs. The SEM image of the GPE without functionalized fillers typically exhibits a porous structure (Figure 5d).

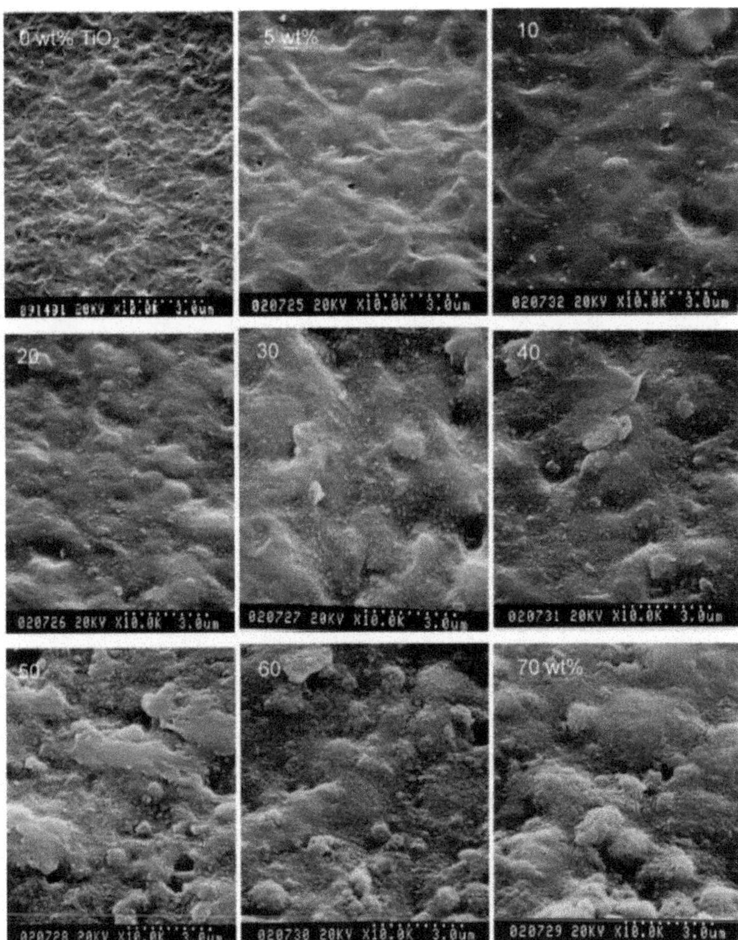

Figure 4. Surface morphology of P poly(vinylidene fluoride-co-hexafluoropropylene) (PVDF-HFP) with varying TiO_2 (rutile) contents from the upper left to the lower right panel. Reprinted with permission from Kim et al. [74]. Copyright 2003 Elsevier B.V.

An experimental investigation by Yahya et al. was performed on proton-conducting GPE nanocomposites based on TiO_2 NP-dispersed cellulose acetate (CA) [83]. The increase in the ionic conductivity of the GPE nanocomposite could be explained by the addition of TiO_2, which increased the total solution/solvent dielectric constant. The dielectric constant of TiO_2 was higher than that of N,N-dimethylformamide (DMF), leading to a reduced Coulombic interaction between the ion aggregates and dissociation of ions from the salt resulting in free NH_4^+ ions. The GPE prepared by adding NH_4BF_4 and TiO_2 to CA was considered a promising material for proton batteries.

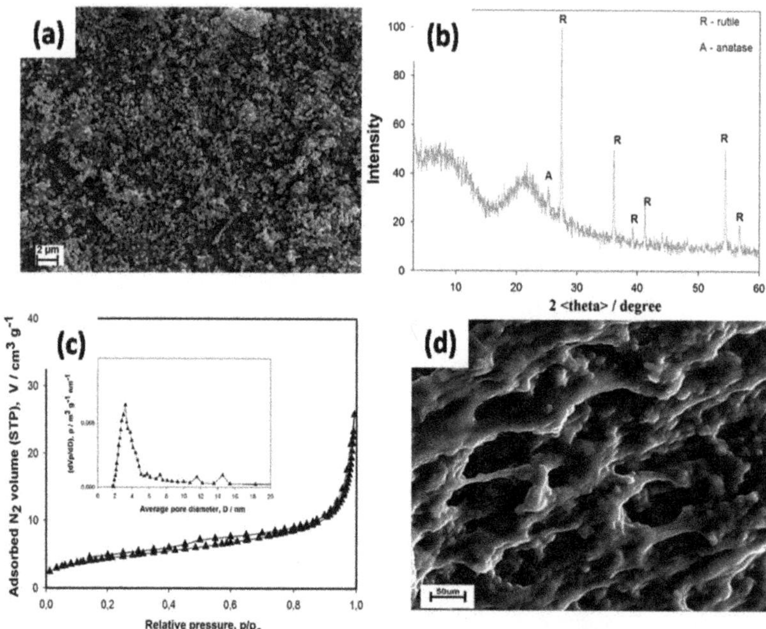

Figure 5. (**a**) SEM image of the TiO_2-SiO_2 hybrid composite, (**b**) wide angle X-ray spectroscopy of the TiO_2-SiO_2 hybrid composite, (**c**) nitrogen adsorption/desorption isotherm and pore size of the TiO_2-SiO_2 hybrid composite, and (**d**) SEM image of the surface of the membrane containing the TiO_2-SiO_2 hybrid composite. Reprinted with permission from Kurc et al. [81]. Copyright 2014 Elsevier Ltd.

Cao et al. [84] demonstrated improvement in the GPE by incorporating TiO_2 in PVDF/PMMA via electrospinning for practical applications in LIBs. The GPE containing 3 wt% TiO_2 showed a highest electrolytic conductance of 3.9×10^{-1} mS cm^{-1} with an electrochemical stability up to 5.1 V vs. Li$^+$/Li at room temperature. The increase in electrolytic conductance with the addition of TiO_2 particles was due to better dispersion through a Lewis acid–base interaction between the polar groups of the electrolytes and the filler and a decrease in the crystallinity of the polymer. Hong Chen et al. [85] investigated the role of nano-TiO_2 dispersibility in a GPE based on a PVDF-HFP polymer for LIBs (schematic is shown in Figure 6a). Figure 6b shows the dispersion of the nanoparticles (pristine, commercial, and modified TiO_2) in DMF. The morphology of nano-TiO_2-PMMA doped PVDF-HFP membrane exhibits a smoother surface with fewer pores compared to the pristine PVDF-HFP membrane and nano-TiO_2 doped PVDF-HFP membrane. Interestingly, the highly dispersed TiO_2-PMMA hybrid membrane has a rougher surface, and the pore size is more uniform (Figure 6c). The effect of TiO_2 dispersion on the C-rate discharge performance is shown in Figure 6d. The addition of nanosized TiO_2 to the PVDF-HFP-based GPE significantly improved the discharge capacity of the cells. In addition, the highly dispersed nano-TiO_2-PMMA doped GPE showed the highest discharge capacity compared to the other electrolytes. This study demonstrated that the dispersion of nanosized TiO_2 is an important factor that influences the performance of the PVDF-HFP-based polymer electrolyte. Sankaranarayanan et al. [86] explained the effect of TiO_2 on the electrochemical performance based on the distribution and aggregation of the fillers, Lewis acid–base interactions, and polymer segment-ion coupling. The addition of TiO_2 to the GPE nanocomposite promoted Lewis acid–base interactions, thereby facilitating the dissolution of $LiClO_4$ salts and increasing the amount of free Li$^+$ ions. Bozkurt et al. [87] produced polymer electrolyte nanocomposites based on borate ester graft copolymer PVA, poly(ethylene glycol) methyl ether (PEGME)), nano-TiO_2, and trifluoromethane sulfonate (CF_3SO_3Li). The conduction pathway for ion transport was improved by the boron-containing PVA backbone and flexible side chains. This study suggested that the presence of inorganic fillers resulted in an increase in ionic conductivity, Li$^+$ ion transfer, interfacial stability between the electrode and electrolyte, and the mechanical strength of the GPEs. Chen et al. [88] showed that after adding TiO$_2$ NPs, the PVDF-HFP/PMMA/TiO$_2$ membrane exhibited enhanced thermal stability and electrolytic conductance. Based on the images of the GPE membrane, its thermal stability was significantly improved after the addition of TiO$_2$ NPs (Figure 7a,b). In particular, the GPE containing 5 wt% TiO$_2$ NPs showed homogeneously interconnected pores which resulted in an excellent performance (Figure 7c,d). The EIS analysis of GPE containing varying concentrations of TiO$_2$, shown in Figure 7c, shows that after the introduction of TiO$_2$, the bulk resistance (R$_b$) of the GPEs significantly decreases for the sample containing the lowest TiO$_2$ content of 5 wt%. The initial discharge capacities were 143.6, 180.5, 188.1, and 163.6 mAh g^{-1} for the different TiO$_2$ contents (0, 2, 5, and 7 wt%), respectively. After 50 cycles, the capacity decreased to 114, 154.3, 173.2, and 139.9 mAh g^{-1}, with a capacity retention of 79.4%, 85.5%, 92.1%, and 85.5%, respectively (Figure 7d). This study showed that the incorporation of TiO$_2$ NPs to the GPE improved its electrochemical stability and ionic conductivity in LIBs. Yamolenko et al. [89] extended this study on the effect of NPs to a polyester-diacrylate (PEDAC)-based network polymer electrolyte. The mechanical strength of the GPE was enhanced after improving the distribution of the NPs in the polymer electrolyte by ultrasonic treatment compared to simple mechanical stirring.

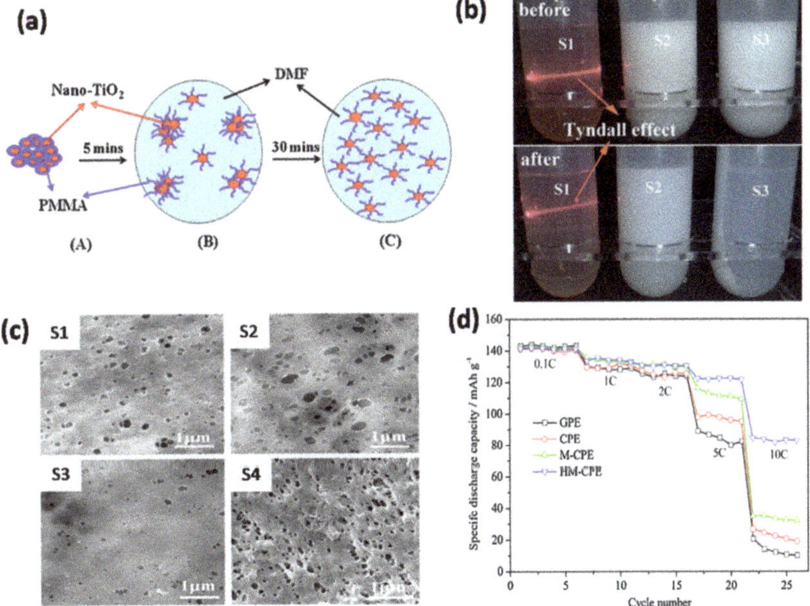

Figure 6. (**a**) Schematic of nano-TiO$_2$- poly(methyl methacrylate) (PMMA) in N,N-dimethylformamide (DMF) (A: mixture of nano-TiO$_2$ and PMMA, B: tethered PMMA on Nano-TiO$_2$, C: self-assembly of PMMA-tethered Nano-TiO$_2$) and (**b**) photographs of the dispersed NPs in DMF before (top panel) and after (bottom panel) 10 min centrifugation at 10,000 rpm. The same TiO$_2$ content (5 wt%) was used in each sample: (S1) highly dispersed nano-TiO$_2$-PMMA, (S2) nano-TiO$_2$-PMMA, and (S3) pristine nano-TiO$_2$. (**c**) SEM images of the GPEs: (S1) pristine PVDF-HFP ("GPE"), (S2) nano-TiO$_2$/PVDF-HFP ("CPE"), (S3) nano-TiO$_2$-PMMA/PVDF-HFP ("M-CPE"), and (S4) highly dispersed nano-TiO$_2$-PMMA/PVDF-HFP ("HM-CPE"). (**d**) The rate capabilities of the different electrolytes shown in (**c**). Reprinted with permission from Chen et al. [85]. Copyright 2013 Elsevier Ltd.

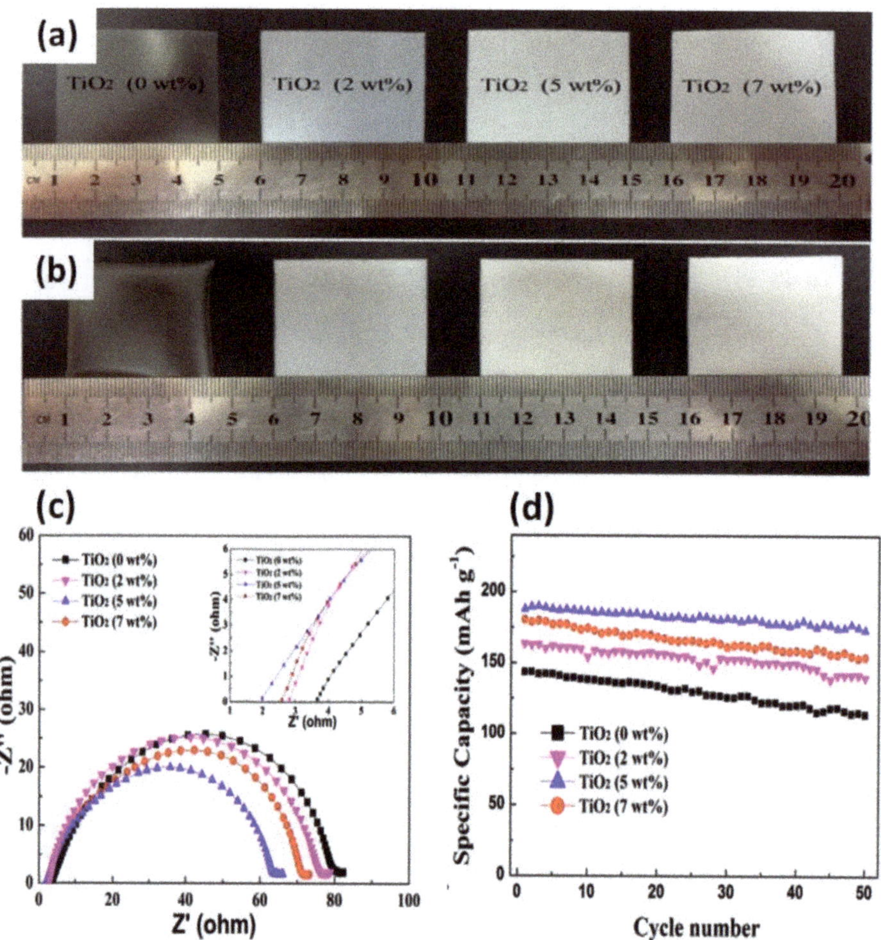

Figure 7. Thermal behavior of the GPE membranes (**a**) before and (**b**) after storing at 130 °C for 1 h, (**c**) EIS results of the GPE membrane containing varying contents of TiO$_2$, and (**d**) cyclic performance of the LiCoO$_2$/Li cells using the GPE at 0.2 C. Reprinted with permission from Chen et al. [88]. Copyright 2015 Elsevier B.V.

Teng et al. [90] reported a GPE prepared by combining poly(acrylonitrile-co-vinyl acetate) (PAV) with PMMA and TiO$_2$ NPs, i.e., PAVM:TiO$_2$ (Figure 8). This study introduced a new concept concerning the association of oxide NPs with the space-charge regimes around the polymer's functional groups to induce 3D conduction pathways for Li$^+$ ions in GPEs applied in LIBs (Figure 8a–c). Based on the impedance spectra shown in Figure 8d, the high-frequency semicircle represents the movement of charge carriers through the SEI layer, middle-frequency semicircle shows charge-transfer resistance, and sloping line is related to the Warburg impedance. The results revealed the superiority of GPE-PAVM:TiO$_2$ in enhancing Li$^+$ ion transport. Figure 8e presents the discharge capacities of the full-cell batteries at a high rate of 20 C. The cell with the GPE-PAVM:TiO$_2$ electrolyte delivered discharge capacities of 152 and 84 mAh g^{-1} at 0.1 and 20 C, respectively, thus outperforming the cell with the SLE electrolyte with capacities of 146 and 40 mAh g^{-1}, respectively. The Li$^+$ ion transport efficiency in the bulk solution as well as at the electrode–electrolyte interface was enhanced through the 3D percolation pathway by immobilizing the PF$_6^-$ anions in the oxide NP framework. Sakunthala et al. performed a comparative study between single-crystalline TiO_2 nanorods and submicronsized TiO_2 fillers in PVDF-HFP/EC/LiClO$_4$. The Li$^+$ ion transfer and tensile strength of the membrane containing 5 wt% TiO_2 nanorods were higher than those of the membrane containing 5 wt% submicronsized TiO_2. This can be explained by the improved interaction of the rod-shaped morphology of the single-crystalline TiO_2 filler with the polymer/salt/plasticizer matrix in the GPE [91].

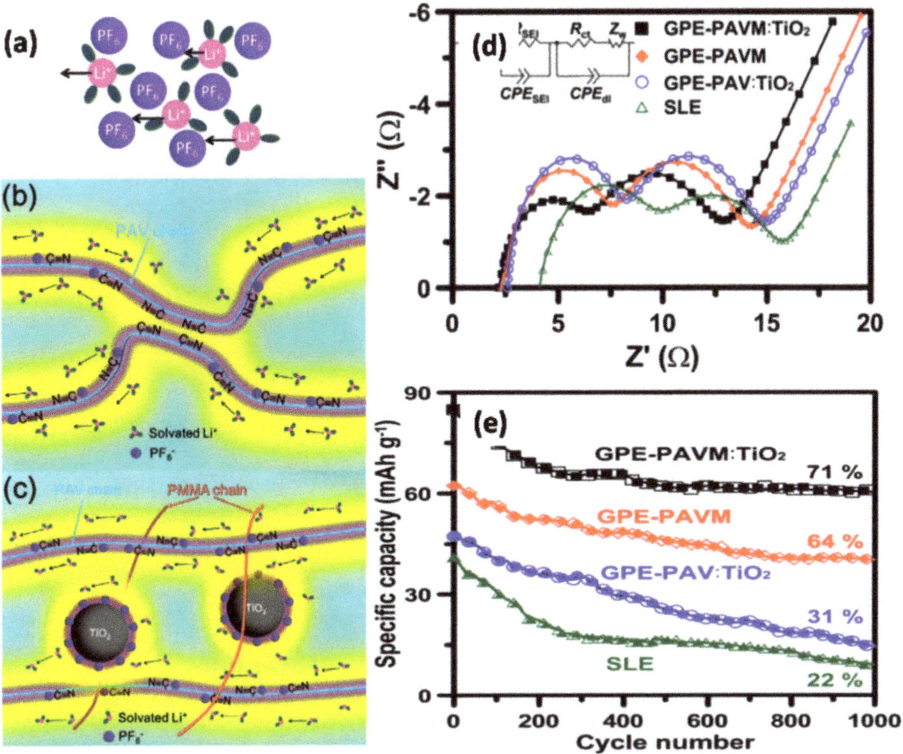

Figure 8. (**a**) Separator-supported LE (SLE): limitation in ionic dissociation. (**b**) GPE- poly(acrylonitrile-co-vinyl acetate) (PAV): a space-charge layer of Li$^+$ ions is formed due to the PF$_6^-$ anion absorbed by the nitrile functional groups on the PAV chain. (**c**) GPE-PAVM:TiO$_2$: the 3D percolation pathway is formed by the space-charged layers surrounding the TiO$_2$ NPs and PAV chain. (**d**) Nyquist plots of the full cell (graphite-GPE-LFP) and (**e**) discharge capacity of the full cell at a 20 C-rate over a voltage range of 2.0–3.8 V. Reprinted with permission from Teng et al. [90]. Copyright 2016 American Chemical Society.

Sivakumar et al. [92] fabricated a GPE containing a PVC-PEMA blend using hydrothermally derived TiO$_2$ NPs as an inorganic filler. The influence of the filler NPs on the surface morphology, thermal stability, and electrochemical properties were studied. The addition of TiO$_2$ NPs reduced the crystallinity of the polymer and enhanced the Li$^+$ ion transport pathways. Similarly, Singh et al. [93] modified the structural properties of PEMA-based plasticized polymer electrolytes by incorporating TiO$_2$ NPs. The addition of TiO$_2$ NPs in the plasticized polymer electrolyte suppressed the crystallinity and enhanced the amorphous nature. Hence, the addition of TiO$_2$ NPs could be a novel approach to enhance the electrolytic conductance in GPEs. The ionic conductivity and temperatures of some significant GPEs containing TiO$_2$ fillers are listed in Table 1.

Table 1. List of GPEs with TiO_2 filler with their conductivity and temperature.

Polymer	Salt/Plasticizers	Solvent	Conductivity (mS cm^{-1})	Temperature (°C)	Reference
PEO	$LiClO_4$	ACN	-	90	[73]
PEO	$LiBF_4$	ACN	7×10^{-4}	30	[67]
PVDF-HFP	$LiPF_6$-EC/DMC	Acetone	10^0	25	[74]
PEG	$LiClO_4$	Dicloro-methane	-	120	[75]
P(VdF-HFP)/P(EO-EC)	$LiCF_3SO_3$	Acetone	4.7×10^0	25	[76]
PEO	$LiClO_4$	THF	1.03×10^{-2}	30	[77]
P(VdF-HFP)/P(EO-EC)	$LiCF_3SO_3$	Acetone	5.1×10^{-2}	25	[78]
PAN/PEGDA	$LiPF_6$/$LiCF_3SO_3$-EC/DMC	-	3.8×10^0	30	[79]
PVDF-HFP	$LiPF_6$-EC/DMC	Acetone + DBP	8.5×10^{-1}	25	[80]
PAN	$LiPF_6$-TMS	DMF	9.8×10^{-1}	25	[81]
Cellulose acetate (CA)	NH_4BF_4	DMF	1.37×10^1	30	[83]
PVDF/PMMA	$LiClO_4$-EC/PC	DMF + acetone	3.9×10^0	30	[84]
PEO-PVC	$LiClO_4$	Cyclohexanone	8.33×10^{-4}	-	[86]
PVA/PEGME	$LiCF_3SO_3$	DMSO	1.58×10^{-1}	30	[87]
PVDF-HFP/PMMA	$LiPF_6$-EC/DMC	DMF + Acetone	2.49×10^0	30	[88]
Polyester diacrylate	$LiClO_4$-EC	Benzoyl peroxide	1.8×10^0	20	[89]
PAN/PVA	$LiPF_6$-EC/DMC/DEC	Dimethylacet-amide	4.5×10^0	30	[90]
PVDF-HFP	$LiClO_4$-EC	Acetone	1.11×10^1	30	[91]
PVC/PEMA	$LiClO_4$-EC/DMC	THF	0.5×10^1	30	[92]
PEMA	NaI-EC	THF	2.42×10^{-1}	30	[93]

3.2. Aluminum Oxide (Al_2O_3)

Li et al. [94] prepared a GPE by combining porous P(VDF-co-HFP) with alumina (Al_2O_3) NPs as the filler. An increase in the Al_2O_3 NP content reduced the level of crystallization in the polymer, thereby increasing the amorphous phase of the membrane. Piotrowska et al. [95] described the effect of oxide fillers on the properties of GPEs with a PVDF-HFP polymer matrix. Modification of the PVDF-HFP membranes with Al_2O_3 NPs led to a decrease in the liquid phase uptake ability owing to the reduction of accessible pore spaces. Egshira et al. [96] assessed the availability of the alumina filler in an imidazolium-based gel electrolyte. The Al_2O_3 filler enhanced Li^+ ion mobility by providing alternative pathways for Li^+ ion movement and changing the interaction between the Li^+ ions and the EO chain. Rai et al. [97] prepared nano-Al_2O_3-filled PVA composite gel electrolytes. As shown in Figure 9a, the PVA membrane exhibits a porous structure, while the addition of 2 wt% Al_2O_3 NPs reduces the porosity of the PVA composite electrolyte as the Al_2O_3 NPs are trapped among the chains in the pores (Figure 9b). Upon the addition of 6 wt% Al_2O_3 NPs (Figure 9c), the PVA chains are fully covered with Al_2O_3 NPs. This indicates complete dispersion of the Al_2O_3 nanofiller in the electrolyte film. Upon further addition of Al_2O_3 NPs (10 wt%), the grain sizes and shapes become irregular resulting in a partially crystalline structure containing Al_2O_3 NPs and PVA electrolyte (Figure 9d). An increase in the Al_2O_3 NP content increased the amorphous phase of pristine PVA. The Li^+ ion transfer capacity of the GPE reached its maximum with the addition of 6 wt% Al_2O_3 NPs.

A novel GPE was prepared by combining a poly(methyl methacrylate-acrylonitrile-ethyl acrylate) (P(MMA-AN-EA)-based polymer electrolyte with nano-SiO_2 and Al_2O_3 as inorganic fillers [98]. The maximum electrolytic conductance of the GPE was achieved when 5 wt% nano-SiO_2 and nano-Al_2O_3 were used. This study showed the different roles of the inorganic fillers: SiO_2 contributed to the enhanced ion conduction due to strong Lewis acid–base interactions, whereas Al_2O_3 improved the structural and thermal stability of the GPE owing to its high stiffness. Wen et al. [99] synthesized a novel GPE with a trilayer structure consisting of polyvinyl formal (PVFM)-4,4-diphenyl-methane diisocyanate (MDI) covered by a PVA-Al_2O_3 solution on both sides to achieve synergistic effects for each layer (Figure 10a). The morphology of the Al_2O_3/PVFM/Al_2O_3 trilayer membrane was determined by FESEM (Figure 10b–e). The thicknesses of the layers were 45.42, 54.27, and 65.77 µm, respectively (Figure 10b). The morphology of the PVFM membrane exhibits sponge-like pores, which are expected to enhance the transport of Li^+ ions (Figure 10c). The porous structure of the inorganic particulate films did in fact increase Li^+ ion transfer

(Figure 10d). The surface morphology of the PVFM membrane shows evenly distributed pores, as shown in Figure 10e. This study indicated that the inorganic layers enhanced the thermal integrity and mechanical properties of the membrane. This multilayer polymer membrane could be a potential material for application in LIBs.

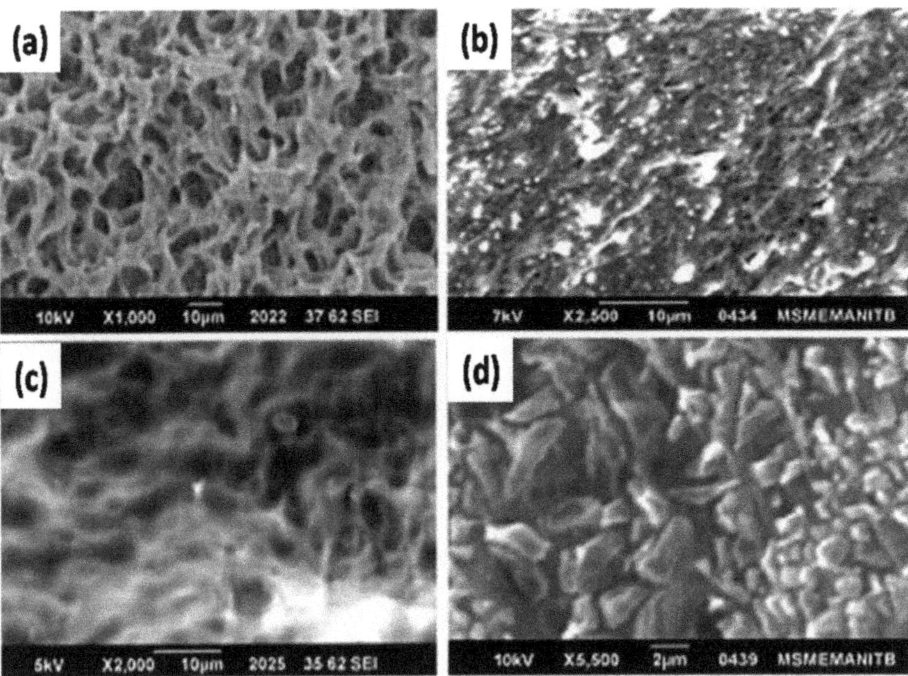

Figure 9. SEM images of the (**a**) PVA:NH$_4$SCN/DMSO GPE containing (**b**) 2, (**c**) 6, and (**d**) 10 wt% Al$_2$O$_3$ NPs. Reprinted with permission from Rat et al. [97]. Copyright 2012 Indian Academy of Sciences.

Kim et al. [100] prepared a GPE containing a PVDF-HFP fibrous matrix with Al$_2$O$_3$ as the inorganic filler by electrospinning. The morphologies of polymer fibrous matrices exhibiting different diameters are shown in Figure 11a–d. The surface morphology of the pure polymer is rougher than that of the Al$_2$O$_3$-composite polymers and exhibits an average diameter of 2.3 µm (Figure 11a,b) compared to 1.2 µm for the Al$_2$O$_3$-composite polymer matrix (Figure 11c,d). The presence of inorganic fillers in the fibrous polymer matrices prevented polymer shrinkage and agglomeration, which was beneficial for the formation of homogeneous pores. Figure 11e shows that the discharge capacities of the NMC and LTO half cells of the GPE-Al$_2$O$_3$ composite are 189.6 and 166.3 mAh g^{-1}, respectively, which are higher than those of the GPE without Al$_2$O$_3$ (168.2 and 146.8 mAh g^{-1}, respectively). The presence of Al$_2$O$_3$ NPs increased the porosity and absorption of free ions, thereby enhancing the electrolytic conductance and electrochemical stability compared to a pure GPE. The Al$_2$O$_3$-composite GPE also showed a better discharge capacity retention of ~96% (initial and final specific capacities of 166.3 and 160.2 mAh g^{-1}, respectively) (Figure 11f). Jain et al. [101] presented efficient conduction pathways constructed in PVdF-based GPEs with Al$_2$O$_3$ and boron nitride (BN) ceramic nano/microparticles. The high dielectric constants of Al$_2$O$_3$ and BN facilitated anion capture in the GPE and the transfer of Li$^+$ without coordination to the anions.

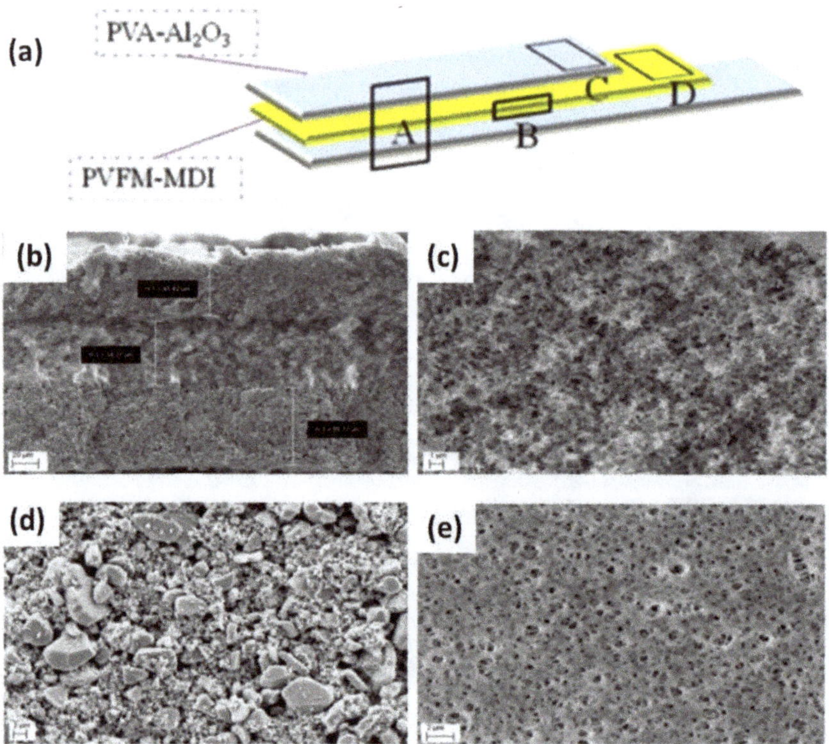

Figure 10. (**a**) Schematic of the membrane (Al$_2$O$_3$/PVFM/Al$_2$O$_3$). SEM images of the trilayer membrane; (**b**) cross-sectional image of the trilayer membrane; (**c**) cross-sectional image of the PVFM membrane; (**d**) the surface morphology of the Al$_2$O$_3$ coating layer; and (**e**) the surface morphology of the PVFM-based membrane. Reprinted with permission from Wen et al. [99]. Copyright 2007 Scientific Research Publishing Inc.

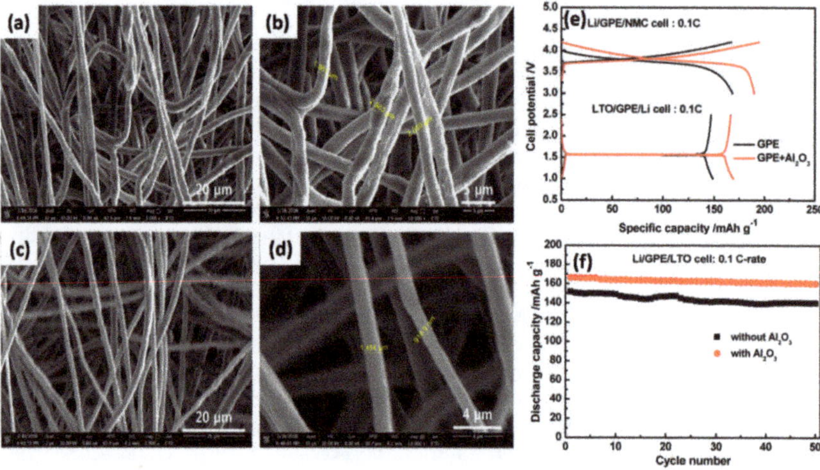

Figure 11. Morphology of (**a,b**) pristine GPE and (**c,d**) Al$_2$O$_3$-GPE membrane. (**e**) Initial charge–discharge curves of the Li$_4$Ti$_5$O$_{12}$ (LTO) and LiNi$_{1/3}$Mn$_{1/3}$Co$_{1/3}$O$_2$ (NMC) half cells and (**f**) the cyclic performance of the GPE membrane. Reprinted with permission from Kim et al. [100]. Copyright 2017 Elsevier Ltd.

Delgado-Rosero et al. [102] synthesized a GPE containing PEO and sodium trifluoroacetate (CF_3COONa) with different contents of Al_2O_3. The addition of inorganic fillers increased the amorphous phase portion surrounding the filler of the $(PEO)_{10}CF_3COONa$ + x wt% Al_2O_3 composite, thereby improving Na^+ ion transport through the pathways of the amorphous phase. Maragani et al. [103] reported a GPE containing a combination of PAN and sodium fluoride (NaF) with Al_2O_3 nanofibers formed through a solution casting technique. With an increase in the Al_2O_3 nanofiber content, the amorphous phase of the GPE increased, resulting in an improvement in ion conduction. Yang et al. [104] designed a novel GPE with uniformly cross-linked β-Al_2O_3 nanowires that compactly covered a P(VDF-co-HFP)-GPE through strong molecular interactions (Figure 12a–c). In this innovative structure, the LEs were immobilized through bonding between the cross-linked Al_2O_3 nanowires and PVDF-HFP (ANs-GPE), thereby creating uniform and continuous Na^+ ion transport channels along the Al_2O_3 nanowires (Figure 12a–c). This innovative structure can significantly improve the density and homogeneity of the Na^+ ion transport channel, resulting in superior electrochemical performance (Figure 12d–f). Mishra et al. [105] studied the effect of Al_2O_3 NP dispersion on a PVdF-HFP/PMMA blend-based nanocomposite GPE system. The electrolytic conductance changed significantly depending on the Al_2O_3 concentration in the PVdF-HFP/PMMA membrane. The maximum electrolytic conductance achieved was ~1.5×10^0 mS cm^{-1} when 6 wt% Al_2O_3 NPs were added to the GPE. The ionic conductivities and temperatures of the GPEs containing Al_2O_3 fillers are summarized in Table 2.

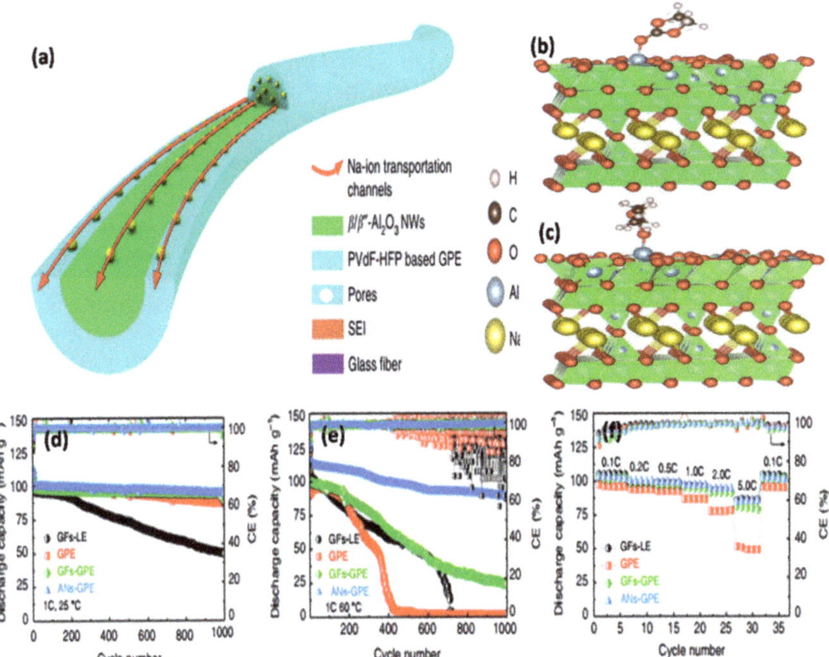

Figure 12. (**a**) Schematic of Na-ion transportation in the Al_2O_3 nanowire (AN)-GPE. (**b,c**) Adsorption of ethylene carbonate (EC) and diethylene carbonate (DEC) on the β-Al_2O_3 (003). (**d,e**) Cyclic performance of the $Na_3V_2(PO_4)_3$ (NVP)/Na cells using a glass fiber (GF)-LE, GPE, GFs-GPE, and ANs-GPE at 1 C under 25 and 60 °C, respectively. (**f**) Rate performance of the NVP/Na cells using the different GPEs. Reprinted with permission from Yang et al. [104]. Copyright 2019 Springer Nature.

Table 2. List of GPEs with Al_2O_3 filler with their conductivity and temperature.

Polymer	Salt/Plasticizers	Solvent	Conductivity (mS cm^{-1})	Temperature (°C)	Reference
PVDF-HFP	LiPF$_6$-EC/DMC DBP	NMP	1.95×10^{-0}	25	[94]
PVDF-HFP	LiPF$_6$-PC/DBP	Acetone	-	20–70	[95]
PEO/PMA	LiTFSI-EC/EMITFSI	-	10^{-1}	25	[96]
PVA	NH$_4$SCN	DMSO	5.81×10^1	30	[97]
P(MMA-AN-EA)	LiPF$_6$-EC/DMC	DMF	2.2×10^0	30	[98]
PVFM	LiPF$_6$-EC/DMC	NMP	4.13×10^{-1}	25	[99]
PVDF-HFP	LiPF$_6$-EC/DMC	Acetone-NMP	4.1×10^0	25	[100]
PVDF	LiPF$_6$-EC/PC	NMP	-	-	[101]
PEO	CF$_3$COONa	ACN	-	-	
PAN	NaF-EC	DMF	4.82×10^0	30	[103]
PVDF-HFP	NaClO$_4$-EC/DEC	Acetone + ethanol	7.13×10^{-1}	25	[104]
PVdF-HFP/PMMA	NaCF$_3$SO$_3$-EC/PC	THF + Acetone	1.5×10^0	25	[105]

3.3. Silicon Dioxide (SiO$_2$)

Wieczorek et al. [106] designed a novel model GPE using a combination of amorphous poly(ethylene oxide) dimethyl ether (PEODME) and LiClO$_4$ with fumed nano-silica. The presence of nanosized fumed silica in the GPE was promising because of the reduction in ion association. Wu et al. [107] prepared a hybrid polymer electrolyte film consisting of PMMA, LiClO$_4$, propylene carbonate (PC), and SiO$_2$ filler using a solvent casting technique. The conductivity was not positively correlated with the increased concentration of SiO$_2$ owing to the aggregation of SiO$_2$, which led to the formation of crystal-like particles on the surface of the membrane. Kim et al. [108] reported novel homogeneous spherical core-shell structured SiO$_2$(Li$^+$) NP fillers that were applied as functional fillers in GPEs (Figure 13). SiO$_2$(Li$^+$) was synthesized by dispersing Li$^+$ ions in the core-shell structure of the SiO$_2$ particles (Figure 13a). Figure 13b shows the charge–discharge curves of the GPE containing 20 wt% SiO$_2$(Li$^+$). The first discharge capacity with LiCoO$_2$ in the cathode was 153 mAh g^{-1}. The GPEs containing the novel filler exhibited unique Li$^+$ ion transport and mechanical strength. Consequently, the battery exhibited low internal resistance, high capacity, and a stable cyclic performance. In addition, the capacity retention was enhanced by increasing the SiO$_2$(Li$^+$) content in the GPEs up to 20 wt% (Figure 13c). The addition of SiO$_2$(Li$^+$) particles resulted in the retention of more LEs in the GPE, thereby improving the electrochemical performance during cycling.

Li et al. [109] obtained similar results and proposed that the addition of SiO$_2$(Li$^+$) increases the amorphous phase and porosity of the polymer film, thus improving the adsorption and gelation of the LE. From the surface morphology of the membranes, as shown in Figure 14, it was established that as the content of SiO$_2$(Li$^+$) increases, the pore size of the film also increases, resulting in a high uptake of liquid electrolyte. The electrolytic conductance of the GPE was improved because of the large number of Li$^+$ ions in SiO$_2$(Li$^+$). However, the GPE became very fragile when the SiO$_2$(Li$^+$) content reached 10 wt%. Manisankar et al. [110] synthesized superhydrophobic PVDF-SiO$_2$ films with different SiO$_2$ contents by electrospinning. When the SiO$_2$ content increased, the surface roughness of the membrane also increased, but the average diameter of the nanofibers was not affected. Kim et al. [111] synthesized a secure and flexible electrolyte through the combination of mesoporous SiO$_2$ NPs containing methacrylate groups and fibrous PAN membrane (Figure 15a,b). The initial discharge capacity delivered was 157.9 mAh g^{-1} after 300 cycles with a capacity retention of 88.0%. In addition, the GPE containing the mesoporous SiO$_2$ particles exhibits better Li$^+$ ion transfer than the GPE containing non-porous SiO$_2$ particles (Figure 15c,d). This study emphasized the role of SiO$_2$ mesoporous NPs compared with non-porous SiO$_2$ NPs when attempting to achieve good electrochemical properties in terms of discharge capacity, capacity retention, rate capability, and cycling stability. GPEs containing a combination of PEO/LiClO$_4$ complex and 1,3 dioxolane (DIOX)/tetraethyleneglycol dimethylether (TEGDME) as plasticizer with a SiO$_2$ filler has also been synthesized [112]. An increase in SiO$_2$ filler content and plasticizer reduced the degree of crystallization of the polymer membrane, thus increasing the ionic conductivity.

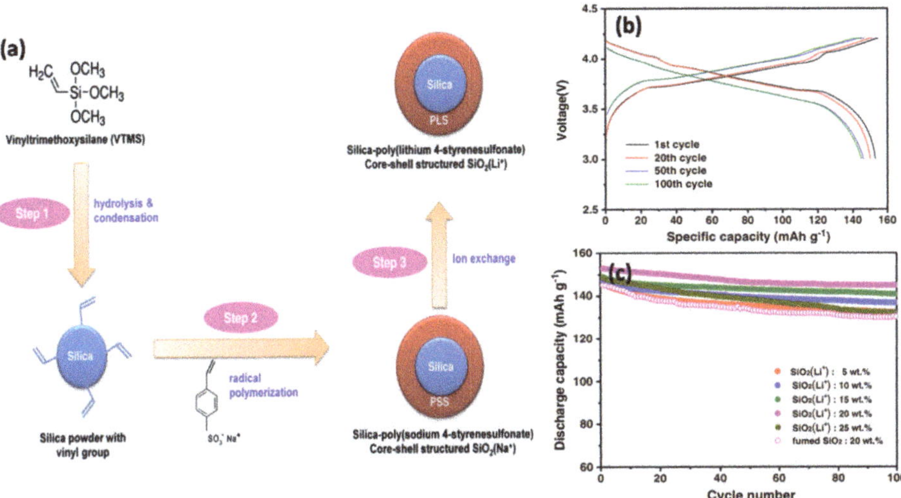

Figure 13. (a) Reaction scheme for the synthesis of the $SiO_2(Li^+)$ particles. (b) Charge-discharge curves of the GPE containing 20 wt.% $SiO_2(Li^+)$ particles. (c) Discharge capacities of the GPE containing different contents of $SiO_2(Li^+)$ particles. Reprinted with permission from Kim et al. [108]. Copyright 2012 Elsevier B.V.

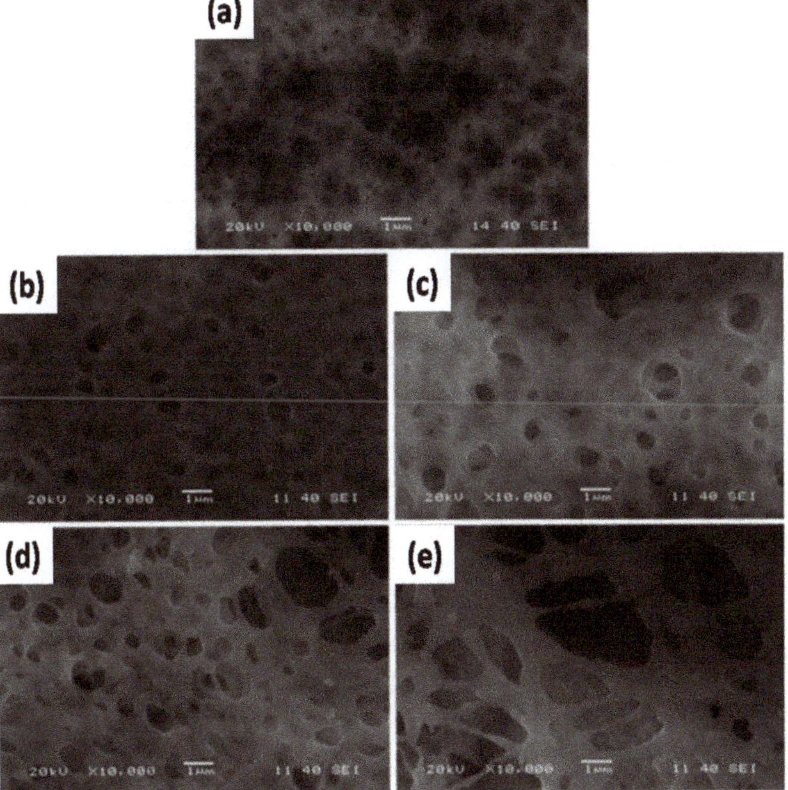

Figure 14. SEM images of the GPE membrane: (a) pristine PVDF, (b) PVDF-1% $SiO_2(Li^+)$, (c) PVDF-2% $SiO_2(Li^+)$, (d) PVDF-5% $SiO_2(Li^+)$, and (e) PVDF-10% $SiO_2(Li^+)$. Reprinted with permission from Li et al. [109]. Copyright 2013 Springer Nature.

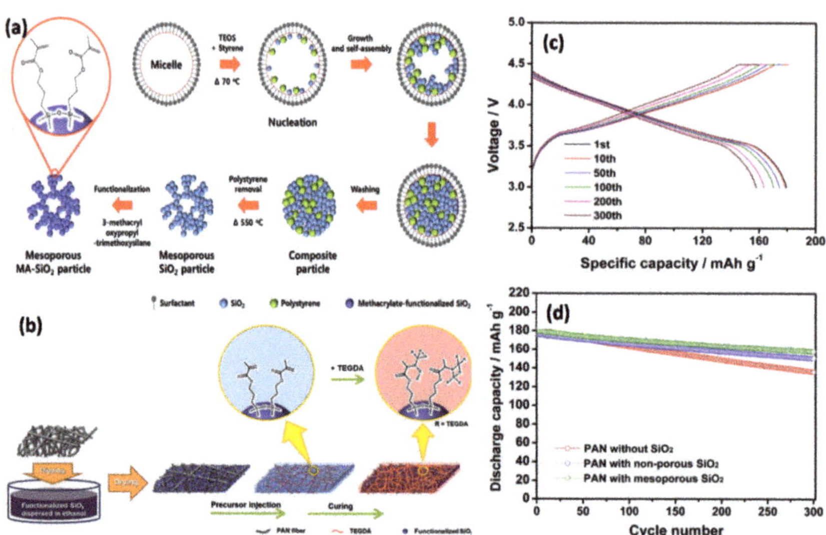

Figure 15. Reaction schemes for the synthesis of (**a**) mesoporous MA-SiO$_2$ particles and (**b**) cross-linked composite GPE. (**c**) Charge-discharge curves of the cell and (**d**) cyclic performance with different electrolytes at 25 °C. Reprinted with permission from Kim et al. [111]. Copyright 2016 Springer Nature.

Wu et al. [22] studied the influence of SiO$_2$ NP content on PVdF-HFP/IL membranes in terms of their ion conduction and discharge capacity in LIBs. The crystallization phase of the membrane was reduced due to the dispersion of SiO$_2$ NPs, which hindered the structural stability of the polymer but improved ion transport because of their interaction with the amorphous phase of the host polymer. The SiO$_2$ NPs acted as multifunctional inorganic fillers with good interfacial stability, which increased the ion conduction and Li$^+$ ion transfer number for the poly(propylene carbonate)-based GPE in Li-S batteries [113].

Hu et al. [114] studied a GPE containing dispersed SiO$_2$ NPs in a PEO matrix. Uniformly dispersed SiO$_2$ NPs were obtained in the polymer matrix owing to the high miscibility of all the precursors. The synthesized GPE nanocomposite membrane significantly improved the electrochemical performance, which suggests a promising strategy for the development of safer and more flexible Li-metal batteries. The ionic conductivities and temperatures of the significant GPEs containing SiO$_2$ fillers are listed in Table 3.

Table 3. List of GPEs with SiO$_2$ filler with their conductivity and temperature.

Polymer	Salt/Plasticizers	Solvent	Conductivity (mS cm^{-1})	Temperature (°C)	Reference
PEO	LiClO$_4$	DMF	3×10^{-3}	30	[106]
PMMA	LiClO$_4$-PC	DMF	5.64×10^{-2}	80	[107]
PVdF-HFP	LiPF$_6$-EC/DEC/DBP	Acetone	-	-	[108]
PVDF	LiPF$_6$-EC/DEC/DBP	DMF	3.87×10^1	30	[109]
PVDF	LiPF6-EC/DMC	DMF	7.73×10^{-4}	30	[110]
PAN	-	DMF	1.8×10^{-3}	25	[111]
PEO	LiCLO$_4$-DIOX/TGDME	ACN	10^{-4}	25	[112]
PVdF-HFP	LiTFSI-SDS	-	1.22×10^{-3}	25	[22]
PPC	LiTFSI + LiNO$_3$	DMAC	1.64×10^{-4}	23	[113]
PEO	LiClO$_4$	ACN	1.1×10^{-4}	30	[114]

3.4. Zirconium Dioxide (ZrO$_2$)

Vickraman et al. [115] studied a novel GPE containing lithium bis(oxalato)borate (LiBOB) as the Li salt and PVdF-PVC as the polymer matrix with varying contents of ZrO$_2$. The high ionic mobility of the GPE is related to the large amorphous phase of the polymer host and its large free volumes that enhanced Li$^+$ ion transfer. Suthanthiraraj et al. [116] investigated the effect of ZrO$_2$ NPs in a new PPG-silver triflate (AgCF$_3$SO$_3$) system in terms of the improvement to ion transport and electrochemical behavior. This study showed that structural modification of the polymer matrix by the addition of ZrO$_2$ NPs increased the ionic mobility and physicochemical properties of the GPE. Sivakumar et al. [117] reported the effect of the different concentrations of dispersed ZrO$_2$ from the perspective of its enhanced ionic conductivity. The ionic conductivity increased upon the introduction of ZrO$_2$ in the bare gel polymer system up to a loading of 6 wt%. However, a further increase in the ZrO$_2$ content reduced the conductivity because of the larger crystalline region present in the matrix, which hindered the ionic mobility. Similar results were observed by Sivakumar et al. [118] demonstrating that the introduction of ZrO$_2$ into a PVDF-HFP-(PC+DEC)-LiClO$_4$ system significantly improved the electrolytic conductance of the GPE. The addition of ZrO$_2$ NPs restricted the reorganization of the polymer chain structure, thereby increasing the amorphous phase of the polymer and improving the electrolytic conductance. Chen et al. [119] synthesized a novel GPE by in situ immobilization of ionic liquids (ILs) and nanoporous ZrO$_2$ in a polymer matrix. This study showed that the ZrO$_2$ skeleton cooperates with Li salts, resulting in improved dissociation of the Li salts and Li$^+$ ion transfer. Therefore, a discharge capacity of 135.9 mAh g^{-1} was obtained after 200 cycles at 30 °C. In addition, the cell operated well in the temperature range of −10 to 90 °C. The good contact and stable interface between the Li-metal electrode and GPE can be attributed to its effective electrochemical performance in Li-metal batteries.

Xiao et al. [120] synthesized a novel GPE by combining synthesized PMMA-ZrO$_2$ (sPZ) hybrid particles with a P(VDF-HFP) polymer matrix. The morphology of the GPE membrane containing homogeneously interconnected micropores is shown in Figure 16a. As shown in Figure 16b, the cell delivers 126.4 mAh g^{-1} at 2.0 C after 150 cycles, corresponding to a capacity retention of 85.2% at 0.1 C. The cell containing a graphite electrode delivers 288.5 mAh g^{-1} at 0.5 C after 80 cycles (Figure 16c). The modified ZrO$_2$ significantly enhanced the properties of the GPE in terms of mechanical strength, ion conduction, and thermal stability.

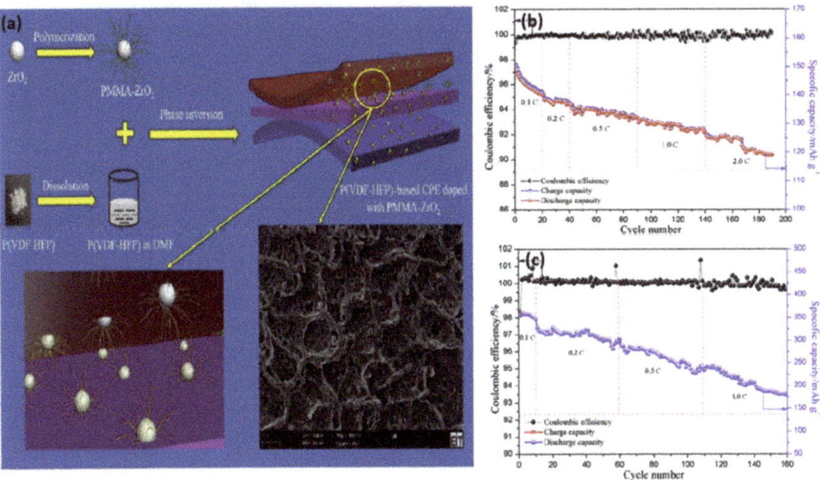

Figure 16. (**a**) Schematic of the GPE-sPZ hybrid particles. (**b,c**) Cyclic performance and Coulombic efficiency of the Li/CPE-sPZ/LiCoO$_2$ and Li/CPE-sPZ/graphite coin cells at different C-rates at room temperature. Reprinted with permission from Xiao et al [120]. Copyright 2018 Elsevier B.V.

Khoon et al. [121] obtained similar results and proposed that the ZrO$_2$-based GPE has the potential to be applied in lithium polymer batteries owing to the improved Li$^+$ ion transfer. The incorporation of ZrO$_2$ NPs into the polymer-salt system resulted in a higher electrolytic conductance than that for the GPE without ZrO_2 NPs. Prasanna et al. [122] prepared a nanocomposite GPE comprising a solution of zinc trifluoromethanesulfonate in a 1-ethyl-3-methylimidazolium bis(trifluoromethylsulfonyl)imide IL entrapped in a PVC/PEMA blend and dispersed ZrO_2 nanofillers via a solution casting method. The GPE film showed the highest ionic conductivity of 3.63×10^{-1} mS cm^{-1} at room temperature when 3 wt% ZrO_2 nanofiller was added. A list of significant GPEs containing ZrO_2 fillers is provided in Table 4 with their ionic conductivities and temperatures.

Table 4. List of GPEs with ZrO_2 filler with their conductivity and temperature.

Polymer	Salt/Plasticizers	Solvent	Conductivity (mS cm^{-1})	Temperature (°C)	Reference
PVDF/PVC	LiBOB-EC/DEC	THF	1.53×10^{0}	70	[115]
PPG	AgCF$_3$SO$_3$	THF	2.9×10^{0}	30	[116]
PVdF-HFP	LiTFSI-PC	THF	4.46×10^{0}	30	[117]
PS/PMMA	LiClO$_4$	-	2.2×10^{0}	30	[118]
PVDF-HFP	LiPF$_6$-EC/DMC/EMC	DMF	3.6×10^{0}	30	[120]
PVDF-HFP/MG49	LiBF$_4$	THF	3.39×10^{0}	30	[121]
PVC/PEMA	Zn(OTf)$_2$	DMF	3.63×10^{-1}	30	[122]

3.5. Cerium Oxide (CeO$_2$)

Rajendran et al. [123] reported a GPE system containing PEO-PMMA-LiClO$_4$-DMP with varying CeO_2 contents. The maximum ionic conductivity achieved was 2.07×10^{-1} mS cm^{-1} when 10 wt% CeO_2 was added to the GPE. Vijayakumar et al. [124] prepared a new GPE membrane containing a PVDF-HFP-based polymer electrolyte and micro/nanosized CeO_2 using a phase inversion technique. The highest ionic conductivity achieved was 2.47×10^{0} mS cm^{-1} at room temperature in the presence of 8 wt% CeO_2. The addition of CeO_2 NPs to the GPE reduced the interfacial resistance and enabled a wider electrochemical window and good cycling performance. Vijayakumar et al. [125] reported similar results in which the incorporation of CeO_2 reduced the ion coupling and increased the charge carrier number, thereby improving the ionic conductivity. However, an excess of CeO_2 led to an increased dilution effect, which resulted in a continuous decrease in the electrolytic conductance. Kumar et al. [126] designed a new class of nanocomposite polymer electrolytes to elucidate the origin and nature of the interactions between the surface of the CeO_2 filler NPs in the polymer chain and the migrating ionic species. The study showed an enhanced percolation network among the CeO_2 particles due to the increased ionic dynamics and surface interactions, which led to an increase in the electrolytic conductance. The surface interactions of the filler played an important role in increasing the charge carrier number and mobility, resulting in the highest ion conductivity of 5.2×10^{-1} mS cm^{-1} being achieved at room temperature. Polu et al. [127] synthesized a GPE based on PEG-Mg(CH$_3$COO)$_2$ as the polymer matrix and CeO_2 NPs as the inorganic filler. The maximum ionic conductivity achieved was 3.40×10^{-3} mS cm^{-1} when 15 wt% CeO_2 was added to the GPE system. This study indicated that the addition of a certain concentration of filler increased the ionic conductivity, but further addition resulted in a decrease in conductivity. A list of GPEs containing CeO_2 fillers is presented in Table 5 along with their ionic conductivities and temperatures.

Table 5. List of GPEs with CeO_2 filler with their conductivity and temperature.

Polymer	Salt/Plasticizers	Solvent	Conductivity (mS cm^{-1})	Temperature (°C)	Reference
PMMA/PEO	LiClO$_4$-DMP	THF	20.65×10^{-2}	30	[123]
PVDF-HFP	LiClO$_4$-EC/DMC	NMP	2.47×10^{0}	30	[124]
PVDF-HFP	LiClO$_4$-EC/DMC	NMP	3.84×10^{0}	30	[125]
PPG	AgCF$_3$SO$_3$	-	5.2×10^{-1}	30	[126]
PEG	Mg(CH$_3$COO)$_2$	Distilled water	3.4×10^{-3}	25	[127]

3.6. Barium Titanate (BaTiO₃)

Kim et al. [128] studied the effect of $BaTiO_3$ nanosized filler content in composite polymer electrolytes (CPEs). The optimum content of 15 wt% $BaTiO_3$ showed reduced crystallinity of the CPEs and high ion conduction with a wide electrochemical window and good thermal stability. The results indicated that the addition of $BaTiO_3$ filler affects the electrochemical properties and the crystallinity of the CPEs. Sivakumar et al. [129] synthesized a GPE by dispersing hydrothermally derived $BaTiO_3$ NPs in PVC-PEMA-EC/DMC-LiClO₄. The use of $BaTiO_3$ NPs as fillers increased the electrolytic conductance owing to the improvement in the polymer fractions and the amorphous phase of the GPE. The presence of dispersed $BaTiO_3$ NPs in the GPE prevented the growth of a passive layer on the surface of the Li-metal anode. Manimuthu et al. [130] discussed the influence of different ratios of $BaTiO_3$ NPs on the PEO/PVDF-HFP-based polymer electrolyte. The presence of a $BaTiO_3$ filler with high polarity reduced the crystallinity of the polymer because of the cooperation between the polymer chain and the filler surface, which increased the ion conduction of the GPE membrane. Table 6 lists $BaTiO_3$ filler-based GPEs with their ionic conductivities and temperatures.

Table 6. List of GPEs with $BaTiO_3$ filler with their conductivity and temperature.

Polymer	Salt/Plasticizers	Solvent	Conductivity (mS cm^{-1})	Temperature (°C)	Reference
PEO-PVDF	LiClO₄-PC	ACN	1.2×10^{-1}	30	[128]
PVC-PEMA	LiClO₄-EC/DMC	THF	0.61×10^1	30	[129]
PEO/PVDF-HFP	LiClO₄-PC	Acetone	6×10^0	30	[130]

4. Mechanism of Li⁺ Transport on the Interface between the Inorganic Filler and Polymer Matrix in GPEs

As GPEs combine the advantages of LEs and SPEs in terms of electrolytic conductance and mechanical stability, they are considered to have more potential for practical applications in LIBs and non-LIBs [13]. During the charge–discharge process, a solid electrolyte interface (SEI) is still formed, similar to LEs, through a reaction between the plasticizers and electrode surface in GPEs. The electrochemical behavior and internal resistance allow the measurement of the ionic conductivity of the GPE at various charge–discharge rates [131]. Generally, LEs have high electrolytic conductance in the range of 10^{-0} to 10^1 mS cm^{-1}; thus, to evaluate the efficient usage of the GPE, its ionic conductivity should be higher than that of LEs (>10^{-1} mS cm^{-1}) [42,132]. In gel-type polymer electrolytes, polymers are used as host matrices to trap the liquid constituent [133]. In this case, Li⁺ ion transfer is not affected by the segmental polymer chain motion but through the swollen gelled phase or liquid phase. In addition, the solvents used for GPEs should have a high dielectric constant and low viscosity. In SPE, polymer chains conduct local segmental motion continuously, leading to the formation of the free volume. Under the effect of the electric field, the lithium ions diffuse to a new coordinating position along the polymer chain or move from chain to chain through these free volumes. In the presence of ceramics, the diffusion of Li⁺ ion is enhanced because the ceramics increase the free volume of the polymer chain. Thus, the mechanism of reduction in the polymer crystallinity is not dependent on the chemical nature of the filler, but on the size, volume fraction, and shape of ceramic filler [134]. Because of the different particle sizes, the optimal content of ceramic fillers to achieve optimum diffusion of Li⁺ ion is different. Inert ceramic oxide fillers such as TiO_2, SiO_2 are often used with only a few percent (ceramic-in-polymer) to achieve optimum ionic conductivity. However, as the inert filler content increases, it decreases the mechanical strength of the polymer membrane. In contrast, active fillers, such as LLZO, LAGP, etc., have high ionic conductivity and can participate in Li⁺ ion transport, thus when the active filler content increases, the conductivity can be increased accordingly. Therefore, depending on the properties and content of the active filler and polymer, the optimal content to achieve maximum electrical conductivity is different [57]. The dissociation of the salt is facilitated by a high dielectric constant, while a low viscosity increases the ionic mobility in the electrolyte, resulting in a high electrolytic conductance [135]. In addition, the functional groups of the polymer matrix should be able to dissolve the Li salt to form the polymer-salt system. This requires that the lattice energy of the salt be relatively low, but the dielectric constant of the polymer matrix is relatively high. More importantly, the properties of the lithium salt will affect the performance of lithium polymer batteries. Because of its high ion conductivity, good electrochemical stability, and corrosion resistance for aluminum current collector, $LiPF_6$ is mainly used in commercial LIBs. However, the halogen-based lithium salts have some disadvantages: $LiClO_4$ is potentially explosive when exposed to an organic substance; $LiAsF_6$ contains toxic As; $LiBF_6$ and $LiPF_4$ are easily decomposed into HF, which is toxic and corrosive in the moist atmosphere. Organic lithium salts, such as $LiCF_3SO_3$, LiTf, etc., are highly resistant to oxidation, thermally stable, and non-toxic, but they have poor ionic conductivity. LiTFSI is the best choice for Li salt in LIBs with high solubility in common solvents, but they are corrosive to aluminum current collectors. Therefore, the search for reliable new lithium salts has remained at the center of research in recent years [136]. Typically, electrolytic conductance is related to the elementary electric charge and ion mobility. The electrolytic conductance of GPEs is mainly dominated by the properties of the trapped liquid electrolyte in the micropores of the membrane; thus, the interconnected micropores of the membrane are the main factor affecting the mobile ions. In contrast, for a homogeneous membrane or a membrane without micropores, the Li^+ ion mobility is primarily determined by the swollen gelled phase. In addition, an incomplete drying during the SPE preparation leads to the presence of moisture and solvent that results in the decomposition reaction of polymer. The standard drying process for the preparation of SPE membrane with a high content of Li salt (polymer-in-salt) will trap a large amount of solvent due to the strong interaction between solvent and ion species, which leads to the high ionic conductivity. The presence of trace water up to a few ppm remained after drying under vacuum was sufficient to trigger the depolymerization process. This process can have a strong impact on the electrochemical performance. Therefore, it is necessary to remove water to avoid the depolymerization of the polymer [137]. The possible mechanism of ion transport is as follows: (i) Li^+ ions are located in the porous structure of the polymer; (ii) through dispersion of the inorganic filler, the porous structure of the polymer is maintained, which assists in the absorption of the LE in the GPE, resulting in alleviation of the leakage problem and an enhanced safety of the applied device; and (iii) Li^+ ions migrate from one coordination site to a new site through the liquid phase or gelled phase (porous structure) via the effect of the inorganic fillers.

A new concept has been recently proposed by Chen et al. [138] who explained the electrolytic conductance is highly affected by the layers of chemical water (immobile ice layer) adsorbed on OH-terminated SiO_2 surface. They proposed the immobilization of 1-butyl-1-methylpyrrolidinium-bis TFSI (BMP TFSI) IL molecules extends the formation of ice layer on the SiO_2 surface (Figure 17). In symmetric O=S=O groups on the TFSI anion, one oxygen can interact with a hydroxyl group on silica surface while other oxygen atom interacts with BMP cation groups. The TFSI anion in LiTFSI also has two groups of O=S=O, ensuring dense adsorption of the TFSI monolayer. However, the molecular interaction between the BMP cation and TFSI (LiTFSI) anion is different because TFSI has free rotation and no polarization from the underlying surface. The TFSI molecules must compensate for the positive bipolar charge due to the formation of H bonds between the O group of the TFSI anion and the OH group of the ice layer. This loosens the bond between the Li^+ cation and the TFSI anion. In this way, the concentration of free Li^+ increases at this interface, resulting in higher ionic conductivity.

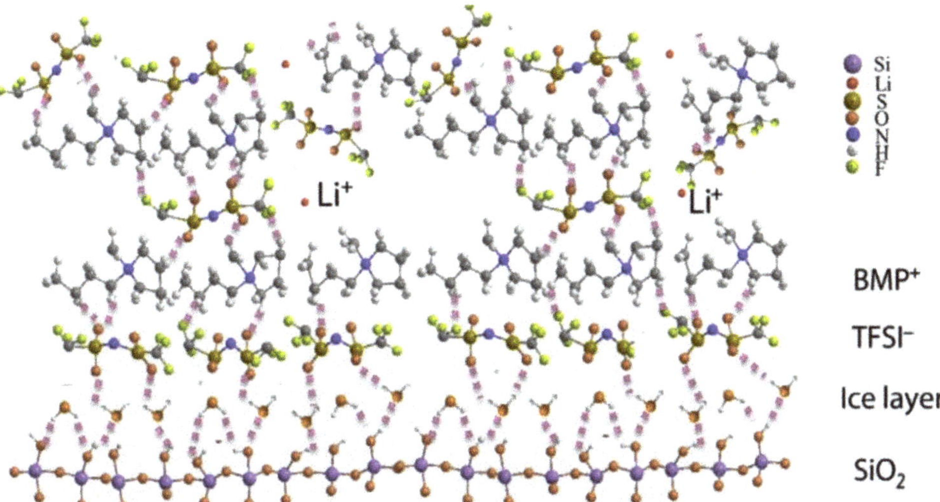

Figure 17. The model of the electrolytic conductance on the inorganic filler (SiO$_2$) interfaces through the layers of immobile ice layer absorbed on OH-terminated silica surface. Reprinted with permission from Chen et al. [138]. 2020 Creative Commons Attribution-Noncommercial liscense.

In addition to the Li$^+$ ions transport mechanism, the interaction of lithium dendrite with the polymer and filler also needs to be considered. The dendrite growth during charge-discharge cycling hinders the utilization of GPE in LIBs, causing detrimental effects related to the safety and electrochemical performance of the battery, although its effect is lower than that of LE [139,140]. The inclusion of inorganic filler into GPE can suppress the formation of lithium dendrite through enhancement of the mechanical properties of the polymer matrix, trapping the LE resulting in a uniform Li$^+$ ions flow at the Li metal-electrolyte interface [141]. Liu et al. studied the growth of Li dendrite through the addition of SiO$_2$ NPs (nano-SiO$_2$ and acid-modified nano-SiO$_2$ filler) into PEO-LiTFSI system at current densities of 0.1 and 0.5 mA cm^{-2} [142]. The study showed that the interfacial resistance of cells is significantly reduced when introducing SiO$_2$ NPs. However, more importantly, the acid-modified nano-SiO$_2$ filler was the most effective to reduce the interfacial resistance. The acid modification on the nano-SiO$_2$ surface immobilized the PEO chains and promoted the movement of Li$^+$ ions. In addition, the Lewis acid–base interaction between the hydroxyl groups of the trace water in polymer electrolyte and the surface of acid-modified nano SiO$_2$ prevented the reaction of the traces water with Li metal, thereby hindering the growth of lithium dendrite. The interaction of Li dendrite with polymer and inorganic filler was also observed by Liu et al. [143]. In this study, they prepared a multi-functional GPE by combining ultraviolet (UV)-cured ethoxylated trimethylolpropane triacrylate (ETPTA) macromer with PEO and Al$_2$O$_3$ NPs as an inorganic filler. The uniform dispersion and high-density of Al$_2$O$_3$ NPs in GPE acted as a filler to stabilize the electrode interface as well as a protective agent to prevent the lithium dendrite growth, thus promoting good cycle stability (Figure 18a). From SEM images (Figure 18b–d), LiFePO$_4$/GPE/Li cell showed large dendrites on the surface of lithium anode after 20 cycles at 0.5 C compared with the smooth surface of pristine lithium anode before cycling. In contrast, in the presence of Al$_2$O$_3$ NPs, no dendrites were observed on the surface of lithium anode in LiFePO$_4$/Al$_2$O$_3$-GPE/Li cell. The addition of inorganic filler can suppress the lithium dendrite growth, which in turn greatly improves the safety of LIBs. Besides, the presence of polymer can prevent direct contact of the electrode and LE.

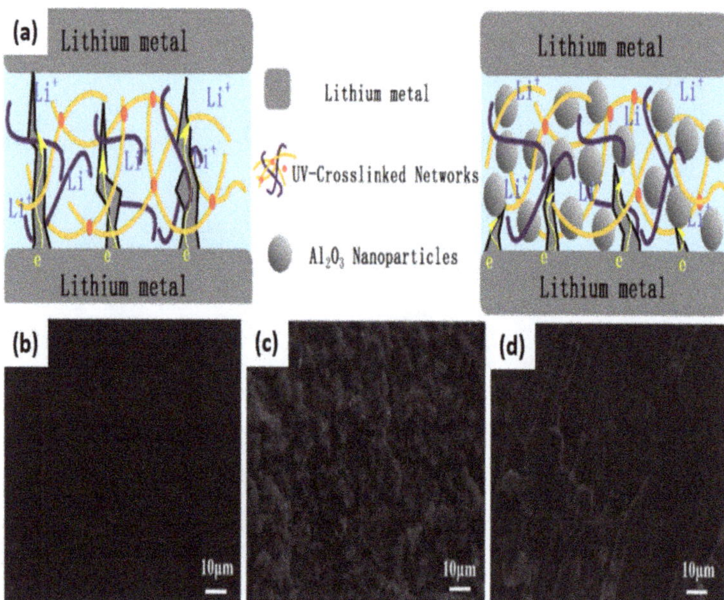

Figure 18. (**a**) Schematic illustrations the prevent of lithium dendrite growth by Al_2O_3-GPE electrolyte; (**b**) pristine of lithium anode before cycling in LFP/GPE/Li cell; (**c**) lithium anode in LFP/GPE/Li cell after 200 cycles at 0.5 C; (**d**) lithium anode in LFP/Al_2O_3-GPE/Li cell after 200 cycles at 0.5 C. Reprinted with permission from Liu et al. [143]. Copyright 2018. The Chinese Ceramic Society. Production and hosting by Elsevier B.V.

For SPE, the dendrite formation in the Li metal anode causes a short circuit of the battery, in which the polymer decomposition is the main failure mechanism of this battery [144]. Golozar et al. studied the mechanism of these dendrite formation [145]. The presence of carbon increases the hardness of the dendrite and facilitates their penetration into the SPE, thereby causing a short circuit of the battery. Li_xC_x is formed through the reduction of SPE throughout the cycle, which produces carbon-rich species. In addition, the decomposition of LiTFSI at grain boundaries has also been observed [146]. The dissolution of the lithium metal into the polyether begins at the grain boundary and continues after many cycles leading to the depletion of lithium. Li_3N initially formed becomes insoluble in the next cycles and further decomposition of salt leads to the formation of Li_2S, LiC_xF_y, Li_xCNF_3, and Li_ySO_x, where LiF acts as a protective layer for the lithium from further dissolution.

5. Synthesis Methods for Inorganic Gel Polymer Electrolytes

5.1. GPEs Based on Physical Preparation Methods

GPE can be separated into two categories, physical and chemical gels, based on their preparation method. In physical gels, LEs are confined to the polymer matrix without significant polymer-solvent bonding. The GPE can be prepared from a dry polymer membrane that undergoes swelling by a liquid electrolyte containing Li salts and plasticizers. The general synthesis methods include conventional solution casting, phase inversion, and electrospinning.

In the solution cast method, the solvent should be able to dissolve both the Li salt and polymer matrix without water molecules absorbed from environment [42]. Forsyth et al. [147] demonstrated a purpose-designed Teflon mold method for the preparation of GPEs which combines an IL electrolyte (consisting of 3.8 m LIFSI in trimethyl(isobutyl)phosphonium bis(fluorosulfonyl)imide (P_{111i4}FSI)), poly(diallyldimethylammonium) bis(trifluoromethanesulfony)imide (PDADMA TFSI), and Al_2O_3 NP inorganic filler. The $P_{111i4}FSI$ showed a wide electrochemical window, and the high concentration of Li salt could significantly enhance the lithium stripping and plating. On the other hand, the introduction of Al_2O_3 increased the mechanical stability, which allowed more IL to interact with the GPE. The optimized composition of GPE with 5 wt% Al_2O_3 NPs and 50% of IL showed the highest ionic conductivity of 0.28×10^0 mS cm^{-1} and enhanced the Li$^+$ ion transport. The solution cast method is a commonly used traditional method due to its ease of fabrication. The obtained GPEs generally show increased electrolytic conductance and good interfacial properties between electrode and electrolyte as well as the good electrochemical performance of the full cell polymer batteries.

Phase inversion method is also commonly used to fabricate highly porous polymeric film through a de-mixing procedure in which an initial homogeneous polymer solution is changed from a liquid state to a solid-state [148]. Liu et al. [149] integrated non-woven fabrics of PVDF-PAN-SiO$_2$-based GPE membrane using a Loeb-Sourirajan (L-S) inverted phase method (i.e., the dry-wet phase inversion technique to prepare cellulose acetate membrane for seawater desalination) to obtain a polymer membrane with uniform pore size. The chemical reaction that occurred between two salts (NaHCO$_3$ and CH$_3$COOH) during the membrane preparation process promotes the formation of porous and interconnected structures in GPE. The GPE membrane showed the highest ionic conductivity of 3.32×10^0 mS cm^{-1} with enhanced the mechanical strength and electrolyte uptake. The combination of L-S inverted phase and chemical reaction process was suitable to prepare polymer matrices in GPE. The facile electrolytic conductance and stable interfacial property between electrode and electrolyte in this GPE resulted in an excellent performance in LIBs.

Electrospinning is another common method to produce polymer fibers with a diameter ranging from tens of nanometers to tens of micrometers through the electrostatic repulsion of the polymer solution [150,151]. A novel PVDF-HFP-based GPE with nanostructured IL and SiO$_2$ NP-tethered 1 methyl-1-propulpiperidinium bis(trifluoromethanesulfonyl) imide (SiO$_2$PPTFSI) was prepared by electrospinning process [152]. The obtained GPE showed good mechanical stability, an increase in the electrolytic conductance as well as Li$^+$ ion transfer. More importantly, the assembled cell showed an initial discharge capacity of 119 mAh g^{-1} and a capacity retention of 92.1% after 460 cycles at 1 C. The use of a nanostructured IL and modified SiO$_2$ presented a potential candidate to improve the cyclic performance and the safety of LIBs. Although the electrospinning method is a cost-effective and simple manufacturing process, the difficulty in controlling the pore structure and the time-consuming process are significant drawbacks that need to be considered.

Generally, physical methods are used to cross-link polymer chains through weak physical interactions in GPEs. Owing to the weak interaction between the constituents, safety issues remain a concern as the polymer matrix can easily swell or dissolve in the LE at high temperatures, which leads to solvent leakage and reduced electrochemical performance. Poor thermal stability is another limitation when applying GPEs synthesized through physical methods to practical applications [131].

5.2. GPEs Based on Chemical Preparation Methods

The chemical methods used to prepare GPEs are also called "in situ synthesis" methods. In chemical gels, the crosslinking agent and functional groups of the polymers form a chemical bond. The precursor solution is prepared by dissolving the crosslinking agent and monomers in a LE in a specific ratio. Subsequently, the GPE is synthesized by polymerization of the monomer, forming a cross-linked network, and the LE is uniformly immobilized in the nanopores. Polyester is the most commonly used material for GPEs owing to the facile interaction between the lithium-ion and ethylene oxide (EO) units. Generally, polyether is formed when using precursors containing methacrylate groups on the surface [153]. Sato et al. [154] designed novel three-dimensional hybrid silica particles (PSiPs) with concentrated polymer brushes (CPB) with trace amount of IL. Initially, the surface of monodisperse silica particle (SiP) was changed by a mixture of ethanol/water/(2-

bromo-2methyl) propionyloxyhexyltriethoxysilane (BHE) to form a new type of colloidal crystal. Then, by surface-initiated living radical polymerization (LRP) process, the PSiPs were successfully synthesized by grafting well-defined polymers on the surface to form the CPB-modified particles. The GPE with CPB-modified SiP provided an advanced ion conduction channel with an orderly and clearly defined structure. A newly designed GPE gave the chance for bipolar LIB device with a good electrochemical performance and made it possible to apply in practical use. Guo et al. [155] fabricated ionic liquid GPEs (ILGPEs) supported by active filler LAGP or inactive filler SiO_2 for enhanced ionic conductivity and electrochemical performance. As for inorganic fillers, the optimized content of nano-SiO_2 could decrease the crystalline phase of the host polymer as well as promote the Li-ion transport. Nevertheless, when the concentration of SiO_2 exceeded the optimal value, the fillers impeded the effective migration of Li^+ ion into the ILGPE. However, in the case of LAGP filler, it not only reduced the crystallinity but also improved the electrolytic conductance of the host polymer even when the concentration was greatly increased because LAGP is also the source of Li^+ ion. The IL-GPE with 10 wt% LAGP showed high thermal stability and no flammability, which suggests it can be a promising electrolyte for the highly safe energy storage devices. Ma et al. [156] improved the ionic conductivity of GPE by LATP ceramic particles, which were dispersed in polymerized ionic liquids (PILs) as a polymer matrix, and LiTFSi as a source of Li^+ ion. The PIL was synthesized using 1-vinyl-3-ethylimidazolium TFSI as a monomer and nonwoven polyethylene terephthalate (PET) as a polymer in GPE. The optimized content of LATP (10 wt%) exhibited good rate performance and capacity retention of 97% after 250 cycles at 60 °C, indicating that PIL-LiTFSI-LATP can yield superior cyclic performance at the high temperature. Wang et al. [157] incorporated two-dimensional (2D) silica NP into PIL to optimize the transport properties of GPE. The desired transport properties were due to the bonding of the grafted PIL and mesoporous structure of 2D silica nano filler with abundant, shorter, and continuous ion transport pathways. In addition, the 2D nanofiller could effectively control the ion transport trajectory through the surface contact orientations, thus leading to higher ion conductivity than the GPEs added with zero-dimensional (0D) or one-dimensional (1D) nanofiller. The assembled cell showed a discharge capacity of 135.8 mAh g^{-1} after 30 cycles at 60 °C, suggesting that their capacity and capacity retention are superior to cells using unmodified PIL/IL PE (50.0 mAh g^{-1}). The ionotropic gelation method is a technique used to prepare micro-and nanoparticles, which are synthesized by adding anion polyelectrolyte solution with drop-by-drop manner to an acidic chitosan solution. Chitosan is ionotropically gelated to be proton conductive membranes, creating proton conductor sites in a single step. In contrast, the pre-formed chitosan in the solution is not highly conductive because of the limited Li^+ diffusion through the polymer matrix [158,159]. Kim et al. prepared a multifunctional binder network by combining chitosan and reduced graphene oxide (rGO) to enhance the electrochemical performance of Li-sulfur batteries [160]. A homogeneous network formed by the reaction of chitosan with GO in an aqueous solution enhanced the redox system by trapping lithium polysulfides, reinforced the mechanical properties, and promoted ion/electron movement. This multifunctional network binder can be used for high-performance Li-S batteries. Zhao et al. used chitosan crosslinked with a carboxylic acid or acrylic acid molecules to form a compatible binder for both silicon and graphite [161]. The crosslinked chitosan lattice can effectively regulate the large volume change of silicon particles throughout the cycles. In this sense, chitosan network can be used as an effective binder in LIBs.

6. Promising Applications of GPEs in Various Battery Systems

The ionic conductivity enhancement and host polymer structure modification realized by the addition of inorganic fillers in GPE can be applied to improve the electrochemical performance of various battery systems.

6.1. Lithium-Ion Batteries

Because GPEs are highly resistant to electrochemical oxidation compared to LEs, GPEs are selected as potential electrolytes for practical applications in LIBs. Many studies have used inorganic fillers such as SiO_2, TiO_2, Al_2O_3, etc., to enhance the mechanical strength and improve the ionic conductivity of GPE in LIBs. Lewis acid–base interaction between a surface group of filler and ions appeared to be responsible for this role. Some new inorganic fillers that have been recently introduced in LIBs are aluminum oxyhydroxide $(AlO(OH)_n)$, graphene oxide (GO), clays, etc., which are expected to modify the structure of polymer matrix and thus enhance the Li^+ ion transport. Stephen et al. [162] prepared a multifunctional GPE by combining a PVdF-HFP as a polymer matrix, lithium bis perfluorosulfonyl imide $(LiN(CF_3SO_2)_2)$ as a lithium salt, and $AlO(OH)_n$ as an inorganic filler. The introduction of $AlO(OH)_n$ not only increases the amorphous domain and acts as "solid plasticizer" to promote the Li^+ ion transfer but also provides a good interfacial property towards Li-metal anode. The full cell with GPE membrane exhibited first discharge capacity of 127 mAh g^{-1} with a capacity retention of 98.4% after 20 cycles at 70 °C. Aravindan et al. [163] continued to employ $AlO(OH)_n$ as an inert filler in PVdF-HFP and indicated that the introduction of 10% $AlO(OH)_n$ in GPE membrane enhanced the ionic conductivity due to the interaction of Lewis acid–base between F atoms in PVdF-HFP and OH^- groups in filler. This interaction increased the amorphous domain by preventing reorganization of polymer chains, thus leading to the enhancement of the electrolyte uptake at ambient temperature condition. In addition, the study showed high cyclic stability with a capacity retention of 97.8% after 10 cycles. The advantages of using $AlO(OH)_n$ over conventional inorganic fillers, such as TiO_2, SiO_2, Al_2O_3, etc., are mainly because of its ability to promote more dissociation of lithium salt, which leads to the enhanced number of charged carriers, ionic conductivity, and electrochemical performance. Chen et al. [164] prepared a high-performance battery using a PVDF-HFP-based GPE co-doped with PEO and GO via weak hydrogen bond interaction. The GPE showed a 3D porous network with superior ionic conductivity up to 2.1×10^0 mS cm^{-1} and excellent cycling stability with a capacity retention of 92% after 2000 cycles at 5 C. The good electrochemical performance was due to the abundant oxygen-functional groups in GO sheets that interact with the copolymer (PVDF-HFP and PEO polymer) to form an amorphous phase and porous structure, which is beneficial for both Li^+ and PF_6^- intercalation/deintercalation kinetics. Zhao et al. [165] used GO as a filler to synthesize the homogeneous GPE for enhancing the electrochemical properties of the cell. The presence of GO enhanced the contact between the GPE and electrodes and formed a more stable SEI layer, leading to the enhancement of electrolytic conductance and the Li^+ ion transfer of GPE. Thus, the discharge capacity and cyclic performance of battery were effectively improved. Liu et al. [166] designed a GPE by incorporating graphene fillers in the presence of PVDF as a host polymer, and $LiPF_6$ as a source of Li^+ ions. This study demonstrated the decrease in crystallinity of porous PVDF due to the homogeneously dispersed graphene in host polymer, resulting in the increased GPE electrolytic conductance from 1.85×10^0 mS cm^{-1} in pure PVDF to 3.61×10^0 mS cm^{-1} in the presence of 0.002 wt% graphene, and enhanced cyclic performance of the cell. Chen et al. [167] studied a GPE that consists of PAVM as a host polymer, $LiPF_6$ as a source of Li^+ ion, and GO quantum dots (GOQD) as an inorganic filler. The GOQD hinders the formation of ion-solvent clusters and immobilizes anions; as a result, the assembled $LiFePO_4$/GPE/Li cell showed a good performance at high rates (up to 20 C) and exhibited capacity retention of 100% after 500 cycles. From these results, it can be expected that layered GO can significantly change the properties of the host polymer even with a very low content due to its highly oxidizing and hydrophilic nature. Compared with conventional ceramic nanofillers, GO has some advantages because of its tunable surface functionalities, high compatibility, and excellent dispersion with the polymer network. Dyartanti et al. [168] prepared a PVDF-PVP-based GPE containing an montmorillonite (MMT) nano-clay as a filler. The addition of MMT clay showed an increased porosity of host polymer and enhanced the uptake of LE, leading to an increased ionic conductivity of 5.61×10^0 mS cm^{-1}. The full cell showed good cyclic stability with a capacity retention of 97.7% after 48 cycles. The use of clay as an inorganic layered filler is beneficial because of its unique characteristics of the length scale (channel width = 16 Å), high cation exchange capacity, suitable interlayer charge, and a substantial specific surface area (~31.82 m^2 g^{-1}). Table 7 summarizes the electrochemical performance of LIBs with inorganic fillers added in GPEs.

Table 7. Lists of GPEs and their electrochemical performance in lithium-ion batteries (LIBs).

Polymer	Salt/Plasticizers/Fillers	Specific Capacity (mA h g^{-1})	Capacity Retention (%)	Long-Term Cycling	Current Density	Reference
PAN/PEGDA	LiPF$_6$ -LiCF$_3$SO$_3$/EC-DMC/TiO$_2$	138.0	-	50	0.2 C	[79]
PAN	LiPF$_6$/TMS/TiO$_2$	345.0	-	20	0.2 C	[81]
PAN	LiPF$_6$/TMS/TiO$_2$-SiO$_2$	182.0	75.8	100	0.2 C	[82]
PVDF-HFP	LiPF$_6$/TiO$_2$	122.0	92.4	100	0.5 C	[85]
PVDF-HFP/PMMA	LiPF$_6$/EC-DMC/TiO$_2$	173.2	92.1	50	0.2 C	[88]
PAN/PVA	LiPF$_6$/EC-DMC-DEC/TiO$_2$	60.0	71.0	1000	20 C	[90]
P(MMA-AN-EA)	LiPF$_6$/EC-DMC/Al$_2$O$_3$	132.0	94.8	100	0.2 C	[98]
PVFM	LiPF$_6$/EC-DMC/Al$_2$O$_3$	140.3	88.6	15	0.2 C	[99]
PVDF-HFP	LiPF$_6$/EC-DMC/Al$_2$O$_3$	160.2	96.0	50	0.1 C	[100]
PVDF-HFP	LiPF$_6$/EC-DEC-DBP/SiO$_2$	149.0	97.4	100	0.5 C	[108]
PAN	SiO$_2$	157.9	88.0	300	0.5 C	[111]
PPC	LiTFSI-LiNO$_3$/SiO$_2$	597.0	85.0	500	0.1 C	[113]
PEO	LiClO$_4$/SiO$_2$	123.5	70.0	90	0.2 C	[114]
PVDF-HFP	LiPF$_6$/EC-DMC-EMC/ZrO$_2$	126.4	85.2	150	2 C	[120]
PVDF-HFP	LiClO$_4$/EC-DMC/CeO$_2$	105.8	86.0	30	0.5 C	[124]
PVDF-HFP	LiClO$_4$/EC-DMC/CeO$_2$	116.4	81.6	50	0.5 C	[125]
PEO/PVDF-HFP	LiClO$_4$/PC/BaTiO$_3$	123.0	-	100	0.3 C	[130]
PVDF-HFP	LiN(CF$_3$SO$_2$)$_2$/AlO(OH)$_n$	125.0	98.4	20	0.1 C	[162]
PVDF-HFP	LiBOB/AlO(OH)$_n$	157.0	97.8	10	0.1 C	[163]
P(VDF-HFP)-co-PEO	LiPF$_6$/GO	103.0	92.0	2000	5 C	[164]
PVDF-HFP	LiTFSI/GO	120.0	94.4	100	0.2 C	[165]
PVDF	LiPF$_6$/Graphene	144.0	96.6	100	2.0 C	[166]
PAVM	LiPF$_6$/GOQD	-	100	500	5 C	[167]
P(OPal-MMA)	LiClO$_4$/Clay	146.4	-	50	0.5 C	[168]

6.2. Sodium-Ion Batteries

Because of their versatility, flexibility, and thermodynamic stability, SPEs have been selected as one of the most promising candidates for high safety sodium-ion batteries (SIBs). However, the low ion mobility in SPEs at room temperature hinders their practical application in SIBs. The addition of inorganic fillers to GPEs is an effective method to improve the ion conduction of electrolytes. Conventional fillers, such as SiO$_2$, TiO$_2$, Al$_2$O$_3$, and BaTiO$_3$, etc., have been used to synthesize composite solid polymer electrolytes for sodium batteries [169–171]. Hwang et al. [172] fabricated a GPE containing a PEO-based polymer electrolyte, NaClO$_4$, and nanosized TiO$_2$ using a solution casting technique. The GPE, with a EO:Na ratio of 20:1 (w:w) and 5 wt% TiO$_2$, showed the highest ionic conductivity of ~2.62 × 10^{-1} mS cm^{-1} at 60 °C and enhanced the stability of cyclic performance of the cell. This result can be explained by the positive effect of the TiO$_2$ filler, which reduced the crystallinity of the polymer, thus enabling faster ionic transport. Zhang et al. [173] prepared a novel GPE based on PMMA, PEG, NaClO$_4$, and α-Al$_2$O$_3$ which contained acidic surface sites. High electrolytic conductance (1.46 × 10^{-1} mS cm^{-1} at 70 °C), wide electrochemical stability window (4.5 V vs. Na$^+$/Na), and good mechanical stability were achieved by the GPE. The reversible capacity reached 85 mAh g^{-1}, corresponding to a capacity retention of 94.1%, even after 350 cycles, when coupled with a Na$_3$V$_2$(PO$_4$)$_3$ cathode. Kumar et al. [174] investigated a GPE containing SiO$_2$ NPs dispersed in PVDF-HFP. This membrane was transparent, flexible, and free-standing, which makes it suitable for flexible SIBs. The material showed a high electrolytic conductance of 4.1 × 10^0 mS cm^{-1} at ambient temperature and good thermal stability owing to the formation of space-charge layers between the SiO$_2$

particles and the gel region. Liu et al. [175] synthesized a PVDF-HFP/PMMA-based GPE membrane containing a suitable number of β-Al_2O_3 NPs. The incorporation of PMMA into PVDF-HFP-based film improved the ionic conductivity due to the amorphous properties of PMMA which can promote the uptake of the LE and enhance the interaction of carbonyl-carbonate groups in MMA monomer and electrolyte, respectively. The GPE membrane showed a high electrolytic conductance of 2.39×10^0 mS cm^{-1} and enhanced the electrochemical stability window up to 5.04 V. Besides, the full cell exhibited good electrochemical performance with the first discharge capacity of 94.1 mAh g^{-1} and capacity retention of 85% after 300 cycles at 0.5 C. Wang et al. [176] employed a sodium ion conductive $Na_3Zr_2Si_2PO_{12}$ (NZSPO) in modified PVDF-HFP/PMMA/polyurethane (TPU)-based GPE to enhance the properties of membrane. The introduction of filler increased the amorphous phase and boosted the porosity of GPE membranes, leading to the increase in LE uptake. Besides, the NZSPO itself was the active filler; thus, it could provide pathways of ions at the interface between the filler and GPE. The GPE film showed ionic conductivity of 2.83×10^0 mS cm^{-1} and a wide electrochemical window of 5.16 V. The full cell exhibited the first discharge capacity of 92.7 mAh g^{-1} with capacity retention of 99.2% after 100 cycles at 0.5 C. Yi et al. [177] prepared a PMMA-based GPE by introducing $Na_3Zr_2Si_2PO_{12}$ and PVDF-HFP to boost the interfacial adhesion between electrode and electrolyte. The GPE membrane showed a high electrolytic conductance of 2.78×10^0 mS cm^{-1} and a wide electrochemical window of 4.9 V. More importantly, the assembled full cell with GPE exhibited the first discharge capacity of 96 mAh g^{-1} with excellent cyclability during 600 cycles.

Recently, inorganic NPs have also been employed to further enhance the electrochemical properties of PEO/Na-salts/ILs in GPEs. Song et al. [178] developed a hybrid GPE consisting of PEO-NaClO$_4$-SiO$_2$ and 1-ethyl-3-methylimidazolium bis(fuorosulfonyl)imide (Emim FSI) for sodium batteries. This GPE demonstrated an integrated structure by redox processes and interactions among the Emim FSI, silicon, and PEO. The GPE showed a high electrolytic conductance of 1.3×10^0 mS cm^{-1} at ambient temperature and stable voltage window of 4.2 V vs. Na/Na$^+$, which is sufficient for most cathode materials in SIBs. These results indicated that the introduction of nanosized inorganic fillers is a good strategy to enhance the electrochemical performance of polymer electrolytes. Table 8 presents the electrochemical performance of SIBs with GPEs containing inorganic fillers.

Table 8. Lists of GPEs and their electrochemical performance in SIBs.

Polymer	Salt/Plasticizers/Fillers	Specific Capacity (mA h g^{-1})	Capacity Retention (%)	Long-Term Cycling	Current Density	Reference
PVDF-HFP	NaClO$_4$/EC-DEC/Al$_2$O$_3$	120.0	95.3	1000	1 C	[104]
PVDF-HFP/PMMA	NaCF$_3$SO$_3$/EC-PC/Al$_2$O$_3$	360.0	90.0	10	-	[105]
PEO	NaClO$_4$/TiO$_2$	45.0	91.5	25	0.1 C	[172]
PMMA-PEG	NaClO$_4$/Al$_2$O$_3$	85.0	94.1	350	0.5 C	[173]
PVDF-HFP	NaCF$_3$SO$_3$/SiO$_2$	21.0	15.0	8	-	[174]
PVDF-HFP/PMMA	NaClO$_4$/β-Al$_2$O$_3$	80.0	85.0	300	0.5 C	[175]
PVDF-HFP/PMMA/TPU	NaClO$_4$/NZSPO	92.0	99.2	100	0.5 C	[176]
PVDF-HFP/PMMA	NaPF$_6$/NZSPO	96.0	-	600	1 C	[177]
PEO	NaClO$_4$/SiO$_2$	46.2	51.0	100	0.5 C	[178]

6.3. Magnesium-Ion Batteries

Magnesium (Mg) is the eighth most abundant element in the Earth's crust, the third most abundant element in seawater, and is geographically widespread [179]. Compared with Li-metal anodes, it is less likely that Mg dendrites would grow as they thermodynamically prefer three dimensional crystal growth rather than one-dimensional growth [180]. Nevertheless, the presence of dendrite Mg has been reported in several LEs and remains the hurdle to overcome with Mg metal anodes. Mg metal interacts more strongly with counter ions or polymer matrices than lithium metal and requires a higher under/overpotential for Mg electrode position/dissolution. Therefore, it is necessary to develop new Mg polymer electrolytes [181]. Kim et al. [182] prepared Mg^{2+} ion-conducting polymer electrolytes containing P(VdF-co-HFP), $Mg(ClO_4)_2$-EC/PC, and SiO_2 filler. This GPE achieved an electrolytic conductance of 3.2×10^0 mS cm^{-1} at room temperature. Hashmi et al. [183] investigated a novel GPE nanocomposite based on PVdF-HFP containing dispersed MgO NPs. The maximum electrolytic conductance was 8×10^0 mS cm^{-1} at room temperature when 3 wt% MgO was introduced. The assembled V_2O_5/GPE/Mg battery showed the low first discharge capacity of 58 mAh g^{-1} and poor cycling performance with capacity retention of 38% after 10 cycles, which are attributed to high interfacial resistance between Mg and GPE. Pandey et al. [184] studied the effect of MgO and SiO_2 particle sizes in a PVDF-HFP-based polymer electrolyte. High conductivities of 1×10^1 mS cm^{-1} for 3 wt% and 9×10^0 mS cm^{-1} for 15 wt% SiO_2 were obtained for the SiO_2 dispersed gel electrolyte. The presence of MgO formed space-charge regions that facilitated the Mg^{2+} ion motion, thereby enhancing the electrolytic conductance. Hashmi et al. [185] investigated the effect of SiO_2 NPs in a PVDF-HFP-based polymer electrolyte. The highest electrolytic conductance achieved was 1.1×10^1 mS cm^{-1} at 25 °C when 3 wt% SiO_2 NPs were added. The assembled full cell with GPE exhibited a first discharge capacity of 175 mAh g^{-1} with poor cyclability after 10 cycles. They also studied the effect of microsized MgO particle dispersion in a PVDF-HFP-based magnesium-ion (Mg^{2+}) conducting GPE [186]. The maximum ionic conductivity reached 6×10^0 mS cm^{-1} at room temperature with the incorporation of 10 wt% MgO particles. Hashmi et al. [187] reported a novel GPE membrane containing PVDF-HFP as the polymer matrix, Mg trifluoromethanesulfonate (Mg-triflate or $Mg(Tf)_2$) in a mixture of EC and PC as the Mg salt and nanosized passive Al_2O_3 filler or active filler Mg aluminate ($MgAl_2O_4$). The presence of the filler increased the porosity of the membrane, thereby increasing the electrolytic conductance of the GPE film. The highest ionic conductivities achieved were 3.3×10^0 and 4.0×10^0 mS cm^{-1} when adding 30 wt% Al_2O_3 and 20 wt% $MgAl_2O_4$ fillers, respectively. To achieve the impressive improvement of electrochemical performance with the inorganic fillers-added GPE, it is thought that high impedance at the interface between Mg metal and GPE needs to be resolved, which requires the finding of suitable electrode materials. Table 9 shows the list of electrochemical performance of MIBs with inorganic filler-added GPEs.

Table 9. Lists of GPEs and their electrochemical performance in MIBs.

Polymer	Salt/Plasticizers/Fillers	Specific Capacity (mA h g^{-1})	Capacity Retention (%)	Long-Term Cycling	Current Density	Reference
PVDF-HFP	$Mg(ClO_4)_2$/EC-PC/SiO_2	24.0	41.4	11	-	[182]
PVDF-HFP	$Mg(ClO_4)_2$/EC-PC/SiO_2	175.0	79.5	10	0.1 C	[185]
PVDF-HFP	$Mg(ClO_4)_2$/EC-PC/MgO	175.0	67.31	10	-	[186]

6.4. Zinc-Ion Batteries

Because of the favorable features of Zn metal, such as low cost, low toxicity, and high natural abundance, Zn has received intensive attention as an anode material in Zn rechargeable batteries. Hashmi et al. [188] prepared a GPE containing PVDF-HFP as the host polymer, a solution of EC-PC-$Zn(Tf)_2$, and nanosized ZnO filler particles. The GPE exhibited good thermal stability and good electrochemical performance achieving an electrolytic conductance of $>10^0$ mS cm^{-1}. LE immobilization and dispersion of the ZnO NPs in the GPE led to appropriate changes in the PVDF-HFP and filler-polymer interactions. Suthanthirarai et al. [189] prepared new GPE using a solution casting technique and evaluated the transport mechanism of a Zn^{2+}-conducting polymer electrolyte system. The highest electrolytic conductance was 3.4×10^{-3} mS cm^{-1} at room temperature when 5 wt% TiO_2 NPs were added. The presence of TiO_2 NPs formed a space-charged region, which significantly enhanced the movement of the Zn^{2+} ions, thereby increasing the ionic conductivity. From good thermal stability and enhanced ionic conductivity achieved by adding inorganic fillers in GPEs, the choice of appropriate inorganic filler in GPE can be seen as a potential approach for the high-performance ZIBs.

7. Conclusions and Perspectives

The incorporation of inorganic fillers into the polymer/salt system has been demonstrated as a promising strategy to enhance the electrochemical performance of GPEs in the last few decades. The introduction of inorganic fillers improves the electrolytic conductance as well as mechanical and thermal stability of gel-state polymer electrolytes. In this review, a historical overview of the developments in GPEs is first provided and subsequently detailed fillers applied in GPEs are discussed. The possible mechanisms behind the conductivity enhancement of inorganic fillers are also briefly discussed. Finally, inorganic filler/polymer GPEs studied for use in various battery systems, including Li-, Na-, Mg-, and Zn-ion batteries, were reviewed.

Although there have been several studies regarding the mechanisms behind the ionic conductivity enhancement and improvement in electrochemical stability with the addition of fillers, further fundamental understanding should be continuously pursued with novel composite polymer electrolyte designs for the successful implementation of GPEs in high-performance lithium batteries, especially for industrial applications. To achieve this, both experimental and theoretical calculation approaches should be synergistically combined. On the experimental side, it is important to find the optimal composite GPE structure, wherein the interfacial volume between the polymer and filler is maximized and agglomeration of the constituents is minimized. This structural feature will be beneficial for ionic conductivity and electrochemical stability, thus improving the cyclic stability in various battery applications. Consideration of the appropriate interfacial structure between the gel electrolyte and the electrode is another important factor. Long and tortuous ion pathways greatly inhibit the utilization of active materials in the performance of Li-polymer batteries. A reduced internal and interfacial resistance at the cathode and electrolyte/electrode interface can significantly enhance the cycling performance and rate capability of LIBs, resulting in a higher energy density. Besides, optimization of the active material loading ($mg \cdot cm^{-2}$) is another important consideration to realize the high energy density of the cell. Although a high thickness film which accompanies the high mass loading can increase the energy storage capacity of LIBs, it does not necessarily result in the high energy density. Since the energy density is proportional to the specific capacity, it is necessary to maximize the specific capacity to obtain the high energy density. Normally, when the film thickness is excessively high, the specific capacitance decreases due to the inefficient electrolyte ion diffusion through the electrode film. If the film thickness is too low, the reproducibility of measured capacity becomes degraded. Therefore, the appropriate film thickness (or mass loading) is required to increase the energy density. In previous studies with GPEs, the best mass loading condition is in the range of 0.35 to 3 mg cm^{-2} [104,114,120,155–157,165,175,176]. In the theoretical calculations, a precise prediction of the overall structure of the constituents in the composite gel will be advantages in the design of GPE components; in particular, it can provide useful information regarding the appropriate content of inorganic filler for a given GPE system.

Finally, the preparation method utilized for GPEs needs to be considered in terms of simplicity and manufacturing costs. To apply newly developed GPEs to batteries in industry, facile and low-cost process should be utilized to prepare the GPEs. In addition, advanced fabrication approaches should be developed to apply the new GPE to micro-batteries and flexible devices.

Author Contributions: Conceptualization, J.H. and V.P.H.H.; writing—original draft preparation, V.P.H.H.; writing—review and editing, S.S. and J.H.; supervision, J.H.; funding acquisition, J.H. All authors have read and agreed to the published version of the manuscript.

Funding: This research was also supported by Basic Science Research Capacity Enhancement Project through Korea Basic Science Institute (National Research Facilities and Equipment Center) grant funded by the Ministry of Education (2019R1A6C1010016) and Korea Institute of Energy Technology Evaluation and Planning (KETEP) and the Ministry of Trade, Industry & Energy (MOTIE) of the Republic of Korea (No. 20194030202290).

Conflicts of Interest: The authors declare no conflict of interest.

References

1. Xu, K. Nonaqueous liquid electrolytes for lithium-based rechargeable batteries. *Chem. Rev.* **2004**, *104*, 4303–4418. [CrossRef]
2. Nguyen, T.A.; Kim, I.T.; Lee, S.W. Chitosan-tethered iron oxide composites as an antisintering porous structure for high-performance Li-ion battery anodes. *J. Am. Ceram. Soc.* **2016**, *99*, 2720–2728. [CrossRef]
3. Mun, Y.S.; Pham, T.N.; Bui, V.K.H.; Tanaji, S.T.; Lee, H.U.; Lee, G.-W.; Choi, J.S.; Kim, I.T.; Lee, Y.-C. Tin oxide evolution by heat-treatment with tin-aminoclay (SnAC) under argon condition for lithium-ion battery (LIB) anode applications. *J. Power Sources* **2019**, *437*, 226946. [CrossRef]
4. Nguyen, T.L.; Park, D.; Kim, I.T. $Fe_xSn_yO_z$ Composites as Anode Materials for Lithium-Ion Storage. *J. Nanosci. Nanotechnol.* **2019**, *19*, 6636–6640. [CrossRef]
5. Mun, Y.S.; Kim, D.; Kim, I.T. Electrochemical performance of $FeSb_2$-P@C composites as anode materials for lithium-ion storage. *J. Nanosci. Nanotechnol.* **2018**, *18*, 1343–1346. [CrossRef]
6. Nguyen, T.L.; Kim, J.H.; Kim, I.T. Electrochemical performance of $Sn/SnO/Ni_3Sn$ composite anodes for lithium-ion batteries. *J. Nanosci. Nanotechnol.* **2019**, *19*, 1001–1005. [CrossRef] [PubMed]
7. Pham, T.N.; Tanaji, S.T.; Choi, J.-S.; Lee, H.U.; Kim, I.T.; Lee, Y.-C. Preparation of Sn-aminoclay (SnAC)-templated Fe_3O_4 nanoparticles as an anode material for lithium-ion batteries. *RSC Adv.* **2019**, *9*, 10536–10545. [CrossRef]
8. Son, S.Y.; Hong, S.-A.; Oh, S.Y.; Lee, Y.-C.; Lee, G.-W.; Kang, J.W.; Huh, Y.S.; Kim, I.T. Crab-shell biotemplated SnO_2 composite anodes for lithium-ion batteries. *J. Nanosci. Nanotechnol.* **2018**, *18*, 6463–6468. [CrossRef] [PubMed]
9. Min, K.; Park, K.; Park, S.Y.; Seo, S.-W.; Choi, B.; Cho, E. Residual Li reactive coating with Co_3O_4 for superior electrochemical properties of $LiNi_{0.91}Co_{0.06}Mn_{0.03}O_2$ cathode material. *J. Electrochem. Soc.* **2018**, *165*, A79. [CrossRef]
10. Park, J.H.; Choi, B.; Kang, Y.S.; Park, S.Y.; Yun, D.J.; Park, I.; Shim, J.H.; Park, J.H.; Han, H.N.; Park, K. Effect of Residual Lithium Rearrangement on Ni-rich Layered Oxide Cathodes for Lithium-Ion Batteries. *Energy Technol.* **2018**, *6*, 1361–1369. [CrossRef]
11. Min, K.; Jung, C.; Ko, D.-S.; Kim, K.; Jang, J.; Park, K.; Cho, E. High-performance and industrially feasible Ni-rich layered cathode materials by integrating coherent interphase. *ACS Appl. Mater. Interfaces* **2018**, *10*, 20599–20610. [CrossRef] [PubMed]
12. Baskoro, F.; Wong, H.Q.; Yen, H.-J. Strategic structural design of a gel polymer electrolyte toward a high efficiency lithium-ion battery. *ACS Appl. Energy Mater.* **2019**, *2*, 3937–3971. [CrossRef]
13. Cheng, X.; Pan, J.; Zhao, Y.; Liao, M.; Peng, H. Gel polymer electrolytes for electrochemical energy storage. *Adv. Energy Mater.* **2018**, *8*, 1702184. [CrossRef]
14. Park, K.; Choi, B. Requirement of high lithium content in Ni-rich layered oxide material for Li ion batteries. *J. Alloy. Compd.* **2018**, *766*, 470–476. [CrossRef]
15. Han, D.; Park, K.; Park, J.-H.; Yun, D.-J.; Son, Y.-H. Selective doping of Li-rich layered oxide cathode materials for high-stability rechargeable Li-ion batteries. *J. Ind. Eng. Chem.* **2018**, *68*, 180–186. [CrossRef]
16. Park, K.; Kim, J.; Park, J.-H.; Hwang, Y.; Han, D. Synchronous phase transition and carbon coating on the surface of Li-rich layered oxide cathode materials for rechargeable Li-ion batteries. *J. Power Sources* **2018**, *408*, 105–110. [CrossRef]
17. Park, J.-H.; Park, K.; Han, D.; Yeon, D.-H.; Jung, H.; Choi, B.; Park, S.Y.; Ahn, S.-J.; Park, J.-H.; Han, H.N. Structure-and porosity-tunable, thermally reactive metal organic frameworks for high-performance Ni-rich layered oxide cathode materials with multi-scale pores. *J. Mater. Chem. A* **2019**, *7*, 15190–15197. [CrossRef]
18. Kang, Y.-S.; Kim, D.Y.; Yoon, J.; Park, J.; Kim, G.; Ham, Y.; Park, I.; Koh, M.; Park, K. Shape control of hierarchical lithium cobalt oxide using biotemplates for connected nanoparticles. *J. Power Sources* **2019**, *436*, 226836. [CrossRef]
19. Xu, K. Electrolytes and interphases in Li-ion batteries and beyond. *Chem. Rev.* **2014**, *114*, 11503–11618. [CrossRef] [PubMed]
20. Meyer, W.H. Polymer electrolytes for lithium-ion batteries. *Adv. Mater.* **1998**, *10*, 439–448. [CrossRef]
21. Zhang, J.; Tan, T.; Zhao, Y.; Liu, N. Preparation of ZnO nanorods/graphene composite anodes for high-performance lithium-ion batteries. *Nanomaterials* **2018**, *8*, 966. [CrossRef]
22. Caimi, S.; Klaue, A.; Wu, H.; Morbidelli, M. Effect of SiO_2 nanoparticles on the performance of PVdF-HFP/ionic liquid separator for lithium-ion batteries. *Nanomaterials* **2018**, *8*, 926. [CrossRef]
23. Wen, L.; Sun, J.; An, L.; Wang, X.; Ren, X.; Liang, G. Effect of conductive material morphology on spherical lithium iron phosphate. *Nanomaterials* **2018**, *8*, 904. [CrossRef] [PubMed]
24. Alonso-Domínguez, D.; Pico, M.P.; Álvarez-Serrano, I.; López, M.L. New Fe_2O_3-Clay@ C nanocomposite anodes for Li-ion batteries obtained by facile hydrothermal processes. *Nanomaterials* **2018**, *8*, 808. [CrossRef]
25. Nacimiento, F.; Cabello, M.; Pérez-Vicente, C.; Alcántara, R.; Lavela, P.; Ortiz, G.F.; Tirado, J.L. On the Mechanism of Magnesium Storage in Micro-and Nano-Particulate Tin Battery Electrodes. *Nanomaterials* **2018**, *8*, 501. [CrossRef] [PubMed]
26. Mauger, A.; Julien, C.M.; Paolella, A.; Armand, M.; Zaghib, K. Building better batteries in the solid state: A review. *Materials* **2019**, *12*, 3892. [CrossRef]
27. Li, A.; Yuen, A.C.Y.; Wang, W.; De Cachinho Cordeiro, I.M.; Wang, C.; Chen, T.B.Y.; Zhang, J.; Chan, Q.N.; Yeoh, G.H. A Review on Lithium-Ion Battery Separators towards Enhanced Safety Performances and Modelling Approaches. *Molecules* **2021**, *26*, 478. [CrossRef] [PubMed]
28. Zhang, B.; Liu, Y.; Liu, J.; Sun, L.; Cong, L.; Fu, F.; Mauger, A.; Julien, C.M.; Xie, H.; Pan, X. "Polymer-in-ceramic" based poly (e-caprolactone)/ceramic composite electrolyte for all-solid-state batteries. *J. Energy Chem.* **2021**, *52*, 318–325. [CrossRef]
29. Chen, L.; Li, Y.; Li, S.-P.; Fan, L.-Z.; Nan, C.-W.; Goodenough, J.B. PEO/garnet composite electrolytes for solid-state lithium batteries: From "ceramic-in-polymer" to "polymer-in-ceramic". *Nano Energy* **2018**, *46*, 176–184. [CrossRef]

30. Osada, I.; de Vries, H.; Scrosati, B.; Passerini, S. Ionic-liquid-based polymer electrolytes for battery applications. *Angew. Chem. Int. Ed.* **2016**, *55*, 500–513. [CrossRef]
31. Park, M.J.; Choi, I.; Hong, J.; Kim, O. Polymer electrolytes integrated with ionic liquids for future electrochemical devices. *J. Appl. Polym. Sci.* **2013**, *129*, 2363–2376. [CrossRef]
32. Wang, C.; Adair, K.R.; Liang, J.; Li, X.; Sun, Y.; Li, X.; Wang, J.; Sun, Q.; Zhao, F.; Lin, X. Solid-state plastic crystal electrolytes: Effective protection interlayers for sulfide-based all-solid-state lithium metal batteries. *Adv. Funct. Mater.* **2019**, *29*, 1900392. [CrossRef]
33. Wang, J.; Huang, G.; Chen, K.; Zhang, X.-B. A Porosity-Adjustable Plastic Crystal Electrolyte Enabled High-Performance All-Solid-State Lithium-Oxygen Batteries. *Angew. Chem. Int. Ed.* **2020**, *132*, 9468–9473. [CrossRef]
34. Tong, Z.; Wang, S.-B.; Jena, A.; Liu, C.-E.; Liao, S.-C.; Chen, J.-M.; Chang, H.; Hu, S.-F.; Guo, X.; Liu, R.-S. Matchmaker of Marriage between a Li Metal Anode and NASICON-Structured Solid-State Electrolyte: Plastic Crystal Electrolyte and Three-Dimensional Host Structure. *ACS Appl. Mater. Interfaces* **2020**, *12*, 44754–44761. [CrossRef]
35. Xue, Z.; He, D.; Xie, X. Poly (ethylene oxide)-based electrolytes for lithium-ion batteries. *J. Mater. Chem. A* **2015**, *3*, 19218–19253. [CrossRef]
36. Ichino, T.; Takeshita, Y.; Yamamoto, F.; Kato, H.; Mushiake, N.; Wani, T. Composite Polymer Electrolyte Membrane. U.S. Patent 5,858,264A, 1 December 1999.
37. Jeon, J.-H.; Ha, S.-H.; Cho, J.-J. Gel Polymer Electrolyte Composition, Gel Polymer Electrolyte and Electrochemical Device Comprising the Same. U.S. Patent 9,130,242B2, 30 May 2012.
38. Oliver, M.; Gies, P.J.; Pendalwar, S.L.; Coalson, C.E.; Eschbach, F.O. Polymer gel Electrolyte. U.S. Patent 5,639,573A, 17 June 1997.
39. Fleischmann, S.; Bunte, C.; Mikhaylik, Y.V.; Viner, V.G. Gel Electrolytes and Electrodes. U.S. Patent 9,755,268B2, 5 September 2017.
40. Park, C.-k.; Zhang, Z.; Sun, L.Y.; Chai, C. High Ionic Conductivity gel Polymer Electrolyte for Rechargeable Polymer Batteries. U.S. Patent 6,841,303B2, 26 September 2002.
41. Balsara, N.P.; Eitouni, H.B.; Gur, I.; Singh, M.; Hudson, W. Gel Polymer Electrolytes for Batteries. U.S. Patent 8,889,301B2, 18 November 2014.
42. Arya, A.; Sharma, A. Polymer electrolytes for lithium ion batteries: A critical study. *Ionics* **2017**, *23*, 497–540. [CrossRef]
43. Long, L.; Wang, S.; Xiao, M.; Meng, Y. Polymer electrolytes for lithium polymer batteries. *J. Mater. Chem. A* **2016**, *4*, 10038–10069. [CrossRef]
44. Dias, F.B.; Plomp, L.; Veldhuis, J.B. Trends in polymer electrolytes for secondary lithium batteries. *J. Power Sources* **2000**, *88*, 169–191. [CrossRef]
45. Marcinek, M.; Syzdek, J.; Marczewski, M.; Piszcz, M.; Niedzicki, L.; Kalita, M.; Plewa-Marczewska, A.; Bitner, A.; Wieczorek, P.; Trzeciak, T. Electrolytes for Li-ion transport—Review. *Solid State Ion.* **2015**, *276*, 107–126. [CrossRef]
46. Das, A.; Thakur, A.K.; Kumar, K. Raman spectroscopic study of ion dissociation effect in clay intercalated polymer blend nano composite electrolyte. *Vib. Spectrosc.* **2017**, *92*, 14–19. [CrossRef]
47. Zhang, Y.; Sheehan, C.J.; Zhai, J.; Zou, G.; Luo, H.; Xiong, J.; Zhu, Y.; Jia, Q. Polymer-embedded carbon nanotube ribbons for stretchable conductors. *Adv. Mater.* **2010**, *22*, 3027–3031. [CrossRef]
48. Hennaoui, F.; Belbachir, M. Green Synthesis of copolymer (PDMS-co-PEO) Catalyzed by an ecocatalyst Maghnite-H. *Arch. Appl. Sci. Res.* **2016**, *8*, 19–26.
49. Liu, X.; Peng, S.; Gao, S.; Cao, Y.; You, Q.; Zhou, L.; Jin, Y.; Liu, Z.; Liu, J. Electric-field-directed parallel alignment architecting 3D lithium-ion pathways within solid composite electrolyte. *ACS Appl. Mater. Interfaces* **2018**, *10*, 15691–15696. [CrossRef] [PubMed]
50. Hassoun, J.; Scrosati, B. Advances in anode and electrolyte materials for the progress of lithium-ion and beyond lithium-ion batteries. *J. Electrochem. Soc.* **2015**, *162*, A2582. [CrossRef]
51. Li, Q.; Chen, J.; Fan, L.; Kong, X.; Lu, Y. Progress in electrolytes for rechargeable Li-based batteries and beyond. *Green Energy Environ.* **2016**, *1*, 18–42. [CrossRef]
52. Meng, Q.; Hu, J. A review of shape memory polymer composites and blends. *Compos. Part A Appl. Sci. Manuf.* **2009**, *40*, 1661–1672. [CrossRef]
53. Park, J.S.; Chung, Y.-C.; Do Lee, S.; Cho, J.W.; Chun, B.C. Shape memory effects of polyurethane block copolymers cross-linked by celite. *Fibers Polym.* **2008**, *9*, 661–666. [CrossRef]
54. Ni, Q.-Q.; Zhang, C.-s.; Fu, Y.; Dai, G.; Kimura, T. Shape memory effect and mechanical properties of carbon nanotube/shape memory polymer nanocomposites. *Compos. Struct.* **2007**, *81*, 176–184. [CrossRef]
55. Mondal, S.; Jinlian, H. Shape memory studies of functionalized MWNT-reinforced polyurethane copolymers. *Iran. Polym. J.* **2006**, *15*, 135–142.
56. Li, Y.; Sun, Y.; Pei, A.; Chen, K.; Vailionis, A.; Li, Y.; Zheng, G.; Sun, J.; Cui, Y. Robust pinhole-free Li3N solid electrolyte grown from molten lithium. *ACS Cent. Sci.* **2018**, *4*, 97–104. [CrossRef]
57. Horowitz, Y.; Lifshitz, M.; Greenbaum, A.; Feldman, Y.; Greenbaum, S.; Sokolov, A.P.; Golodnitsky, D. Polymer/Ceramic Interface Barriers: The Fundamental Challenge for Advancing Composite Solid Electrolytes for Li-Ion Batteries. *J. Electrochem. Soc.* **2020**, *167*, 160514. [CrossRef]
58. Xu, B.; Zhai, H.; Liao, X.; Qie, B.; Mandal, J.; Gong, T.; Tan, L.; Yang, X.; Sun, K.; Cheng, Q. Porous insulating matrix for lithium metal anode with long cycling stability and high power. *Energy Storage Mater.* **2019**, *17*, 31–37. [CrossRef]
59. Park, K.; Goodenough, J.B. Dendrite-Suppressed Lithium Plating from a Liquid Electrolyte via Wetting of Li3N. *Adv. Energy Mater.* **2017**, *7*, 1700732. [CrossRef]

60. Weston, J.; Steele, B. Effects of inert fillers on the mechanical and electrochemical properties of lithium salt-poly (ethylene oxide) polymer electrolytes. *Solid State Ion.* **1982**, *7*, 75–79. [CrossRef]

61. Kumar, B.; Scanlon, L.G. Polymer-ceramic composite electrolytes. *J. Power Sources* **1994**, *52*, 261–268. [CrossRef]

62. Zhou, J.; Fedkiw, P.S. Ionic conductivity of composite electrolytes based on oligo (ethylene oxide) and fumed oxides. *Solid State Ion.* **2004**, *166*, 275–293. [CrossRef]

63. Wang, X.-L.; Cai, Q.; Fan, L.-Z.; Hua, T.; Lin, Y.-H.; Nan, C.-W. Gel-based composite polymer electrolytes with novel hierarchical mesoporous silica network for lithium batteries. *Electrochim. Acta* **2008**, *53*, 8001–8007. [CrossRef]

64. Yang, C.-C.; Lian, Z.-Y.; Lin, S.; Shih, J.-Y.; Chen, W.-H. Preparation and application of PVDF-HFP composite polymer electrolytes in LiNi$_{0.5}$Co$_{0.2}$Mn$_{0.3}$O$_2$ lithium-polymer batteries. *Electrochim. Acta* **2014**, *134*, 258–265. [CrossRef]

65. Itoh, T.; Miyamura, Y.; Ichikawa, Y.; Uno, T.; Kubo, M.; Yamamoto, O. Composite polymer electrolytes of poly (ethylene oxide)/BaTiO$_3$/Li salt with hyperbranched polymer. *J. Power Sources* **2003**, *119*, 403–408. [CrossRef]

66. Pitawala, H.; Dissanayake, M.; Seneviratne, V.; Mellander, B.-E.; Albinson, I. Effect of plasticizers (EC or PC) on the ionic conductivity and thermal properties of the (PEO) 9 LiTf: Al$_2$O$_3$ nanocomposite polymer electrolyte system. *J. Solid State Electrochem.* **2008**, *12*, 783–789. [CrossRef]

67. Liu, Y.; Lee, J.Y.; Hong, L. Morphology, crystallinity, and electrochemical properties of in situ formed poly (ethylene oxide)/TiO$_2$ nanocomposite polymer electrolytes. *J. Appl. Polym. Sci.* **2003**, *89*, 2815–2822. [CrossRef]

68. Dey, A.; Karan, S.; De, S. Effect of nanofillers on thermal and transport properties of potassium iodide–polyethylene oxide solid polymer electrolyte. *Solid State Commun.* **2009**, *149*, 1282–1287. [CrossRef]

69. Ali, T.M.; Padmanathan, N.S. Effect of nanofiller CeO$_2$ on structural, conductivity, and dielectric behaviors of plasticized blend nanocomposite polymer electrolyte. *Ionics* **2015**, *21*, 829–840.

70. Osińska, M.; Walkowiak, M.; Zalewska, A.; Jesionowski, T. Study of the role of ceramic filler in composite gel electrolytes based on microporous polymer membranes. *J. Membr. Sci.* **2009**, *326*, 582–588. [CrossRef]

71. Kumar, B. Heterogeneous electrolytes: Variables for and uncertainty in conductivity measurements. *J. Power Sources* **2008**, *179*, 401–406. [CrossRef]

72. Wieczorek, W.; Such, K.; Chung, S.; Stevens, J. Comparison of Properties of Composite Polymeric Electrolytes Based on the Oxymethylene-Linked Poly (ethylene oxide) NaClO$_4$ Electrolyte with Polyacrylamide or. alpha.-Al$_2$O$_3$ Additives. *J. Phys. Chem.* **1994**, *98*, 9047–9055. [CrossRef]

73. Chung, S.; Wang, Y.; Persi, L.; Croce, F.; Greenbaum, S.; Scrosati, B.; Plichta, E. Enhancement of ion transport in polymer electrolytes by addition of nanoscale inorganic oxides. *J. Power Sources* **2001**, *97*, 644–648. [CrossRef]

74. Kim, K.M.; Ko, J.M.; Park, N.-G.; Ryu, K.S.; Chang, S.H. Characterization of poly (vinylidenefluoride-co-hexafluoropropylene)-based polymer electrolyte filled with rutile TiO$_2$ nanoparticles. *Solid State Ion.* **2003**, *161*, 121–131. [CrossRef]

75. Karlsson, C.; Best, A.; Swenson, J.; Howells, W.; Börjesson, L. Polymer dynamics in 3PEG–LiClO$_4$–TiO$_2$ nanocomposite polymer electrolytes. *J. Chem. Phys.* **2003**, *118*, 4206–4212. [CrossRef]

76. Jeon, J.-D.; Kim, M.-J.; Chung, J.W.; Kwak, S.-Y. Effects of the Addition of TiO$_2$ Nanoparticles for Polymer Electrolytes Based on Porous Poly (vinylidene fluoride-co-hexafluoropropylene)/Poly (ethylene oxide-co-ethylene carbonate) Membranes. *TechConnect Briefs* **2005**, *2*, 581–584.

77. Lin, C.; Hung, C.; Venkateswarlu, M.; Hwang, B. Influence of TiO$_2$ nano-particles on the transport properties of composite polymer electrolyte for lithium-ion batteries. *J. Power Sources* **2005**, *146*, 397–401. [CrossRef]

78. Ahmad, S.; Saxena, T.; Ahmad, S.A. The effect of nanosized TiO$_2$ addition on poly (methylmethacrylate) based polymer electrolytes. *J. Power Sources* **2006**, *159*, 205–209. [CrossRef]

79. Cho, B.-W.; Kim, D.H.; Lee, H.-W.; Na, B.-K. Electrochemical properties of gel polymer electrolyte based on poly (acrylonitrile)-poly (ethylene glycol diacrylate) blend. *Korean J. Chem. Eng.* **2007**, *24*, 1037–1041. [CrossRef]

80. Walkowiak, M.; Osińska, M.; Jesionowski, T.; Siwińska-Stefańska, K. Synthesis and characterization of a new hybrid TiO$_2$/SiO$_2$ filler for lithium conducting gel electrolytes. *Cent. Eur. J. Chem.* **2010**, *8*, 1311–1317. [CrossRef]

81. Kurc, B. Gel electrolytes based on poly (acrylonitrile)/sulpholane with hybrid TiO$_2$/SiO$_2$ filler for advanced lithium polymer batteries. *Electrochim. Acta* **2014**, *125*, 415–420. [CrossRef]

82. Kurc, B.; Jesionowski, T. Modified TiO$_2$-SiO$_2$ ceramic filler for a composite gel polymer electrolytes working with LiMn$_2$O$_4$. *J. Solid State Electrochem.* **2015**, *19*, 1427–1435. [CrossRef]

83. Johari, N.; Kudin, T.; Ali, A.; Yahya, M. Effects of TiO$_2$ on conductivity performance of cellulose acetate based polymer gel electrolytes for proton batteries. *Mater. Res. Innov.* **2011**, *15*, s229–s231. [CrossRef]

84. Zhou, L.; Wu, N.; Cao, Q.; Jing, B.; Wang, X.; Wang, Q.; Kuang, H. A novel electrospun PVDF/PMMA gel polymer electrolyte with in situ TiO$_2$ for Li-ion batteries. *Solid State Ion.* **2013**, *249*, 93–97. [CrossRef]

85. Cao, J.; Wang, L.; Shang, Y.; Fang, M.; Deng, L.; Gao, J.; Li, J.; Chen, H.; He, X. Dispersibility of nano-TiO$_2$ on performance of composite polymer electrolytes for Li-ion batteries. *Electrochim. Acta* **2013**, *111*, 674–679. [CrossRef]

86. Jayanthi, S.; Kulasekarapandian, K.; Arulsankar, A.; Sankaranarayanan, K.; Sundaresan, B. Influence of nano-sized TiO$_2$ on the structural, electrical, and morphological properties of polymer-blend electrolytes PEO–PVC–LiClO4. *J. Compos. Mater.* **2015**, *49*, 1035–1045. [CrossRef]

87. Aydin, H.; Bozkurt, A. Nanocomposite polymer electrolytes comprising PVA-graft-PEGME/TiO$_2$ for Li-ion batteries. *J. Mater. Res.* **2014**, *29*, 625. [CrossRef]

88. Song, D.; Xu, C.; Chen, Y.; He, J.; Zhao, Y.; Li, P.; Lin, W.; Fu, F. Enhanced thermal and electrochemical properties of PVDF-HFP/PMMA polymer electrolyte by TiO$_2$ nanoparticles. *Solid State Ion.* **2015**, *282*, 31–36. [CrossRef]

89. Yarmolenko, O.; Yudina, A.; Marinin, A.; Chernyak, A.; Volkov, V.; Shuvalova, N.; Shestakov, A. Nanocomposite network polymer gel-electrolytes: TiO$_2$-and Li$_2$TiO$_3$-nanoparticle effects on their structure and properties. *Russ. J. Electrochem.* **2015**, *51*, 412–420. [CrossRef]

90. Wang, S.-H.; Lin, Y.-Y.; Teng, C.-Y.; Chen, Y.-M.; Kuo, P.-L.; Lee, Y.-L.; Hsieh, C.-T.; Teng, H. Immobilization of anions on polymer matrices for gel electrolytes with high conductivity and stability in lithium ion batteries. *ACS Appl. Mater. Interfaces* **2016**, *8*, 14776–14787. [CrossRef]

91. Kumar, P.S.; Sakunthala, A.; Govindan, K.; Reddy, M.; Prabu, M. Single crystalline TiO$_2$ nanorods as effective fillers for lithium ion conducting PVdF-HFP based composite polymer electrolytes. *RSC Adv.* **2016**, *6*, 91711–91719. [CrossRef]

92. Jagadeesan, A.; Sasikumar, M.; Krishna, R.H.; Raja, N.; Gopalakrishna, D.; Vijayashree, S.; Sivakumar, P. High electrochemical performance of nano TiO$_2$ ceramic filler incorporated PVC-PEMA composite gel polymer electrolyte for Li-ion battery applications. *Mater. Res. Express* **2019**, *6*, 105524. [CrossRef]

93. Sachdeva, A.; Singh, P.K. Modification of properties of polymer electrolyte by incorporation of titanium dioxide nanoparticles. *Mol. Cryst. Liq. Cryst.* **2019**, *693*, 97–106. [CrossRef]

94. Li, Z.; Su, G.; Gao, D.; Wang, X.; Li, X. Effect of Al$_2$O$_3$ nanoparticles on the electrochemical characteristics of P (VDF-HFP)-based polymer electrolyte. *Electrochim. Acta* **2004**, *49*, 4633–4639. [CrossRef]

95. Piotrowska, K.; Zalewska, A.; Syzdek, J.S.; Niedzicki, L.; Marcinek, M. Properties of PVdF/HFP Inorganic Al$_2$O$_3$ Filler Modified Gel Electrolytes. *Ecs Trans.* **2010**, *25*, 221. [CrossRef]

96. Egashira, M.; Yoshimoto, N.; Morita, M. The Effect of Alumina Filler on the Properties of Imidazolium Ionic Liquid Gel Electrolyte. *Electrochem. Commun.* **2010**, *78*, 423–426. [CrossRef]

97. Chand, N.; Rai, N.; Agrawal, S.; Patel, S. Morphology, thermal, electrical and electrochemical stability of nano aluminium-oxide-filled polyvinyl alcohol composite gel electrolyte. *Bull. Mater. Sci.* **2011**, *34*, 1297–1304. [CrossRef]

98. Sun, P.; Liao, Y.; Luo, X.; Li, Z.; Chen, T.; Xing, L.; Li, W. The improved effect of co-doping with nano-SiO$_2$ and nano-Al$_2$O$_3$ on the performance of poly (methyl methacrylate-acrylonitrile-ethyl acrylate) based gel polymer electrolyte for lithium ion batteries. *RSC Adv.* **2015**, *5*, 64368–64377. [CrossRef]

99. Wen, Y.; Li, G.; Zhang, P.; Xiong, G. Performance Improvement of Polyvinyl Formal Based Gel Polymer Electrolyte for Lithium-Ion Batteries by Coating Al$_2$O$_3$. *J. Appl. Math. Phys.* **2016**, *4*, 189–194. [CrossRef]

100. Kim, K.-W.; Kim, H.W.; Kim, Y.; Kim, J.-K. Composite gel polymer electrolyte with ceramic particles for LiNi$_{1/3}$Mn$_{1/3}$Co$_{1/3}$O$_2$-Li$_4$Ti$_5$O$_{12}$ lithium ion batteries. *Electrochim. Acta* **2017**, *236*, 394–398. [CrossRef]

101. Vishwakarma, V.; Jain, A. Enhancement of thermal transport in Gel Polymer Electrolytes with embedded BN/Al$_2$O$_3$ nano-and micro-particles. *J. Power Sources* **2017**, *362*, 219–227. [CrossRef]

102. Jurado-Meneses, N.M.; Delgado-Rosero, M.I.; Meléndez-Lira, M.A. Structural and vibrational studies on composites polymer electrolytes (PEO) 10CF3COONa+ x wt.% Al$_2$O$_3$. *Rev. Fac. Ing. Univ. Antioq.* **2017**, *83*, 43–49. [CrossRef]

103. Maragani, N.; Vijaya Kumar, K. Structural and Conductivity Studies of PAN-based Al$_2$O$_3$ Nano Composite Gel Polymer Electrolytes. *Iran. J. Mater. Sci. Eng.* **2018**, *15*, 11–18.

104. Lei, D.; He, Y.-B.; Huang, H.; Yuan, Y.; Zhong, G.; Zhao, Q.; Hao, X.; Zhang, D.; Lai, C.; Zhang, S. Cross-linked beta alumina nanowires with compact gel polymer electrolyte coating for ultra-stable sodium metal battery. *Nat. Commun.* **2019**, *10*, 1–11. [CrossRef]

105. Mishra, K.; Arif, T.; Kumar, R.; Kumar, D. Effect of Al$_2$O$_3$ nanoparticles on ionic conductivity of PVdF-HFP/PMMA blend-based Na+-ion conducting nanocomposite gel polymer electrolyte. *J. Solid State Electrochem.* **2019**, *23*, 2401–2409. [CrossRef]

106. Świerczyński, D.; Zalewska, A.; Wieczorek, W. Composite polymeric electrolytes from the PEODME– LiClO$_4$– SiO$_2$ System. *Chem. Mater.* **2001**, *13*, 1560–1564. [CrossRef]

107. Kuo, C.-W.; Li, W.-B.; Chen, P.-R.; Liao, J.-W.; Tseng, C.-G.; Wu, T.-Y. Effect of plasticizer and lithium salt concentration in PMMA-based composite polymer electrolytes. *Int. J. Electrochem. Sci.* **2013**.

108. Lee, Y.-S.; Ju, S.H.; Kim, J.-H.; Hwang, S.S.; Choi, J.-M.; Sun, Y.-K.; Kim, H.; Scrosati, B.; Kim, D.-W. Composite gel polymer electrolytes containing core-shell structured SiO$_2$ (Li+) particles for lithium-ion polymer batteries. *Electrochem. Commun.* **2012**, *17*, 18–21. [CrossRef]

109. Li, W.-l.; Tang, J.-j.; Li, B.-t. Preparation and Characterization of Composite Microporous Gel Polymer Electrolytes Containing SiO$_2$ (Li+). *J. Inorg. Organomet. Polym. Mater.* **2013**, *23*, 831–838. [CrossRef]

110. Sethupathy, M.; Sethuraman, V.; Manisankar, P. Preparation of PVDF/SiO$_2$ composite nanofiber membrane using electrospinning for polymer electrolyte analysis. *Soft Nanocsci. Lett.* **2013**, *3*, 37–43. [CrossRef]

111. Shin, W.-K.; Cho, J.; Kannan, A.G.; Lee, Y.-S.; Kim, D.-W. Cross-linked composite gel polymer electrolyte using mesoporous methacrylate-functionalized SiO$_2$ nanoparticles for lithium-ion polymer batteries. *Sci. Rep.* **2016**, *6*, 1–10. [CrossRef]

112. Nagajothi, A.; Kannan, R.; Rajashabala, S. Lithium ion conduction in plasticizer based composite gel polymer electrolytes with the addition of SiO$_2$. *Mater. Res. Innov.* **2018**, *22*, 226–230. [CrossRef]

113. Huang, H.; Ding, F.; Zhong, H.; Li, H.; Zhang, W.; Liu, X.; Xu, Q. Nano-SiO$_2$-embedded poly (propylene carbonate)-based composite gel polymer electrolyte for lithium–sulfur batteries. *J. Mater. Chem. A* **2018**, *6*, 9539–9549. [CrossRef]

114. Tan, X.; Wu, Y.; Tang, W.; Song, S.; Yao, J.; Wen, Z.; Lu, L.; Savilov, S.V.; Hu, N.; Molenda, J. Preparation of Nanocomposite Polymer Electrolyte via In Situ Synthesis of SiO_2 Nanoparticles in PEO. *Nanomaterials* **2020**, *10*, 157. [CrossRef]
115. Aravindan, V.; Vickraman, P.; Kumar, T.P. ZrO_2 nanofiller incorporated PVC/PVdF blend-based composite polymer electrolytes (CPE) complexed with LiBOB. *J. Membr. Sci.* **2007**, *305*, 146–151. [CrossRef]
116. Suthanthiraraj, S.A.; Kumar, R.; Paul, B.J. Impact Of Zro 2 Nanoparticles On Ionic Transport And Electrochemical Properties Of Nanocomposite Gel Polymer Electrolyte: Ppg (4000)–Agcf3so3: ZrO_2. *Int. J. Nanosci.* **2011**, *10*, 241–246. [CrossRef]
117. Sivakumar, M.; Subadevi, R.; Muthupradeepa, R. Studies on the effect of dispersoid (ZrO_2) in PVdF-co-HFP based gel polymer electrolytes. *Am. Inst. Phys.* **2013**, *1536*, 857–858.
118. Ramachandran, M.; Subadevi, R.; Wang, F.-M.; Liu, W.-R.; Sivakumar, M. Structural, morphology and ionic conductivity studies on composite P (S-MMA)-ZrO_2 Polymer electrolyte for Lithium Polymer battery. In Proceedings of the International Conference on Materials and Characterization Techniques, Vellore, India, 10–12 March 2014; pp. 1687–1689.
119. Chen, R.; Qu, W.; Qian, J.; Chen, N.; Dai, Y.; Guo, C.; Huang, Y.; Li, L.; Wu, F. Zirconia-supported solid-state electrolytes for high-safety lithium secondary batteries in a wide temperature range. *J. Mater. Chem. A* **2017**, *5*, 24677–24685. [CrossRef]
120. Xiao, W.; Wang, Z.; Zhang, Y.; Fang, R.; Yuan, Z.; Miao, C.; Yan, X.; Jiang, Y. Enhanced performance of P (VDF-HFP)-based composite polymer electrolytes doped with organic-inorganic hybrid particles PMMA-ZrO_2 for lithium ion batteries. *J. Power Sources* **2018**, *382*, 128–134. [CrossRef]
121. Khoon, L.T.; Fui, M.-L.W.; Hassan, N.H.; Su'ait, M.S.; Vedarajan, R.; Matsumi, N.; Kassim, M.B.; Shyuan, L.K.; Ahmad, A. In situ sol–gel preparation of ZrO_2 in nano-composite polymer electrolyte of PVDF-HFP/MG49 for lithium-ion polymer battery. *J. Sol-Gel Sci. Technol.* **2019**, *90*, 665–675. [CrossRef]
122. Sai Prasanna, C.M.; Austin Suthanthiraraj, S. PVC/PEMA-based blended nanocomposite gel polymer electrolytes plasticized with room temperature ionic liquid and dispersed with nano-ZrO_2 for zinc ion batteries. *Polym. Compos.* **2019**, *40*, 3402–3411. [CrossRef]
123. Rajendran, S.; Mahendran, O.; Krishnaveni, K. Effect of CeO_2 on Conductivity of PMMA/PEO Polymer Blend Electrolytes. *J. New Mater. Electrochem. Syst.* **2003**, *6*, 25–28.
124. Vijayakumar, G.; Karthick, S.; Priya, A.S.; Ramalingam, S.; Subramania, A. Effect of nanoscale CeO_2 on PVDF-HFP-based nanocomposite porous polymer electrolytes for Li-ion batteries. *J. Solid State Electrochem.* **2008**, *12*, 1135–1141. [CrossRef]
125. Vijayakumar, G.; Karthick, S.; Subramania, A. A new class of P (VdF-HFP)-CeO_2-LiClO4-based composite microporous membrane electrolytes for Li-ion batteries. *Int. J. Electrochem.* **2011**, *2011*. [CrossRef]
126. Kumar, R.; Suthanthiraraj, S.A. Ion dynamics and segmental relaxation of CeO_2 nanoparticles loaded soft-matter like gel polymer electrolyte. *J. Non-Cryst. Solids* **2014**, *405*, 76–82. [CrossRef]
127. Polu, A.R.; Kumar, R. Preparation and characterization of PEG–Mg $(CH_3COO)_2$–CeO_2 composite polymer electrolytes for battery application. *Bull. Mater. Sci.* **2014**, *37*, 309–314. [CrossRef]
128. Lee, L.; Park, S.-J.; Kim, S. Effect of nano-sized barium titanate addition on PEO/PVDF blend-based composite polymer electrolytes. *Solid State Ion.* **2013**, *234*, 19–24. [CrossRef]
129. Jagadeesan, A.; Sasikumar, M.; Jeevani, R.; Therese, H.; Ananth, N.; Sivakumar, P. Fabrication of $BaTiO_3$ ceramic filler incorporated PVC-PEMA based blend nanocomposite gel polymer electrolytes for Li ion battery applications. *J. Mater. Sci. Mater. Electron.* **2019**, *30*, 17181–17194. [CrossRef]
130. Prabakaran, P.; Manimuthu, R.P.; Gurusamy, S. Influence of barium titanate nanofiller on PEO/PVdF-HFP blend-based polymer electrolyte membrane for Li-battery applications. *J. Solid State Electrochem.* **2017**, *21*, 1273–1285. [CrossRef]
131. Zhou, D.; Shanmukaraj, D.; Tkacheva, A.; Armand, M.; Wang, G. Polymer electrolytes for lithium-based batteries: Advances and prospects. *Chem. Rev.* **2019**, *5*, 2326–2352. [CrossRef]
132. Goodenough, J.B.; Kim, Y. Challenges for rechargeable Li batteries. *Chem. Mater.* **2010**, *22*, 587–603. [CrossRef]
133. Aziz, S.B.; Woo, T.J.; Kadir, M.; Ahmed, H.M. A conceptual review on polymer electrolytes and ion transport models. *J. Sci. Adv. Mater. Devices* **2018**, *3*, 1–17. [CrossRef]
134. Commarieu, B.; Paolella, A.; Daigle, J.-C.; Zaghib, K. Toward high lithium conduction in solid polymer and polymer–ceramic batteries. *Curr. Opin. Electrochem.* **2018**, *9*, 56–63. [CrossRef]
135. Kumar, M. A conceptual review on polymer gel electrolytes and its conduction mechanism. *J. Pharm. Innov.* **2018**, *7*, 194–198.
136. Mauger, A.; Julien, C.; Paolella, A.; Armand, M.; Zaghib, K. A comprehensive review of lithium salts and beyond for rechargeable batteries: Progress and perspectives. *Mater. Sci. Eng. R Rep.* **2018**, *134*, 1–21. [CrossRef]
137. Commarieu, B.; Paolella, A.; Collin-Martin, S.; Gagnon, C.; Vijh, A.; Guerfi, A.; Zaghib, K. Solid-to-liquid transition of polycarbonate solid electrolytes in Li-metal batteries. *J. Power Sources* **2019**, *436*, 226852. [CrossRef]
138. Chen, X.; Put, B.; Sagara, A.; Gandrud, K.; Murata, M.; Steele, J.A.; Yabe, H.; Hantschel, T.; Roeffaers, M.; Tomiyama, M. Silica gel solid nanocomposite electrolytes with interfacial conductivity promotion exceeding the bulk Li-ion conductivity of the ionic liquid electrolyte filler. *Sci. Adv.* **2020**, *6*, eaav3400. [CrossRef] [PubMed]
139. Liu, H.; Cheng, X.B.; Xu, R.; Zhang, X.Q.; Yan, C.; Huang, J.Q.; Zhang, Q. Plating/stripping behavior of actual lithium metal anode. *Adv. Energy Mater.* **2019**, *9*, 1902254. [CrossRef]
140. Sun, Y.; Liu, N.; Cui, Y. Promises and challenges of nanomaterials for lithium-based rechargeable batteries. *Nat. Energy* **2016**, *1*, 1–12. [CrossRef]

141. Ye, F.; Zhang, X.; Liao, K.; Lu, Q.; Zou, X.; Ran, R.; Zhou, W.; Zhong, Y.; Shao, Z. A smart lithiophilic polymer filler in gel polymer electrolyte enables stable and dendrite-free Li metal anode. *J. Mater. Chem. A* **2020**, *8*, 9733–9742. [CrossRef]

142. Liu, S.; Imanishi, N.; Zhang, T.; Hirano, A.; Takeda, Y.; Yamamoto, O.; Yang, J. Effect of nano-silica filler in polymer electrolyte on Li dendrite formation in Li/poly (ethylene oxide)–Li (CF$_3$SO$_2$)$_2$N/Li. *J. Power Sources* **2010**, *195*, 6847–6853. [CrossRef]

143. Liu, R.; Wu, Z.; He, P.; Fan, H.; Huang, Z.; Zhang, L.; Chang, X.; Liu, H.; Wang, C.-a.; Li, Y. A self-standing, UV-cured semi-interpenetrating polymer network reinforced composite gel electrolytes for dendrite-suppressing lithium ion batteries. *J. Mater.* **2019**, *5*, 185–194. [CrossRef]

144. Kaboli, S.; Demers, H.; Paolella, A.; Darwiche, A.; Dontigny, M.; Clément, D.; Guerfi, A.; Trudeau, M.L.; Goodenough, J.B.; Zaghib, K. Behavior of solid electrolyte in Li-Polymer battery with NMC cathode via in-situ scanning electron microscopy. *Nano Lett.* **2020**, *20*, 1607–1613. [CrossRef]

145. Golozar, M.; Hovington, P.; Paolella, A.; Bessette, S.p.; Lagacé, M.; Bouchard, P.; Demers, H.; Gauvin, R.; Zaghib, K. In situ scanning electron microscopy detection of carbide nature of dendrites in Li–polymer batteries. *Nano Lett.* **2018**, *18*, 7583–7589. [CrossRef]

146. Golozar, M.; Paolella, A.; Demers, H.; Bessette, S.; Lagacé, M.; Bouchard, P.; Guerfi, A.; Gauvin, R.; Zaghib, K. In situ observation of solid electrolyte interphase evolution in a lithium metal battery. *Commun. Chem.* **2019**, *2*, 1–9. [CrossRef]

147. Wang, X.; Zhu, H.; Girard, G.M.; Yunis, R.; MacFarlane, D.R.; Mecerreyes, D.; Bhattacharyya, A.J.; Howlett, P.C.; Forsyth, M. Preparation and characterization of gel polymer electrolytes using poly (ionic liquids) and high lithium salt concentration ionic liquids. *J. Mater. Chem. A* **2017**, *5*, 23844–23852. [CrossRef]

148. Purkait, M.K.; Sinha, M.K.; Mondal, P.; Singh, R. Introduction to membranes. In *Interface Science and Technology*; Elsevier: Amsterdam, The Netherlands, 2018; Volume 25, pp. 1–37.

149. Liu, L.; Wang, Z.; Zhao, Z.; Zhao, Y.; Li, F.; Yang, L. PVDF/PAN/SiO$_2$ polymer electrolyte membrane prepared by combination of phase inversion and chemical reaction method for lithium ion batteries. *J. Solid State Electrochem.* **2016**, *20*, 699–712. [CrossRef]

150. Dahlin, R.L.; Kasper, F.K.; Mikos, A.G. Polymeric nanofibers in tissue engineering. *Tissue Eng. Part B Rev.* **2011**, *17*, 349–364. [CrossRef]

151. Sun, G.; Sun, L.; Xie, H.; Liu, J. Electrospinning of nanofibers for energy applications. *Nanomaterials* **2016**, *6*, 129. [CrossRef] [PubMed]

152. Cheng, Y.; Zhang, L.; Xu, S.; Zhang, H.; Ren, B.; Li, T.; Zhang, S. Ionic liquid functionalized electrospun gel polymer electrolyte for use in a high-performance lithium metal battery. *J. Mater. Chem. A* **2018**, *6*, 18479–18487. [CrossRef]

153. Li, H.; Ma, X.-T.; Shi, J.-L.; Yao, Z.-K.; Zhu, B.-K.; Zhu, L.-P. Preparation and properties of poly (ethylene oxide) gel filled polypropylene separators and their corresponding gel polymer electrolytes for Li-ion batteries. *Electrochim. Acta* **2011**, *56*, 2641–2647. [CrossRef]

154. Sato, T.; Morinaga, T.; Marukane, S.; Narutomi, T.; Igarashi, T.; Kawano, Y.; Ohno, K.; Fukuda, T.; Tsujii, Y. Novel solid-state polymer electrolyte of colloidal crystal decorated with ionic-liquid polymer brush. *Adv. Mater.* **2011**, *23*, 4868–4872. [CrossRef] [PubMed]

155. Guo, Q.; Han, Y.; Wang, H.; Xiong, S.; Sun, W.; Zheng, C.; Xie, K. Flame retardant and stable Li$_{1.5}$Al$_{0.5}$Ge$_{1.5}$(PO$_4$)$_3$-supported ionic liquid gel polymer electrolytes for high safety rechargeable solid-state lithium metal batteries. *J. Phys. Chem. C* **2018**, *122*, 10334–10342. [CrossRef]

156. Ma, F.; Zhang, Z.; Yan, W.; Ma, X.; Sun, D.; Jin, Y.; Chen, X.; He, K. Solid polymer electrolyte based on polymerized ionic liquid for high performance all-solid-state lithium-ion batteries. *ACS Sustain. Chem. Eng.* **2019**, *7*, 4675–4683. [CrossRef]

157. Wang, S.; Shi, Q.X.; Ye, Y.S.; Xue, Y.; Wang, Y.; Peng, H.Y.; Xie, X.L.; Mai, Y.W. Constructing desirable ion-conducting channels within ionic liquid-based composite polymer electrolytes by using polymeric ionic liquid-functionalized 2D mesoporous silica nanoplates. *Nano Energy* **2017**, *33*, 110–123. [CrossRef]

158. Bucchella, P. Ionotropic gelation of chitosan for next-generation composite proton conducting flat structures. *Molecules* **2020**, *25*, 1632. [CrossRef]

159. Sacco, P.; Pedroso-Santana, S.; Kumar, Y.; Joly, N.; Martin, P.; Bocchetta, P. Ionotropic Gelation of Chitosan Flat Structures and Potential Applications. *Molecules* **2021**, *26*, 660. [CrossRef] [PubMed]

160. Kim, S.; Cho, M.; Lee, Y. Multifunctional chitosan–RGO network binder for enhancing the cycle stability of Li–S batteries. *Adv. Funct. Mater.* **2020**, *30*, 1907680. [CrossRef]

161. Zhao, X.; Yim, C.-H.; Du, N.; Abu-Lebdeh, Y. Crosslinked chitosan networks as binders for silicon/graphite composite electrodes in li-ion batteries. *J. Electrochem. Soc.* **2018**, *165*, A1110. [CrossRef]

162. Stephan, A.M.; Nahm, K.S.; Kulandainathan, M.A.; Ravi, G. Poly (vinylidene fluoride-hexafluoropropylene)(PVdF-HFP) based composite electrolytes for lithium batteries. *Eur. Polym. J.* **2006**, *42*, 1728–1734. [CrossRef]

163. Aravindan, V.; Vickraman, P. Nanoparticulate AlO(OH)$_n$ filled polyvinylidenefluoride-co-hexafluoropropylene based microporous membranes for lithium ion batteries. *J. Renew. Sustain. Energy* **2009**, *1*, 023108. [CrossRef]

164. Chen, G.; Zhang, F.; Zhou, Z.; Li, J.; Tang, Y. A flexible dual-ion battery based on PVDF-HFP-modified gel polymer electrolyte with excellent cycling performance and superior rate capability. *Adv. Energy Mater.* **2018**, *8*, 1801219. [CrossRef]

165. Zhao, X.; Tao, C.-a.; Li, Y.; Chen, X.; Wang, J.; Gong, H. Preparation of gel polymer electrolyte with high lithium ion transference number using GO as filler and application in lithium battery. *Ionics* **2020**, *26*, 4299–4309. [CrossRef]

166. Liu, J.; Wu, X.; He, J.; Li, J.; Lai, Y. Preparation and performance of a novel gel polymer electrolyte based on poly (vinylidene fluoride)/graphene separator for lithium ion battery. *Electrochim. Acta* **2017**, *235*, 500–507. [CrossRef]

167. Chen, Y.M.; Hsu, S.T.; Tseng, Y.H.; Yeh, T.F.; Hou, S.S.; Jan, J.S.; Lee, Y.L.; Teng, H. Minimization of Ion–Solvent Clusters in Gel Electrolytes Containing Graphene Oxide Quantum Dots for Lithium-Ion Batteries. *Small* **2018**, *14*, 1703571. [CrossRef]
168. Dyartanti, E.R.; Purwanto, A.; Widiasa, I.N.; Susanto, H. Ionic conductivity and cycling stability improvement of PVdF/nano-clay using PVP as polymer electrolyte membranes for LiFePO4 batteries. *Membranes* **2018**, *8*, 36. [CrossRef] [PubMed]
169. Gebert, F.; Knott, J.; Gorkin, R.; Chou, S.-L.; Dou, S.-X. Polymer electrolytes for sodium-ion batteries. *Energy Storage Mater.* **2021**, *36*, 10–30. [CrossRef]
170. Iturrondobeitia, A.; Goñi, A.; Gil de Muro, I.; Lezama, L.; Rojo, T. Physico-chemical and electrochemical properties of nanoparticulate NiO/C composites for high performance lithium and sodium ion battery anodes. *Nanomaterials* **2017**, *7*, 423. [CrossRef] [PubMed]
171. Ayandele, E.; Sarkar, B.; Alexandridis, P. Polyhedral oligomeric silsesquioxane (POSS)-containing polymer nanocomposites. *Nanomaterials* **2012**, *2*, 445–475. [CrossRef] [PubMed]
172. Ni'mah, Y.L.; Cheng, M.-Y.; Cheng, J.H.; Rick, J.; Hwang, B.-J. Solid-state polymer nanocomposite electrolyte of TiO_2/PEO/NaClO4 for sodium ion batteries. *J. Power Sources* **2015**, *278*, 375–381. [CrossRef]
173. Zhang, X.; Wang, X.; Liu, S.; Tao, Z.; Chen, J. A novel PMA/PEG-based composite polymer electrolyte for all-solid-state sodium ion batteries. *Nano Res.* **2018**, *11*, 6244–6251. [CrossRef]
174. Kumar, D.; Suleman, M.; Hashmi, S. Studies on poly (vinylidene fluoride-co-hexafluoropropylene) based gel electrolyte nanocomposite for sodium–sulfur batteries. *Solid State Ion.* **2011**, *202*, 45–53. [CrossRef]
175. Liu, Z.; Wang, X.; Chen, J.; Tang, Y.; Mao, Z.; Wang, D. Gel Polymer Electrolyte Membranes Boosted with Sodium-Conductive β-Alumina Nanoparticles: Application for Na-Ion Batteries. *ACS Appl. Energy Mater.* **2021**, *4*, 623–632. [CrossRef]
176. Wang, X.; Liu, Z.; Tang, Y.; Chen, J.; Mao, Z.; Wang, D. PVDF-HFP/PMMA/TPU-based gel polymer electrolytes composed of conductive $Na_3Zr_2Si_2PO_{12}$ filler for application in sodium ions batteries. *Solid State Ion.* **2021**, *359*, 115532. [CrossRef]
177. Yi, Q.; Zhang, W.; Li, S.; Li, X.; Sun, C. Durable Sodium Battery with a Flexible $Na_3Zr_2Si_2PO_{12}$–PVDF–HFP Composite Electrolyte and Sodium/Carbon Cloth Anode. *ACS Appl. Mater. Interfaces* **2018**, *10*, 35039–35046. [CrossRef]
178. Song, S.; Kotobuki, M.; Zheng, F.; Xu, C.; Savilov, S.V.; Hu, N.; Lu, L.; Wang, Y.; Li, W.D.Z. A hybrid polymer/oxide/ionic-liquid solid electrolyte for Na-metal batteries. *J. Mater. Chem. A* **2017**, *5*, 6424–6431. [CrossRef]
179. Yoo, H.D.; Shterenberg, I.; Gofer, Y.; Gershinsky, G.; Pour, N.; Aurbach, D. Mg rechargeable batteries: An on-going challenge. *Energy Environ. Sci.* **2013**, *6*, 2265–2279. [CrossRef]
180. Jäckle, M.; Helmbrecht, K.; Smits, M.; Stottmeister, D.; Groß, A. Self-diffusion barriers: Possible descriptors for dendrite growth in batteries? *Energy Environ. Sci.* **2018**, *11*, 3400–3407. [CrossRef]
181. Park, B.; Schaefer, J.L. Polymer Electrolytes for Magnesium Batteries: Forging Away from Analogs of Lithium Polymer Electrolytes and Towards the Rechargeable Magnesium Metal Polymer Battery. *J. Electrochem. Soc.* **2020**, *167*, 070545. [CrossRef]
182. Oh, J.-S.; Ko, J.-M.; Kim, D.-W. Preparation and characterization of gel polymer electrolytes for solid state magnesium batteries. *Electrochim. Acta* **2004**, *50*, 903–906. [CrossRef]
183. Pandey, G.; Agrawal, R.; Hashmi, S. Magnesium ion-conducting gel polymer electrolytes dispersed with nanosized magnesium oxide. *J. Power Sources* **2009**, *190*, 563–572. [CrossRef]
184. Pandey, G.; Agrawal, R.; Hashmi, S. Electrical and electrochemical properties of magnesium ion conducting composite gel polymer electrolytes. *J. Phys. D Appl. Phys.* **2010**, *43*, 255501. [CrossRef]
185. Pandey, G.; Agrawal, R.; Hashmi, S. Magnesium ion-conducting gel polymer electrolytes dispersed with fumed silica for rechargeable magnesium battery application. *J. Solid State Electrochem.* **2011**, *15*, 2253–2264. [CrossRef]
186. Pandey, G.; Agrawal, R.; Hashmi, S. Performance studies on composite gel polymer electrolytes for rechargeable magnesium battery application. *J. Phys. Chem. Solids* **2011**, *72*, 1408–1413. [CrossRef]
187. Sharma, J.; Hashmi, S. Magnesium ion-conducting gel polymer electrolyte nanocomposites: Effect of active and passive nanofillers. *Polym. Compos.* **2019**, *40*, 1295–1306. [CrossRef]
188. Hashmi, S. Enhanced zinc ion transport in gel polymer electrolyte: Effect of nano-sized ZnO dispersion. *J. Solid State Electrochem.* **2012**, *16*, 3105–3114.
189. Johnsi, M.; Suthanthiraraj, S.A. Preparation, zinc ion transport properties, and battery application based on poly (vinilydene fluoride-co-hexa fluoro propylene) polymer electrolyte system containing titanium dioxide nanofiller. *High. Perform. Polym.* **2015**, *27*, 877–885. [CrossRef]

Article

Self-Assembled Few-Layered MoS$_2$ on SnO$_2$ Anode for Enhancing Lithium-Ion Storage

Thang Phan Nguyen and Il Tae Kim *

Department of Chemical and Biological Engineering, Gachon University,
Seongnam-si, Gyeonggi-do 13120, Korea; phanthang87@gmail.com
* Correspondence: itkim@gachon.ac.kr

Received: 2 December 2020; Accepted: 18 December 2020; Published: 20 December 2020

Abstract: SnO$_2$ nanoparticles (NPs) have been used as reversible high-capacity anode materials in lithium-ion batteries, with reversible capacities reaching 740 mAh·g^{-1}. However, large SnO$_2$ NPs do not perform well in charge–discharge cycling. In this work, we report the incorporation of MoS$_2$ nanosheet (NS) layers with SnO$_2$ NPs. SnO$_2$ NPs of ~5 nm in diameter synthesized by a facile hydrothermal precipitation method. Meanwhile, MoS$_2$ NSs of a few hundreds of nanometers to a few micrometers in lateral size were produced by top-down chemical exfoliation. The self-assembly of the MoS$_2$ NS layer on the gas–liquid interface was first demonstrated to achieve up to 80% coverage of the SnO$_2$ NP anode surface. The electrochemical properties of the pure SnO$_2$ NPs and MoS$_2$-covered SnO$_2$ NP anodes were investigated. The results showed that the SnO$_2$ electrode with a single-layer MoS$_2$ NS film exhibited better electrochemical performance than the pure SnO$_2$ anode in lithium storage applications.

Keywords: SnO$_2$; self-assembly; MoS$_2$; nanosheets; lithium-ion battery

---

1. Introduction

The use of lithium-ion batteries (LIBs) helps to solve issues of energy and powering devices in modern life and industries. The prevalence of electronic devices in life and the rapid development of industry have led to a demand for higher-quality energy storage systems with improvements such as higher capacity, greater stability, and improved safety. Currently, commercialized LIBs use lithium–nickel–manganese–cobalt oxide cathodes, which are generally stable but difficult to improve, and graphite anodes, which have a low capacity of ~372 mAh·g^{-1} [1,2]. The issues requiring attention in the development of LIBs are low anode capacity, degradation, and mechanical instability in LIB structures due to anode material expansion by the insertion and extraction of lithium ions. To improve LIB characteristics, several anode materials have been considered, such as metal alloys, metal oxides, and transition metal chalcogenides (TMCs) [3–5]. Recently, the use of metal oxide materials such as TiO$_2$, SnO$_2$, Fe$_2$O$_3$, and V$_2$O$_5$ for lithium-ion storage anodes has attracted much attention owing to their chemical stability and high reversible capacity [6–15]. Among them, SnO$_2$ nanoparticles (NPs) have shown promising lithium-ion storage capability with very high capacity reaching 740 mAh·g^{-1} [16,17]. It has been reported that the required SnO$_2$ material size is approximately 3 nm to achieve a high reversible capacity [16]. However, the lithiation process of SnO$_2$ leads to large volume expansion, which results in poor reversible cycling performance [18]. Many methods have been employed to enhance the reaction stability of SnO$_2$ materials by using carbon-based support materials such as amorphous carbon or graphene, as well as MoS$_2$. For example, Lou et al. used carbon-supported SnO$_2$ nanocolloids prepared by a facile hydrothermal method and carbonization [19]. Wang et al. developed a SnO$_2$–graphene composite material for high-capacity reversible LIBs, which had a capacity of ~440 mAh·g^{-1} after 100 cycles [20]. Chen et al. reported a composite of SnO$_2$ and MoS$_2$

nanosheets (NS) as a LIB anode to diminish the volume expansion during lithiation in SnO_2 NPs [21]. However, the appropriate working conditions and optimization of the SnO_2 morphologies still require investigation and remain challenging in realizing commercial batteries.

Along with the development of metal oxide materials, 2D materials such as graphene and TMCs have drawn great attention owing to their superior properties and flexibility [22–26]. Single- and few-layered TMCs, including MoS_2 and WS_2, have revealed superior electronic and mechanical properties. They have also been utilized in many optical and electrical devices such as solar cells, light-emitting diodes, and transistors, as well as applications in catalysts for hydrogen generation and energy-storage applications in LIBs or sodium-ion batteries [27–30]. MoS_2 in 1T phase has been used to overcome charge–discharge decay and to effectively store lithium or sodium ions with the aid of carbon derivatives such as graphite, carbon nanotubes (CNTs), and graphene. For instance, a combination of CNTs and 1T-structured MoS_2 reported by Nguyen et al. [14] was introduced to develop 3D MoS_2@graphite-CNT for a long-term stable anode material with a very high lithium-ion storage capacity. The 3D-structured MoS_2@gaphite-CNT was prepared by a scalable ball-milling method. This 3D structure allowed a high anodic charge–discharge rate and showed a high capacity of ~1200 mAh·g^{-1} after 450 cycles. Moreover, Lane et al. reported the computed electronic structure of lithium- and sodium-intercalated MoS_2. They concluded that the 1T-MoS_2 could be a high-capacity and high-conductivity anode material [31]. Li et al. also discovered that the growth of 3D bulky MoS_2 in the 1T phase supported prolonged anodic cycling, where the structure helped to release strain during cycling and led to improved capacity at high current rates [32]. Therefore, the use of 1T-MoS_2, which has high conductivity and flexibility, can help traditional anode materials exhibit superior performance in lithium storage applications.

In this work, we first demonstrated the use of a 1T-MoS_2 self-assembling layer as protection on the SnO_2 surface. The MoS_2 NS layer was formed by using the gas–liquid interface method, which is well-known for the thin-film preparation of monolayer colloidal crystals such as polystyrene NPs [33–35]. SnO_2 NPs were fabricated by a facile hydrothermal method. The presence of the 1T-MoS_2 layer greatly enhanced the cycling performance of SnO_2 anodes in LIBs. Moreover, the effect of different numbers of MoS_2 layers was also investigated.

2. Experimental

2.1. Chemical Materials

Molybdenum(VI) sulfide (MoS_2, powder), tin(II) chloride dihydrate ($SnCl_2 \cdot 2H_2O$, powder), polyvinylidene fluoride (PVDF, M_W 534,000), Sodium dodecylbenzensulfonate (technical grade) and a 2.5-M solution of *n*-butyllithium in hexane were purchased from Sigma-Aldrich Inc. (St. Louis, MO, USA). Super-P amorphous carbon black (C, approximately 40 nm, 99.99%) and absolute ethanol (C_2H_5OH) were purchased from Alpha Aesar Inc. (Ward Hill, MA, USA). All chemicals were used as delivered without any purification.

2.2. Exfoliation of MoS$_2$ Nanosheets (NSs)

MoS_2 NSs were synthesized using the liquid chemical exfoliation method [36–38]. In brief, 1 g of MoS_2 powder was placed in a 10-mL vessel. Then, 3 mL of butyllithium in hexane was added to reach the powder level. The solution was kept for 2 days to allow lithium intercalation into the MoS_2 to form Li_xMoS_2 ($x > 1$) [36,39]. For the exfoliation of MoS_2, Li_xMoS_2 was washed with hexane to remove excess butyllithium by centrifugation at 5000 rpm for 5 min. The Li_xMoS_2 obtained was exfoliated in 100 mL deionized (DI) water in a sonication bath for 1 h. The floating large-sized MoS_2 in the solution was removed. Then, 1T-MoS_2 was centrifuged with DI water four times and kept in DI water for further use or dried at 60 °C in a vacuum oven for characterization.

2.3. SnO$_2$ Nanoparticle (NP) Synthesis

SnO$_2$ NPs were prepared by a facile hydrothermal method. In a typical synthesis, 0.9 g of SnCl$_2$·2H$_2$O was dissolved in 100 mL of DI water under stirring for 30 min. Then, 10 mL of 10% ammonia solution (NH$_4$OH) was dropped slowly to obtain a gel form. The solution was then transferred to a Teflon autoclave line and heated at 200 °C for 2 h. The white precipitate, SnO$_2$, was collected via centrifugation and washing for 3 times, then dried and calcinated in air at 600 °C for 1 h.

2.4. Self-Assembled MoS$_2$ NS Layer

The self-assembled MoS$_2$ was prepared based on the wettability of the MoS$_2$ NSs. MoS$_2$ NSs can be dispersed in DI water; however, an NS floats on the surface of DI water if only one NS surface is wetted. This behavior is similar to that of graphene or polystyrene, both of which can assemble as 2D layers floating on a solution. Here, MoS$_2$ NSs were dispersed in a 1:1 volumetric DI water–ethanol mixture at a concentration of 2 mg·mL^{-1}. A water bath and half-dipped glass substrate are shown in Figure 1. The prepared solution was then dropped slowly to allow evaporation of the DI water–ethanol solution as the drop met the water surface. The MoS$_2$ NSs remained randomly floating at the water–air interface. To consolidate the single-layer MoS$_2$ NSs, a ~10-µL drop of 1% sodium dodecylbenzensulfonate solution was spread on the surface to increase the surface tension. This formed a semi-transparent layer of MoS$_2$ on the side of the water bath. The dried anode material was dipped in DI water before it was dipped in the assembled MoS$_2$ NS layer bath. The MoS$_2$ NS layers were easily deposited on the electrode surface. Finally, the anode was obtained by drying in a vacuum oven at 70 °C.

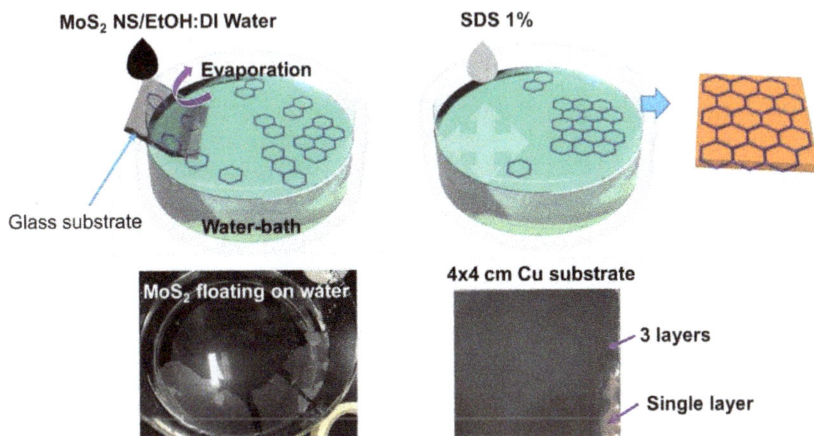

Figure 1. Illustration and photographs of self-assembled MoS$_2$ nanosheet (NS) on Cu substrate.

2.5. Characterization

X-ray diffraction (XRD) (D/MAX-2200 Rigaku, Tokyo, Japan) was used to investigate the structures of the powder samples. The XRD patterns of the samples were performed over the 2θ range of 10–70°. The structures, morphologies, and sizes of the materials were analyzed by scanning electron microscopy (SEM) (Hitachi S4700, Tokyo, Japan) and transmission electron microscopy (TEM) integrated with energy-dispersive X-ray spectroscopy (EDS) (TECNAI G2F30, FEI Corp., OR, USA).

2.6. Electrochemical Measurements

The SnO$_2$ NP materials were employed to assemble a half-cell LIB using coin-type cells (CR 2032, Rotech Inc., Gwangju, Korea). The assembly process follows a typical LIB construction method. The working electrode was prepared by pasting a slurry of 70% active material (SnO$_2$ NP powder),

15% PVDF, and 15% Super P in *N*-methyl-2-pyrrolidinone on a Cu foil by doctor blading. The electrode was then dried in a vacuum oven at 70 °C for at least 12 h before use. The MoS_2 NS layer was lifted from the water surface, as mentioned before, and dried before cell assembly. Electrodes with one, two, and three MoS_2 NS layers were denoted as $M1SnO_2$, $M2SnO_2$, and $M3SnO_2$, respectively. The anodes were punched into circular discs of 12 mm in diameter. The areal loading of active materials was 0.84–1.05 mg cm^{-2}. The battery half-cell structures were assembled under a neutral gas of Ar in a glovebox with positive pressure to ambient air conditions. The reference electrode, separator, and electrolyte were lithium foil, polyethylene, and 1-M $LiPF_6$ in ethylene carbonate–diethylene carbonate (1:1 by volume), respectively. The galvanostatic electrochemical charge–discharge performances of the different cells were measured using a battery cycle tester (WBCS3000, WonAtech, Seocho-gu, Seoul) over the voltage range of 0.01–3.0 V versus Li/Li$^+$. Cyclic voltammetry (CV) tests and electrochemical impedance spectroscopy (EIS) were performed using a ZIVE MP1 (WonAtech, Seocho-gu, Seoul) over the voltage range of 0.01–3.0 V at a scanning rate of 0.1 mV·s^{-1} and over the frequency range of 100 kHz–0.1 Hz.

3. Results and Discussion

Figure 2 shows the SEM images of the SnO_2 NP, MoS_2 NS, and SnO_2 anode covered with the MoS_2 NS layer. To prepare the SnO_2 NPs, an easy hydrothermal method was selected to obtain a uniform and highly crystalline material [40–42]. $Sn(OH)_x$ precipitation using NH_4OH solution and hydrothermal processing were applied to develop the crystalline structures of the SnO_2 NPs. It was observed that the size of the NPs was <100 nm. However, the detailed size distribution of the NPs cannot be determined by SEM measurement, as illustrated in Figure 2a,b. The diameters of the SnO_2 NPs are discussed later via TEM analyses. Meanwhile, the MoS_2 NSs were well prepared with lateral sizes ranging from a few hundred nanometers to a few micrometers, as shown in Figure 2c,d. The size of the NSs is larger than that of MoS_2 fabricated by ultrasonication or ball milling [43,44]. The low-magnification SEM image shows that the MoS_2 NSs uniformly cover the Cu electrode. Figure 2e,f shows a single-layer MoS_2 NS thin film on the surface of the SnO_2 electrode. It was calculated that the MoS_2 NSs covered approximately 80% of the SnO_2 electrode surface. The semi-transparent yellow indicates the uncovered electrode surface (Figure 2f), while the dark colors indicate the thin MoS_2 layer covering the surface. From analysis of the SEM images, a novel SnO_2 electrode uniformly covered with MoS_2 NSs layers was prepared, from which better electrochemical performance is expected, compared to that of a pure SnO_2 electrode.

Figure 2. Scanning electron microscope (SEM) images of (**a,b**) SnO_2 nanoparticle (NP) powder, (**c,d**) single-layer MoS_2 NS thin film on Cu electrode, and (**e,f**) $M1SnO_2$ anode.

To confirm the crystalline nature of the SnO_2 NPs, powder XRD measurements were performed in the range of 10–70°, as shown in Figure 3a. In comparison with PDF #41-1445, the XRD pattern of the SnO_2 NPs was in good agreement with the $P4_2/mnm$ space group [16]. No impurity peaks appear in the pattern. This confirmed the high crystallinity of the SnO_2 NPs. Furthermore, the average sizes of the crystals (D) can be calculated using the Scherrer equation:

$$D = 0.9\lambda/\beta\cos\theta \tag{1}$$

where λ is the X-ray wavelength of the XRD measurement, θ is the Bragg angle, and β is the full width at half maximum of the peak. Based on this information, the calculated crystal size of the SnO_2 NPs was ~2–5 nm. The structure of the assembled MoS_2 layer on SnO_2 anode was also analyzed by XRD as shown in Figure 3a. The peaks of MoS_2 NS, SnO_2 and Cu were clearly revealed, indicating successful coverage of MoS_2 on the SnO_2 anode. Figure 3b shows the XRD patterns of Li_xMoS_2 and the MoS_2 NSs. The intercalation of lithium in the interface of each layer MoS_2 introduced a peak at ~15.4°. After exfoliation in DI water, only the peak of (002) MoS_2 remained, indicating the 2D structure of the MoS_2 NSs. The size of MoS_2 based on the (002) plane was calculated to be ~1 nm, illustrating the formation of thin and single- or few-layered MoS_2 NSs [37,38]. The nanoscale sizes of the SnO_2 NPs are promising for achieving the critical size of SnO_2 that yields the best electrochemical performance. Moreover, high-quality single- and few-layer MoS_2 with large lateral sizes effectively cover the electrode surface.

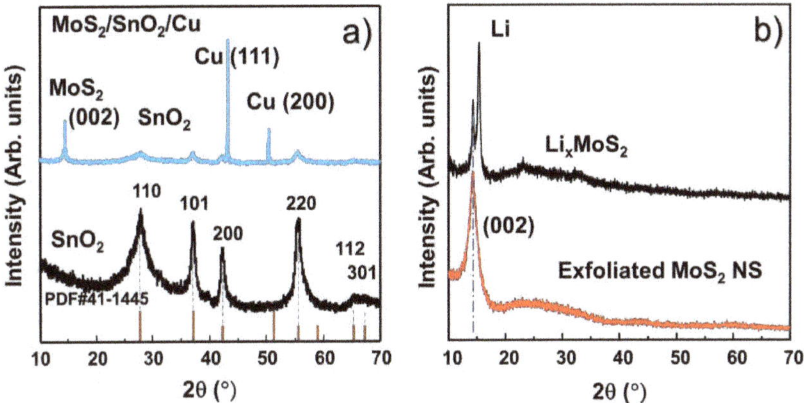

Figure 3. X-ray diffraction (XRD) patterns of (**a**) SnO_2 NPs and M3SnO$_2$ anode and (**b**) lithium-ion intercalated and exfoliated MoS_2 nanosheet.

To further confirm the sizes and structures of the SnO_2 NPs, the SnO_2 powder was observed by TEM. Figure 4a–c shows the SnO_2 EDS mapping images, where clear contrasts indicating Sn and O atoms were detected, indicating the high purity of the synthesized materials. In addition, the high-resolution TEM (HRTEM) image (Figure 4d) clearly shows the lattices of SnO_2 NPs with (101), (110), and (220) planes with sizes of ~5 nm. This result agrees well with the calculation of the SnO_2 crystal size from the XRD patterns. Furthermore, the selected-area electron diffraction (SAED) pattern in Figure 4d also confirms the lattice planes from the crystal structures, as marked with semi-transparent yellow circles. No other elements were found as contaminants. Therefore, pure SnO_2 NPs were well prepared with a diameter of ~5 nm, which would contribute to a high performance in LIBs. Meanwhile, the morphology of MoS_2 NS was also confirmed by HRTEM (Figure 4e), where a lattice spacing of 0.27 nm can be assigned to (100) plane. The spacing between lattice fringes in HRTEM image were to be 1.92 and 3.19 nm for 3 and 5 layers of MoS_2, respectively. Furthermore, the SAED as an inset of

Figure 4e also confirmed the high crystallinity of the MoS$_2$ layer. Therefore, it can be considered that the MoS$_2$ NS was successfully exfoliated to single- and few-layer MoS$_2$.

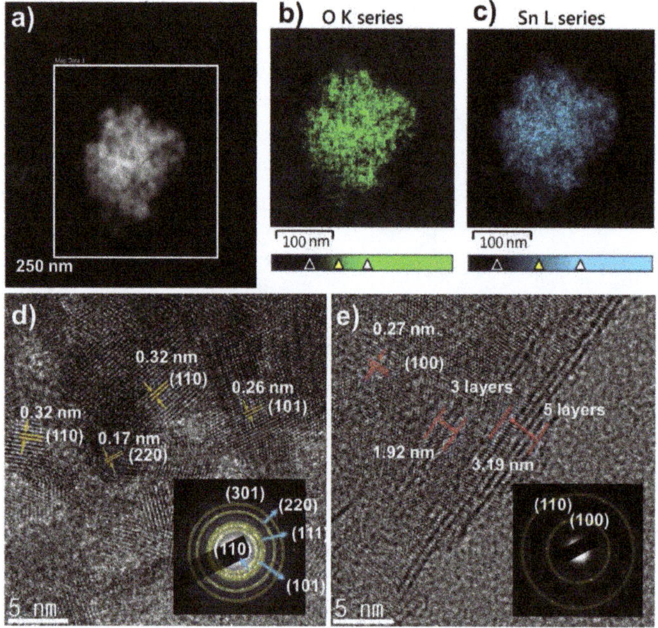

Figure 4. (**a**) Transmission electron microscope (TEM) image of SnO$_2$ NP; energy-dispersive X-ray spectroscopy (EDS) mapping of the elements (**b**) O and (**c**) Sn; (**d**) high-resolution TEM (HRTEM) image of SnO$_2$ NP with inset selected-area electron diffraction (SAED) pattern; (**e**) exfoliated MoS$_2$ NS with inset SAED pattern.

To understand the effect of the MoS$_2$ NS layer on the electrochemical properties of the SnO$_2$ anode, CV tests were performed between 0.01 and 3.00 V (vs. Li/Li$^+$) at a scanning rate of 0.1 mV·s^{-1}. The electrochemical process in the anode can be expressed by the following reaction equations:

$$Li^+ + e^- + electrolyte \rightarrow SEI\ layer \tag{2}$$

$$SnO_2 + 4Li^+ + 4e^- \rightarrow Sn + 2Li_2O \tag{3}$$

$$Sn + xLi^+ + xe^- \leftrightarrow Li_xSn\ (0 \leq x \leq 4.4) \tag{4}$$

Equation (2) relates to the formation of a solid electrolyte (SEI) layer formed from the first lithium-ion insertion. Meanwhile, Equation (3) is the conversion reaction of SnO$_2$ to Sn. Both reactions (2) and (3) contribute to the irreversible capacity of the anode. Reaction (4) represents the reversible reaction of Sn with lithium ions. Figure 5a shows the electrochemical performance of the pure SnO$_2$ anode in the initial three cycles. The peak at ~0.84 V in the first cathodic cycle is attributed to the reduction of SnO$_2$ to Sn metal and the formation of the SEI layer, as illustrated in Equations (2) and (3). In the 2nd and 3rd cycles, these peaks showed reduced intensity and were shifted to the lower potential of ~0.75 V. In the oxidation process, two peaks appear at approximately 0.67 and 1.30 V. Interestingly, the oxidation peak at 0.67 V increases in intensity as the cycle number was increased. This phenomenon could be explained by the activation of the reversible reaction that occurred in the electrode materials [17]. Meanwhile, the oxidation peak at 1.30 V was ascribed to the oxidation of metallic Sn to SnO_2, which is the reversible case of reaction (2) [17,20,45]. When the MoS_2 NS layer was added to the SnO_2 anode surface, the CV profiles changed depending on the number of layers, as shown in Figure 5b–d, respectively. First, it is easily observed that the oxidation peaks remain stable, with two peaks at 0.67 and 1.30 V related to the reversible reduction and oxidation reaction of Sn metal. With the $M1SnO_2$ electrode, additional peaks appeared at ~0.42 V and 0.45 V, arising from the conversion reaction as: $Li_xMoS_2 + (4 − x) Li^+ + (4 − x) e^− → Mo + 2 Li_2S$ and from SEI layer formation on the MoS_2 materials [46–49]. The $M2/M3SnO_2$ electrodes also have peaks at ~0.42 V; however, the intensities are weaker than that of the $M1SnO_2$ electrode. This can be explained by the energy barrier of the single-layer MoS_2 for lithium-ion intercalation, which is in the range of ~0.42 to ~0.16 eV. Meanwhile, bulk or multilayer MoS_2 has an intercalation energy barrier between 0.73 and 0.59 eV. Therefore, the peaks in the $M2/M3SnO_2$ electrodes were divided by a small peak at 0.42 V and a joined peak in the range of 0.75–0.84 V of SnO_2 [50]. In addition, the peak intensity at ~0.8 V related to the reduction of SnO_2 increased compared to those of the pure SnO_2 and $M1SnO_2$ electrodes because of the joined peak of the MoS_2 multilayer. Therefore, it is thought that the addition of MoS_2 NS (two and three layers) might lead to the formation of a large SEI layer. Further, the direct contact of each NS layer can increase the possibility of restacking of the MoS_2 NSs, leading to thicker and bulky MoS_2 NS structures.

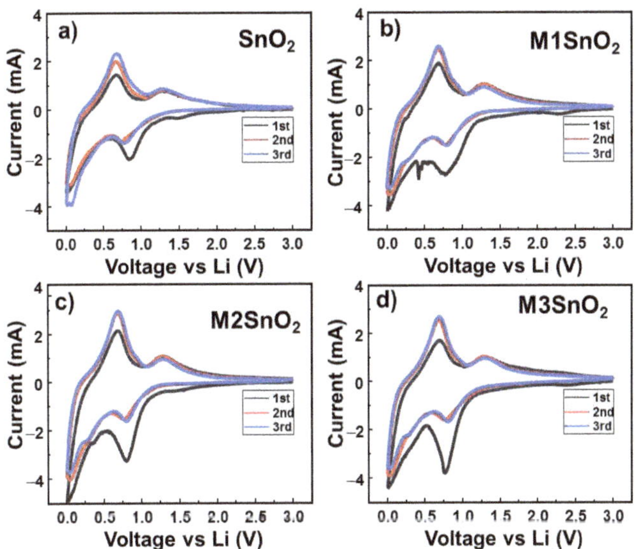

Figure 5. Cyclic voltammetry (CV) profiles of (**a**) SnO_2 NS and (**b–d**) $M1/M2/M3SnO_2$ anodes over three cycles.

The initial voltage profiles of the SnO_2 and $M1/M2/M3SnO_2$ electrodes at 100 $mA·g^{-1}$ from 0.01 to 3.0 V are shown in Figure 6. The initial discharge capacity of SnO_2 showed a very high value of ~1760 $mAh·g^{-1}$. This was dramatically reduced to ~880 and ~760 $mAh·g^{-1}$ for the 2nd and 3rd discharges, respectively, indicating a large irreversible reaction due to the formation of an SEI layer and an initial coulombic efficiency (ICE) of 42.3%. However, in the case of the $M1SnO_2$ electrode (Figure 6b), the 1st, 2nd, and 3rd cycles exhibited slow decreases in the discharge/charge capacities from 1740/760 $mAh·g^{-1}$ to 932/755 $mAh·g^{-1}$ and 835/710 $mAh·g^{-1}$, respectively. Figure 6c shows the voltage profile of the $M2SnO_2$ electrode. It shows the 1st discharge/charge capacity of 1730/823 $mAh·g^{-1}$, corresponding to an ICE of 47.6%. Therefore, it is thought that the better coverage of the MoS_2 NS double layer improved the charge capacity from 760 to 823 $mAh·g^{-1}$ compared to that of the

M1SnO$_2$ electrode in the first cycle. However, from the next cycle, the capacity was reduced rapidly to 580 mAh·g^{-1} in the 3rd cycle. Similarly, the additional coating layer on SnO$_2$ (M3SnO$_2$) caused a significant reduction in the discharge/charge capacity in the 1st cycle to 1640/701 mAh·g^{-1}, which was decreased to ~700/580 mAh·g^{-1} in the 3rd cycle. In summary, the addition of MoS$_2$ layers to the SnO$_2$ electrode improved the electrochemical performance. However, thicker MoS$_2$ NSs in excess of two layers decreased the lithium storage capability. This could be explained by the restacking of the MoS$_2$ NS layers, leading to the formation of bulky or thicker layers. Moreover, TMC materials such as MoS$_2$ and WS$_2$ have been reported to improve the electronic properties when present as very thin single- or few-layer structures [31]. This information is in good agreement with the single-layer MoS$_2$ NS-coated SnO$_2$ meeting the requirements of enhancing the storage ability, whereas multilayer MoS$_2$ NS-coated SnO$_2$ showed faster degradation in battery performance.

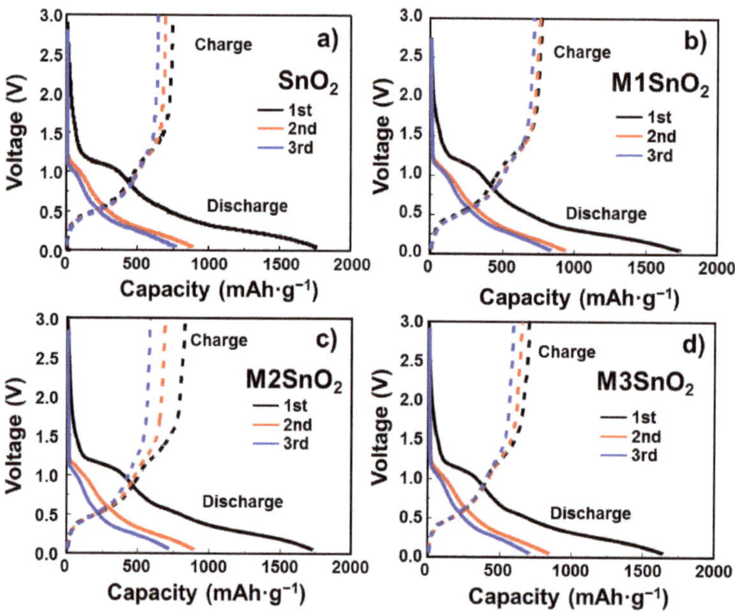

Figure 6. Galvanostatic charge–discharge profiles of (**a**) SnO$_2$ NS and (**b–d**) M1/M2/M3SnO$_2$ anodes for the initial three cycles.

To evaluate the long-term cyclability, the LIB half-cell with and without the MoS$_2$ NS layer on the SnO$_2$ surface were subjected to 100 charge–discharge cycles at a current rate of 100 mA·g^{-1}, as shown in Figure 7a–c. The mass in all measurements was calculated as the real mass of active materials, including SnO$_2$ and MoS$_2$, in the electrode. The pure SnO$_2$ anode showed a substantial decrease in the discharge/charge capacity from 1750/745 mAh·g^{-1} to 373/345 mAh·g^{-1} at the 10th cycle and to 74.4/73.9 mAh·g^{-1} at the 100th cycle, as illustrated in Figure 7a. With the addition of the MoS$_2$ NS layer, the cyclic stability of lithium storage was much higher than that of the bare electrode. With a similar initial discharge/charge capacity of 1740/760 mAh·g^{-1}, at the 10th cycle, they exhibited those of 593/551 mAh·g^{-1} and retained those of 281/277 mAh·g^{-1} at the 100th cycle. It is well observed that the coulombic efficiency retained a high value of ~99%, as shown in Figure 7b. The electrode with >2 layers of MoS$_2$ NSs showed the worst cyclic performance, retaining a discharge/charge capacity of only 103/102 mAh·g^{-1} at the 100th cycle (Figure 7c). Furthermore, the charge-transfer resistance of the cells was measured from the EIS analysis, as depicted in Figure 7d. The modified Randles equivalent circuit was determined to contain a series resistance (R_s), a charge-transfer resistance (R_{ct}), an SEI resistance (R_{SEI}), and a Warburg impedance element (W) (Figure 7d inset) [4]. The extracted values of these resistances are shown in Table 1. It is easy to observe that the bare SnO_2 has the lowest charge-transfer resistance of ~23.86 Ω, while an increasing number of MoS_2 NS layers corresponded to increased charge-transfer resistance in the electrodes. In contrast, the R_{SEI} was dramatically reduced from 500.2 Ω in the SnO_2 NP electrode to 182.7 Ω in the $M1SnO_2$ electrode, and then slightly increased to 284.8 Ω and 305.1 Ω for the M2 and $M3SnO_2$ electrodes, respectively. The MoS_2 layer provides a large uniform surface compared to that of bare SnO_2, thus reducing R_{SEI}. Moreover, as discussed above, the electrode with single-layer MoS_2 has a low lithium intercalation energy barrier of ~0.42 eV, while the electrodes with thicker layers have the higher energy barrier of ~0.73 eV. This could lead to an increase in the R_{SEI}. Based on the aforementioned results, the single layer of MoS_2 NS is important in reducing the SEI resistance, thus helping the long-term cycling stability of the SnO_2 anode for lithium storage. This study suggests that coverage of SnO_2 NPs with a thin and complete large-scale MoS_2 layer can significantly enhance the electrode performance in LIBs.

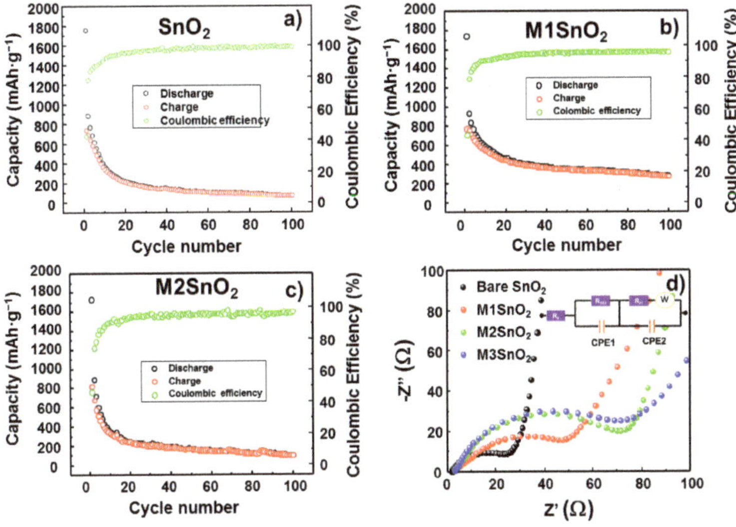

Figure 7. Cyclic performances and Nyquist plots of (**a**) SnO_2 NPs and (**b–d**) M1/M2/M3SnO_2 anodes.

Table 1. Resistance values extracted from the equivalent circuit of M1/M2/M3SnO_2 anodes.

Sample	R_s (Ω)	R_{ct} (Ω)	R_{SEI} (Ω)
SnO_2 NPs	1.9	23.8	500.2
$M1SnO_2$	2.4	59.3	182.7
$M2SnO_2$	2.8	76.2	284.8
$M3SnO_2$	2.8	86.1	305.1

The rate cycling performance of the bare SnO_2 and $M1SnO_2$ electrodes was also investigated, as shown in Figure 8. Performance was recorded for 20 cycles each at charge rates of 100, 500, and 1000 mA·g^{-1} before the charge rate returning to 100 mA·g^{-1}. The SnO_2 electrode shows greater degradation in specific capacity retention when the scan rate is increased. The retentions after 5 cycles at 500 and 1000 mA·g^{-1} were 47% and 26%, respectively. In contrast, the $M1SnO_2$ electrode shows a smaller decrease in specific capacity retention compared to that of pure SnO2, where the retentions at 500 and 1000 mA·g^{-1} were 70% and 47%, respectively, thus demonstrating much better rate performance than the SnO_2 electrode did. The restoration of the $M1SnO_2$ electrode when the current rate was decreased from 1000 to 100 mA·g^{-1} was also remarkable, corresponding to 95%, far exceeding that

(55%) of the SnO_2 electrode. These results suggest that the introduction of a single MoS_2 layer on SnO_2 electrodes could further enhance the electrochemical performance.

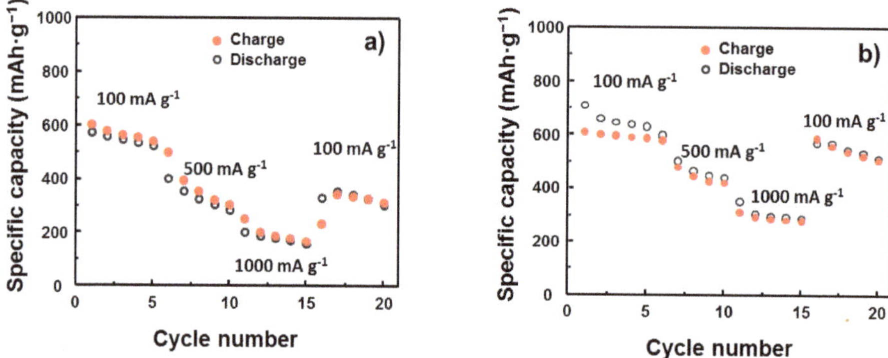

Figure 8. Rate cycling performance of (**a**) bare SnO_2 electrode and (**b**) $M1SnO_2$ electrode at different current rates.

To further confirm the effect of the MoS_2 layer, the surface of the electrode was characterized by ex situ SEM, as shown in Figure 9. The assembled half-cells were run for 10 cycles, and then they were disassembled, washed carefully with dimethyl carbonate and acetone, and dried. Figure 9a,d clearly show that the bare SnO_2 electrode was easily broken, creating micro-holes on the surface. However, the $M1SnO_2$ electrode was protected from washing, as illustrated in Figure 9b,e. This indicates that the MoS_2 layer can facilitate the formation of a stable and smooth SEI layer. In Figure 9c,f, the $M3SnO_2$ electrode after 10 cycles shows a thick MoS_2 surface that is much rougher than that of the $M1SnO_2$ surface. This would be due to the restacking of the MoS_2 NS into a bulky multilayer state. Based on the results, a thin MoS_2 coating layer acted as a protecting layer and formed a uniform SEI layer. The multiple MoS_2 NS layers generated thick and rough surfaces, leading to inferior electrochemical performance for lithium-ion storage.

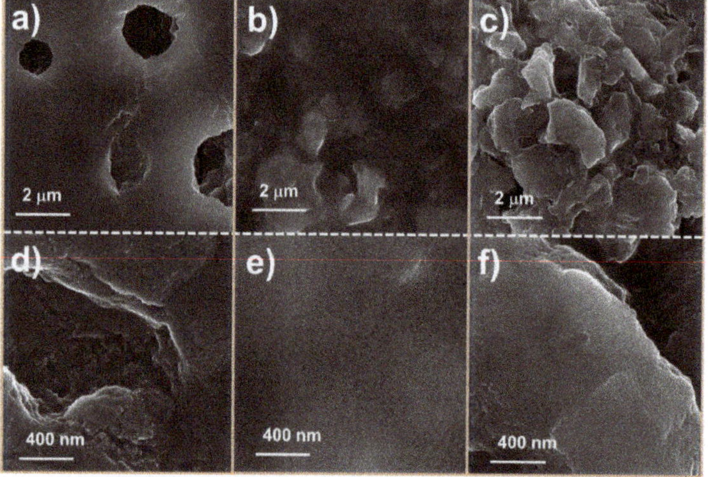

Figure 9. SEM images of (**a,d**) SnO_2 electrode, (**b,e**) $M1SnO_2$ electrode and (**c,f**) $M3SnO_2$ electrode after 10 cycles at different magnifications.

4. Conclusions

In summary, this study successfully prepared SnO_2 NPs by a facile hydrothermal method and MoS_2 NS using a liquid chemical exfoliation method. The materials were characterized by XRD, SEM, and TEM measurements. The size of the SnO_2 NPs was about 5–10 nm, and the MoS_2 NSs were hundreds of nanometers to a few micrometers in size. A new coating method utilizing the self-assembly of MoS_2 NS thin films was successfully developed based on a rigid water–air interface. The addition of single-layer MoS_2 NSs enhanced the cyclic stability of the SnO_2 anodes for lithium-ion storage with a high coulombic efficiency of ~99% and high charge/discharge capacity of 281/277 mAh·g^{-1} compared to those with multiple MoS_2 layers after 100 cycles. These results also suggest that single-layer MoS_2 in large-scale fabrication methods, such as chemical vapor deposition, could be applied to further enhance the cyclic stability of anodes in LIBs.

Author Contributions: T.P.N. Conceptualization, Methodology, Validation, Visualization, Writing—review and editing. I.T.K. Project administration, Funding acquisition, Review & editing. All authors have read and agreed to the published version of the manuscript.

Funding: This work was supported by the National Research Foundation of Korea (NRF) grant funded by the Korean government (MSIT) (NRF-2020R1F1A1048335). This research was also supported by the Basic Science Research Capacity Enhancement Project through a Korea Basic Science Institute (National Research Facilities and Equipment Center) grant funded by the Ministry of Education (2019R1A6C1010016).

Conflicts of Interest: The authors declare no conflict of interest.

References

1. Mo, R.; Tan, X.; Li, F.; Tao, R.; Xu, J.; Kong, D.; Wang, Z.; Xu, B.; Wang, X.; Wang, C.; et al. Tin-graphene tubes as anodes for lithium-ion batteries with high volumetric and gravimetric energy densities. *Nat. Commun.* **2020**, *11*, 1374. [CrossRef] [PubMed]

2. Asenbauer, J.; Eisenmann, T.; Kuenzel, M.; Kazzazi, A.; Chen, Z.; Bresser, D. The success story of graphite as a lithium-ion anode material—Fundamentals, remaining challenges, and recent developments including silicon (oxide) composites. *Sustain. Energy Fuels* **2020**, *4*, 5387–5416. [CrossRef]

3. Nguyen, T.P.; Kim, I.T. W_2C/WS_2 Alloy Nanoflowers as Anode Materials for Lithium-Ion Storage. *Nanomaterials* **2020**, *10*, 1336. [CrossRef] [PubMed]

4. Vo, T.N.; Kim, D.S.; Mun, Y.S.; Lee, H.J.; Ahn, S.K.; Kim, I.T. Fast charging sodium-ion batteries based on Te-P-C composites and insights to low-frequency limits of four common equivalent impedance circuits. *Chem. Eng. J.* **2020**, *398*, 125703. [CrossRef]

5. Zhu, J.; Lu, Y.; Chen, C.; Ge, Y.; Jasper, S.; Leary, J.D.; Li, D.; Jiang, M.; Zhang, X. Porous one-dimensional carbon/iron oxide composite for rechargeable lithium-ion batteries with high and stable capacity. *J. Alloys Compd.* **2016**, *672*, 79–85. [CrossRef]

6. Song, T.; Paik, U. TiO_2 as an active or supplemental material for lithium batteries. *J. Mater. Chem. A* **2016**, *4*, 14–31. [CrossRef]

7. Nguyen, T.L.; Hur, J.; Kim, I.T. Facile Synthesis of quantum dots SnO_2/Fe_3O_4 hybrid composites for superior reversible lithium-ion storage. *J. Ind. Eng. Chem.* **2019**, *72*, 504–511. [CrossRef]

8. Wu, Q.L.; Li, J.C.; Deshpande, R.D.; Subramanian, N.; Rankin, S.E.; Yang, F.Q.; Cheng, Y.T. Aligned TiO_2 Nanotube Arrays As Durable Lithium-Ion Battery Negative Electrodes. *J. Phys. Chem. C* **2012**, *116*, 18669–18677. [CrossRef]

9. Liu, H.; Li, W.; Shen, D.; Zhao, D.; Wang, G. Graphitic Carbon Conformal Coating of Mesoporous TiO_2 Hollow Spheres for High-Performance Lithium Ion Battery Anodes. *J. Am. Chem. Soc.* **2015**, *137*, 13161–13166. [CrossRef]

10. Jiang, T.; Bu, F.; Feng, X.; Shakir, I.; Hao, G.; Xu, Y. Porous Fe_2O_3 Nanoframeworks Encapsulated within Three-Dimensional Graphene as High-Performance Flexible Anode for Lithium-Ion Battery. *ACS Nano* **2017**, *11*, 5140–5147. [CrossRef]

11. He, P.; Ding, Z.; Zhao, X.; Liu, J.; Yang, S.; Gao, P.; Fan, L.Z. Single-Crystal alpha-Fe_2O_3 with Engineered Exposed (001) Facet for High-Rate, Long-Cycle-Life Lithium-Ion Battery Anode. *Inorg. Chem.* **2019**, *58*, 12724–12732. [CrossRef] [PubMed]

12. Xu, N.; Liang, J.Q.; Qian, T.; Yang, T.Z.; Yan, C.L. Half-cell and full-cell applications of horizontally aligned reduced oxide graphene/V_2O_5 sheets as cathodes for high stability lithium-ion batteries. *RSC Adv.* **2016**, *6*, 98581–98587. [CrossRef]

13. Nguyen, T.-A.; Lee, S.-W. Bulky carbon layer inlaid with nanoscale Fe_2O_3 as an excellent lithium-storage anode material. *J. Ind. Eng. Chem.* **2018**, *68*, 140–145. [CrossRef]

14. Nguyen, Q.H.; Choi, J.S.; Lee, Y.C.; Kim, I.T.; Hur, J. 3D hierarchical structure of MoS_2@G-CNT combined with post-film annealing for enhanced lithium-ion storage. *J. Ind. Eng. Chem.* **2019**, *69*, 116–126. [CrossRef]

15. Nhung Pham, T.; Ko, J.; Khac Hoang Bui, V.; So, S.; Uk Lee, H.; Hur, J.; Lee, Y.-C. Facile two-step synthesis of innovative anode design from tin-aminoclay (SnAC) and rGO for Li-ion batteries. *Appl. Surf. Sci.* **2020**, *532*, 147435. [CrossRef]

16. Kim, C.; Noh, M.; Choi, M.; Cho, J.; Park, B. Critical size of a nano SnO_2 electrode for Li-secondary battery. *Chem. Mater.* **2005**, *17*, 3297–3301. [CrossRef]

17. Li, J.; Zhao, Y.; Wang, N.; Guan, L. A high performance carrier for SnO_2 nanoparticles used in lithium ion battery. *Chem. Commun.* **2011**, *47*, 5238–5240. [CrossRef]

18. Zhao, S.; Sewell, C.D.; Liu, R.; Jia, S.; Wang, Z.; He, Y.; Yuan, K.; Jin, H.; Wang, S.; Liu, X.; et al. SnO_2 as Advanced Anode of Alkali-Ion Batteries: Inhibiting Sn Coarsening by Crafting Robust Physical Barriers, Void Boundaries, and Heterophase Interfaces for Superior Electrochemical Reaction Reversibility. *Adv. Energy Mater.* **2019**, *10*, 1902657. [CrossRef]

19. Lou, X.W.; Chen, J.S.; Chen, P.; Archer, L.A. One-Pot Synthesis of Carbon-Coated SnO_2 Nanocolloids with Improved Reversible Lithium Storage Properties. *Chem. Mater.* **2009**, *21*, 2868–2874. [CrossRef]

20. Wang, X.Y.; Zhou, X.F.; Yao, K.; Zhang, J.G.; Liu, Z.P. A SnO_2/graphene composite as a high stability electrode for lithium ion batteries. *Carbon* **2011**, *49*, 133–139. [CrossRef]

21. Chen, Y.; Lu, J.; Wen, S.; Lu, L.; Xue, J.M. Synthesis of SnO_2/MoS_2 composites with different component ratios and their applications as lithium ion battery anodes. *J. Mater. Chem. A* **2014**, *2*, 17857–17866. [CrossRef]

22. Kim, C.; Nguyen, T.P.; Le, Q.V.; Jeon, J.M.; Jang, H.W.; Kim, S.Y. Performances of Liquid-Exfoliated Transition Metal Dichalcogenides as Hole Injection Layers in Organic Light-Emitting Diodes. *Adv. Funct. Mater.* **2015**, *25*, 4512–4519. [CrossRef]

23. Nguyen, T.P.; Le, Q.V.; Choi, K.S.; Oh, J.H.; Kim, Y.G.; Lee, S.M.; Chang, S.T.; Cho, Y.H.; Choi, S.; Kim, T.Y.; et al. MoS_2 Nanosheets Exfoliated by Sonication and Their Application in Organic Photovoltaic Cells. *Sci. Adv. Mater.* **2015**, *7*, 700–705. [CrossRef]

24. Voiry, D.; Goswami, A.; Kappera, R.; e Silva, C.D.C.C.; Kaplan, D.; Fujita, T.; Chen, M.; Asefa, T.; Chhowalla, M. Covalent functionalization of monolayered transition metal dichalcogenides by phase engineering. *Nat. Chem.* **2015**, *7*, 45–49. [CrossRef]

25. Castellanos-Gomez, A.; Poot, M.; Steele, G.A.; van der Zant, H.S.J.; Agrait, N.; Rubio-Bollinger, G. Elastic Properties of Freely Suspended MoS_2 Nanosheets. *Adv. Mater.* **2012**, *24*, 772–775. [CrossRef]

26. Nguyen, T.P.; Kim, S.Y.; Lee, T.H.; Jang, H.W.; Le, Q.V.; Kim, I.T. Facile synthesis of W_2C@WS_2 alloy nanoflowers and their hydrogen generation performance. *Appl. Surf. Sci.* **2020**, *504*, 144389. [CrossRef]

27. Coleman, J.N.; Lotya, M.; O'Neill, A.; Bergin, S.D.; King, P.J.; Khan, U.; Young, K.; Gaucher, A.; De, S.; Smith, R.J.; et al. Two-dimensional nanosheets produced by liquid exfoliation of layered materials. *Science* **2011**, *331*, 568–571. [CrossRef]

28. Yu, Y.; Zhang, Y.T.; Song, X.X.; Zhang, H.T.; Cao, M.X.; Che, Y.L.; Dai, H.T.; Yang, J.B.; Zhang, H.; Yao, J.Q. PbS-Decorated WS_2 Phototransistors with Fast Response. *ACS Photonics* **2017**, *4*, 950–956. [CrossRef]

29. Le, Q.V.; Nguyen, T.P.; Jang, H.W.; Kim, S.Y. The use of UV/ozone-treated MoS_2 nanosheets for extended air stability in organic photovoltaic cells. *Phys. Chem. Chem. Phys.* **2014**, *16*, 13123–13128. [CrossRef]

30. Nguyen, T.P.; Le, Q.V.; Choi, S.; Lee, T.H.; Hong, S.-P.; Choi, K.S.; Jang, H.W.; Lee, M.H.; Park, T.J.; Kim, S.Y. Surface extension of MeS_2 (Me = Mo or W) nanosheets by embedding MeS_x for hydrogen evolution reaction. *Electrochim. Acta* **2018**, *292*, 136–141. [CrossRef]

31. Lane, C.; Cao, D.X.; Li, H.Y.; Jiao, Y.C.; Barbiellini, B.; Bansil, A.; Zhu, H.L. Understanding Phase Stability of Metallic 1T-MoS_2 Anodes for Sodium-Ion Batteries. *Condens. Matter* **2019**, *4*, 53. [CrossRef]

32. Li, S.C.; Liu, P.; Huang, X.B.; Tang, Y.G.; Wang, H.Y. Reviving bulky MoS_2 as an advanced anode for lithium-ion batteries. *J. Mater. Chem. A* **2019**, *7*, 10988–10997. [CrossRef]

33. Li, C.; Hong, G.S.; Qi, L.M. Nanosphere Lithography at the Gas/Liquid Interface: A General Approach toward Free-Standing High-Quality Nanonets. *Chem. Mater.* **2010**, *22*, 476–481. [CrossRef]

34. Rybczynski, J.; Ebels, U.; Giersig, M. Large-scale, 2D arrays of magnetic nanoparticles. *Colloids Surf. A-Physicochem. Eng. Asp.* **2003**, *219*, 1–6. [CrossRef]

35. Li, C.; Hong, G.S.; Wang, P.W.; Yu, D.P.; Qi, L.M. Wet Chemical Approaches to Patterned Arrays of Well-Aligned ZnO Nanopillars Assisted by Monolayer Colloidal Crystals. *Chem. Mater.* **2009**, *21*, 891–897. [CrossRef]

36. Joensen, P.; Frindt, R.F.; Morrison, S.R. Single-layer MoS_2. *Mater. Res. Bull.* **1986**, *21*, 457–461. [CrossRef]

37. Eda, G.; Yamaguchi, H.; Voiry, D.; Fujita, T.; Chen, M.; Chhowalla, M. Photoluminescence from chemically exfoliated MoS_2. *Nano Lett.* **2011**, *11*, 5111–5116. [CrossRef]

38. Park, M.; Nguyen, T.P.; Choi, K.S.; Park, J.; Ozturk, A.; Kim, S.Y. MoS_2-nanosheet/graphene-oxide composite hole injection layer in organic light-emitting diodes. *Electron. Mater. Lett.* **2017**, *13*, 344–350. [CrossRef]

39. Yang, D.; Frindt, R.F. Li-intercalation and exfoliation of WS_2. *J. Phys. Chem. Solids* **1996**, *57*, 1113–1116. [CrossRef]

40. Akhir, A.M.; Rezan, S.A.; Mohamed, K.; Arafat, M.M.; Haseeb, A.S.M.A.; Lee, H.L. Synthesis of SnO_2 Nanoparticles via Hydrothermal Method and Their Gas Sensing Applications for Ethylene Detection. *Mater. Today Proc.* **2019**, *17*, 810–819. [CrossRef]

41. Patil, G.E.; Kajale, D.D.; Gaikwad, V.B.; Jain, G.H. Preparation and characterization of SnO_2 nanoparticles by hydrothermal route. *Int. Nano Lett.* **2012**, *2*, 17. [CrossRef]

42. Dias, J.S.; Batista, F.R.M.; Bacani, R.; Triboni, E.R. Structural characterization of SnO nanoparticles synthesized by the hydrothermal and microwave routes. *Sci. Rep.* **2020**, *10*, 9446. [CrossRef] [PubMed]

43. Han, C.; Zhang, Y.; Gao, P.; Chen, S.; Liu, X.; Mi, Y.; Zhang, J.; Ma, Y.; Jiang, W.; Chang, J. High-Yield Production of MoS_2 and WS_2 Quantum Sheets from Their Bulk Materials. *Nano Lett.* **2017**, *17*, 7767–7772. [CrossRef] [PubMed]

44. Nguyen, T.P.; Sohn, W.; Oh, J.H.; Jang, H.W.; Kim, S.Y. Size-Dependent Properties of Two-Dimensional MoS_2 and WS_2. *J. Phys. Chem. C* **2016**, *120*, 10078–10085. [CrossRef]

45. Reddy, M.V.; Andreea, L.Y.T.; Ling, A.Y.; Hwee, J.N.C.; Lin, C.A.; Admas, S.; Loh, K.P.; Mathe, M.K.; Ozoemena, K.I.; Chowdari, B.V.R. Effect of preparation temperature and cycling voltage range on molten salt method prepared SnO_2. *Electrochim. Acta* **2013**, *106*, 143–148. [CrossRef]

46. Wu, H.; Hou, C.Y.; Shen, G.Z.; Liu, T.; Shao, Y.L.; Xiao, R.; Wang, H.Z. MoS_2/C/C nanofiber with double-layer carbon coating for high cycling stability and rate capability in lithium-ion batteries. *Nano Res.* **2018**, *11*, 5866–5878. [CrossRef]

47. Gong, Y.J.; Yang, S.B.; Zhan, L.; Ma, L.L.; Vajtai, R.; Ajayan, P.M. A Bottom-Up Approach to Build 3D Architectures from Nanosheets for Superior Lithium Storage. *Adv. Funct. Mater.* **2014**, *24*, 125–130. [CrossRef]

48. Gao, S.N.; Yang, L.T.; Shao, J.; Qu, Q.T.; Wu, Y.P.; Holze, R. Construction of Hierarchical Hollow MoS_2/Carbon Microspheres for Enhanced Lithium Storage Performance. *J. Electrochem. Soc.* **2020**, *167*, 100525. [CrossRef]

49. Cui, C.Y.; Li, X.; Hu, Z.; Xu, J.T.; Liu, H.K.; Ma, J.M. Growth of MoS2@C nanobowls as a lithium-ion battery anode material. *RSC Adv.* **2015**, *5*, 92506–92514. [CrossRef]

50. Cha, E.; Patel, M.D.; Park, J.; Hwang, J.; Prasad, V.; Cho, K.; Choi, W. 2D MoS_2 as an efficient protective layer for lithium metal anodes in high-performance Li-S batteries. *Nat. Nanotechnol.* **2018**, *13*, 337–344. [CrossRef]

Publisher's Note: MDPI stays neutral with regard to jurisdictional claims in published maps and institutional affiliations.

MDPI
St. Alban-Anlage 66
4052 Basel
Switzerland
Tel. +41 61 683 77 34
Fax +41 61 302 89 18
www.mdpi.com

Nanomaterials Editorial Office
E-mail: nanomaterials@mdpi.com
www.mdpi.com/journal/nanomaterials